Respiratory Biology of Animals

Respiratory Biology of Animals

Evolutionary and Functional Morphology

Steven F. Perry
Institut für Zoologie,
Rheinische Friedrich-Wilhelms-Universität Bonn, Germany

Markus Lambertz
Institut für Zoologie,
Rheinische Friedrich-Wilhelms-Universität Bonn, Germany
Sektion Herpetologie,
Zoologisches Forschungsmuseum Alexander Koenig, Germany

Anke Schmitz
Institut für Zoologie,
Rheinische Friedrich-Wilhelms-Universität Bonn, Germany

OXFORD
UNIVERSITY PRESS

Great Clarendon Street, Oxford, OX2 6DP,
United Kingdom

Oxford University Press is a department of the University of Oxford.
It furthers the University's objective of excellence in research, scholarship,
and education by publishing worldwide. Oxford is a registered trade mark of
Oxford University Press in the UK and in certain other countries

First Edition published in 2019

Published in the United States of America by Oxford University Press
198 Madison Avenue, New York, NY 10016, United States of America

British Library Cataloguing in Publication Data
Data available

Library of Congress Control Number: 2019945437

ISBN 978–0–19–923846–0 (hbk.)
ISBN 978–0–19–923847–7 (pbk.)

DOI: 10.1093/oso/9780199238460.001.0001

We wish to dedicate this book to the memory of Steve Morris (1956–2009), whose untimely death was a great loss, not only for respiratory biology.

Preface

This book is meant to be a supplementary text for advanced courses in metabolic physiology, evolutionary biology, or related areas. As prerequisites we suggest courses in general biology, general zoology, animal physiology, and comparative anatomy. In addition, it may be of interest to students of medicine, biophysics, biochemistry, or related areas of molecular biology. To this end, we do not delve into the nuts and bolts of biochemistry, biophysics, and molecular biology, but rather emphasize the overarching evolutionary implications and basic principles that the study of respiratory biology can illustrate. But since special courses in respiratory physiology are not offered everywhere, we perhaps go into this area in a bit more detail than would be expected in a book on functional and evolutionary morphology. The same applies to the control of breathing, and some basic information must be supplied in order to understand the more detailed descriptions that follow. As we will discover again and again: 'Before you can do what you want to do, you always have to do something else'.

In general, it was our intention to provide an overview of the big picture of the evolution of respiration in animals. We consequently regularly cite relatively recent overview or review articles on certain subjects rather than referencing all of the primary sources. This book was never intended to serve as an encyclopaedic reference volume. We therefore frequently selected interesting examples to illustrate a given phenomenon as a case in point, rather than providing a complete list of all known details in each and every species that has been examined. Furthermore and most importantly, you may not find all the answers you are interested in, but instead regularly will be confronted with persisting questions. This is in keeping with the primary intention of this book, which was to expose respiratory biology as a fertile ground for further research, in fact on all levels from molecular to evolutionary considerations. We hope that you will enjoy the read and in turn look forward to reading new advancements in this field from your end as well, some of which may even be stimulated by critically reading the present volume.

Bonn,
January 2019
The authors

Acknowledgements

First of all we would like to thank our publisher Oxford University Press for their incredible patience with us before this book eventually became finished. We furthermore wish to express our sincerest thanks to those colleagues who read preliminary drafts of individual chapters and provided valuable criticism that helped us to improve the contents. These are, in alphabetical order: Thorsten Burmester (Hamburg, Germany), Jon F. Harrison (Tempe, AZ, USA), John B. West (San Diego, CA, USA), Jürgen Markl (Mainz, Germany), Richard J.A. Wilson (Calgary, Canada). M.L. gratefully acknowledges the support of Michael H. Hofmann (Bonn, Germany) for providing the facilities and latitude for pursuing this and other projects.

Contents

'The necessity of respiration to the support of life, and the evident injuries arising from any impediment in this function, induced the earliest medical philosophers to make it a subject of enquiry; and, from that time to this, it has afforded a continued subject of admiration, discussion, and dispute: not, indeed, of dispute, as whether it was necessary or not, but why it was necessary? what where the advantages the animal economy derived from thence? and by what means it was carried on?'

Robert Townson (1799). ***Tracts and Observations in Natural History and Physiology*****. London, printed for the author: p. 7.**

CHAPTER 1

Prolegomena

Probably no process epitomizes life more than respiration. By respiration we mean the cascade of energy-producing biochemical reactions called oxidative phosphorylation, powered by a gradient of oxidation. If you are reading this book you probably already know that the terminal oxidant in most cases is atmospheric oxygen (O_2), but a variety of micro-organisms also show respiration powered by other terminal molecules such as sulphur compounds or even carbon dioxide (CO_2) or perchlorate (e.g. Liebensteiner et al., 2016 and references therein). The end products of respiration are phosphorylated nucleotides such as adenosine triphosphate (ATP) or guanosine triphosphate (GTP) and, as waste products, carbon dioxide and water. As a side effect of respiration, many of the reactions involved in oxidative phosphorylation are exothermic. The heat released can aid in the maintenance of a high, constant body temperature (homeothermy), but in extremely large animals such as sauropod dinosaurs the sheer amount of heat produced may even result in substantial physiological problems (gigantothermy).

However, the 'cellular respiration' just described represents only the final phase (here called phase 5). In many organisms, respiration also includes four other phases, which, now working back from the cells to the outside world are (4) gas consumption and production at the tissue and cellular level, (3) the transport of these gases between the gas exchanger and tissues via body fluids (e.g. a circulatory system, if present), (2) diffusive gas exchange between the medium (air or water) and the organism, and finally (1) the convection of oxygen and carbon dioxide to and from a gas exchanger (e.g. lung, gill, skin, or tracheae). Phase 4 is also referred to as 'aerobic metabolism', phase 3 as 'circulation' or 'perfusion', phase 2 is 'gas exchange', and phase 1 is 'breathing' or 'ventilation'.

The focus of this book is the evolution of respiratory systems in animals. But in order to really understand this, we must first deal with some basic concepts of respiration, including the properties of the gases involved and the historical sequence of events that led to the development of respiratory biology as we know it today. We then move on to some basics of functional morphology and physiology as they relate to respiration. But no organ system can be studied on its own, since all of them interact with each other to comprise an organism. So, to be more precise, we will deal in detail with the evolution of the structure–function interaction in respiratory physiology and anatomy. The term 'faculty' as defined by Bock and von Wahlert (1965) describes exactly this structure–function interaction, whereby every structure can have multiple functions and every function can involve numerous structures.

The 'faculty' concept is based on the assumption that a given function—in this case respiration—is related to a variety of anatomical components, thereby composing a functional unit (Figure 1.1). This idea is not new and was probably best expressed by the French comparative and functional morphologist and founder of reconstructive palaeontology, Georges Cuvier (1769–1832), in his so-called law of the correlation of the parts (Cuvier, 1812). There he

Respiratory Biology of Animals: Evolutionary and Functional Morphology. Steven F. Perry, Markus Lambertz, Anke Schmitz, Oxford University Press (2019).
DOI: 10.1093/oso/9780199238460.001.0001

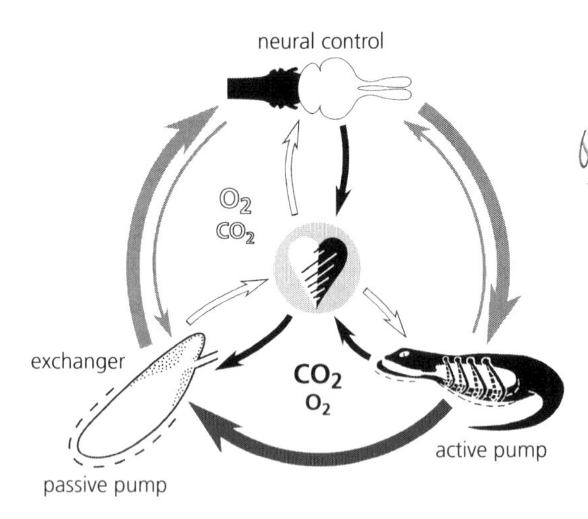

Figure 1.1 A schematic representation of the concept of a respiratory faculty, exemplified in a lizard. Brainstem shown in black. Active respiratory pump illustrating only costal breathing and a simplified rib cage. Gray arrows indicate direct biomechanical (dark shading) and neuronal (light shading) connections, whereby the thinner arrows indicate feedback. Black or white arrows indicate an indirect loop via the circulatory sytem, here with an intercardiac shunt possibility: Black arrows and print, deoxygenated blood; white arrows and print, oxygenated blood. The size and colour of the print (O_2 or CO_2) indicates the predominance of this gas in the circulatory loop. Original M.L. and S.F.P in Hsia et al. (2013).

states that an animal is a closed unit in which all component parts mutually correspond and unite in reciprocal reaction, such that no one part can be changed without changing all of the others. Focusing on the cardiorespiratory faculty, Weibel and Taylor (1981) coined the term 'symmorphosis'. In a series of publications over at least the next 10 years (see Weibel et al., 1991) they applied physiological and quantitative morphological methods to determine just how close structure and function are coupled with regard to oxygen supply and demand in mammals. Although within closely related phylogenetic groups symmorphosis appeared to generally apply, detailed independent statistical analysis revealed weaknesses in the general validity of this principle (Garland and Huey, 1987). The latter authors concluded: 'Although symmorphosis has heuristic value as a working hypothesis, it should not presently be considered an established principle; available data appear largely contradictory'. It is just this heuristic value that causes symmorphosis to surface as a hypothesis even in recent publications (Gifford et al., 2016).

As you read along, you will notice two overriding principles that appear and reappear at all organizational levels from molecular to organismic. The first of these can be summarized as 'Getting it right the first time'. Basically, you can't kill a good idea! A couple of examples illustrate how this works. When we look more closely at the evolution of the respiratory faculty it will become evident that sometimes functions change while structures are conserved. At the molecular level, for example, we see nucleotides such as adenosine appearing with multiple functions ranging from a building block of nucleic acids to energy transport, and they can even assume new functions, such as influencing oxygen affinity of oxygen-carrying molecules such as haemoglobin. An example at the organ level is the lung. This primary gas-exchange organ also assumes crucial physiological functions such as the conversion of angiotensin I to angiotensin II, or as a site of detoxification and immune activity. But sometimes function is conserved or improved upon while the structures around it radically change. This can be recognized at any organizational level and is not limited to respiratory biology. As an example at the cellular physiological level, we see a plethora of so-called respiratory proteins that serve in oxygen transport. At the organ level we find, for example, countercurrent gas exchange being conserved in fish gills in spite of radically different ventilatory mechanisms and gill structures in hagfish, lampreys, sharks, and bony fish. And sometimes a primary function becomes secondary, while an ancestral secondary function becomes primary. An example for such an exaptation at the organ level is the pseudobranch in fishes. What used to be a gas-exchange organ (gill) that contained oxygen receptors, lost its primary gas-exchange function but retained the oxygen receptors as an important monitoring organ for oxygen tension in the blood.

The second reoccurring principle is the evolutionary cascade. This positive feedback phenomenon dovetails in an evolutionary context with exaptation when the exaptive faculty complements with others synergistically. One example is the avian respiratory faculty. The essential structural elements were probably already present in non-avian dinosaurs or even in ancestral archosaurs, but it could only become exploited to its limits in actively flying birds, an evolutionary quirk that could not have been predicted.

Evolution leaves tracks. This will become evident in every chapter of this book. But certainly the most far-reaching of these tracks is the realization that each plant and animal, including us humans, carries within every cell a molecular memory of the conditions under which life originated and which are still necessary. To understand what we are talking about, let us journey back to a time about 4–3.5 thousand million years ago (mya). We must imagine a hot atmosphere consisting mostly of water vapour and gases such as nitrogen (N_2), carbon dioxide, hydrogen sulphide, sulphur dioxide, and traces of ammonia and methane, but no free oxygen. Sort of like living in the mouth of a volcano. The world was spinning much faster than today, meaning that the days and nights were much shorter than they are now. But more importantly, this must have resulted in tremendous winds. As the Earth cooled, water vapour condensed to water and the seas formed. The Moon was much closer to the Earth than it is now, resulting in gigantic tides. As the fog dissipated to form clouds, the Sun's rays could reach the surface unfiltered by an ozone (O_3) layer, since oxygen was lacking in the atmosphere. The sunlight contained high-energy ultraviolet rays that would certainly have been lethal to any life form exposed to them. It is difficult to imagine the origin of life under these violent conditions.

A recent scenario for the origin of life suggests that it all began in the deep sea, near sulphur vents (Lane and Martin, 2012). This hypothesis holds that life began from organic macromolecules reacting with each other in a closed, sulphur-rich environment that was relatively stable compared with what was going on at the surface. We speculate that this primordial life was repeatedly brought to the surface by upwelling activity and washed into inland seas by the immense tides. Under these conditions, a sort of 'training' could have taken place by which proto-life forms were exposed under reasonably calm conditions to relatively stable molecular oxygen and highly unstable oxygen and hydroxyl radicals, generated during photolysis of the water by the ionizing radiation. The life forms that were able to utilize oxygen rather than sulphur as a terminal oxidant would have had a great advantage over those which were not, simply due to the immense surplus of oxygen at the surface relative to sulphur. In other words, we envision a two-step origin of life: the primary origin of cellular organisms in the deep sea and the modification of these organisms for survival through the development of circadian gene activity programmes and the ability to detoxify or utilize oxygen and oxygen radicals.

Rainfall and filling of the inland seas together with the high tides could have brought life back to the sea. This means that the capability to deal with oxygen in a way that is not only compatible with but also advantageous to life could have originated long before the photosynthetic origin of atmospheric oxygen (see later in this chapter) and served both inland sea and marine organisms. In the course of this book you will discover the tracks of many of these stages conserved within the evolutionary history of respiration in animals. Then something happened.

About 2 thousand mya, the most complex forms of life were bacteria: among them, photosynthetic cyanobacteria. They were the cutting edge of life back then since they could use the sun's energy to power the assimilation of carbon dioxide and its conversion into metabolic intermediates and structural macromolecules. This process poured out molecular oxygen, which began to accumulate in the atmosphere. By somewhere between 900 and 600 mya it reached about 1 per cent of the present level. This is called the 'Pasteur point of evolution', since it was the famous French chemist and microbiologist Louis Pasteur (1822–1895) who discovered that oxygen paradoxically decreases the first steps of metabolism while at the same time stimulating growth of yeast cells in culture. But this low level of atmospheric oxygen was still sufficient for many organisms to switch to the highly efficient aerobic form of metabolism. As we shall see later, it is just this level of oxygen that is maintained in our tissues and cells to this day: we are carrying with us an ancient environment dating back to the early days of aerobic life.

Among the ancient eubacteria were also ones that had been internalized by archaea. These endosymbionts could utilize not only molecular oxygen but also acetyl coenzyme A: an adenosine-based substance produced by the host. These endosymbiotic bacteria, now called mitochondria, later became assimilated by nucleated cells (eukaryotes) and gave them the ability to produce about 16 times as much ATP from each mole of glucose as could those that didn't have mitochondria. ATP is produced by phosphorylation of adenosine diphosphate (ADP) either anaerobically by glycolysis in the cytosol or aerobically in the mitochondria. Although cyanobacteria initially changed the atmosphere to an oxidizing one, the large amounts of atmospheric oxygen—reaching peak levels of nearly double present-day levels—were later the result of photosynthesis of green plants, which through independent endosymbiotic processes had acquired chloroplasts.

Mitochondria produce ATP as we have already seen, but equally importantly, they remove oxygen from the cytosol and keep the level similar to that prevailing when oxygen began to accumulate in the atmosphere. This is important, because more oxygen results in higher levels of ATP, which in turn puts the brakes on metabolism right at the glycolytic level (Pasteur effect). So these aerobic eukaryotes were exapted for high energy production in a high-oxygen atmosphere that would come later, by keeping the oxygen in the cells at a low, 'ancient' level!

Later, the actual timing is still highly debated, multicellular eukaryotic organisms evolved and what we nowadays call animals (Metazoa) entered the game. And this, in a nutshell, takes us up to the point where the story of the structure, function, and evolution of respiratory faculties in major animal groups begins. So join us on this journey!

We shall be travelling on the evolutionary biology ship, so before embarking we need to have a look at some of the guidelines for a safe and unencumbered journey. First of all we need to use a vocabulary that may not be familiar to everyone and may even appear cumbersome. For example, nothing ever evolves 'for' a specific purpose, so even implications of such teleological thinking are categorically avoided. Having said this, we realize that epigenetics does exist and that many adaptive phenomena appear to conform to teleological guidelines. But even adaptability is genetically anchored and does not escape evolutionary selection.

Another minefield that needs to be briefly touched on here is the species problem. The specific identity of the organisms in question remains uncertain in many cases. We thus frequently also just refer to a certain genus instead of a specific species, well aware that the details in fact may differ between congeneric species. We refer to this as the 'taxonomic uncertainty relation'. The examples provided within the present book are meant to illustrate the diversity of the respiratory faculty exhibited among taxa. In other words, we focus on the big picture rather than being comprehensive with regard to every species-level specialization.

This leads us to the topic of ranks. Many textbooks still use 'phyla' or 'orders' for groupings such as craniotes or turtles, respectively. We completely avoid referring to ranks and adhere to a more phylogenetic framework, which also makes it inappropriate to speak of 'primitive' species or taxa. For example, hagfish are not primitive craniotes, but a basally branching radiation of them. Speaking of craniotes: most commonly this group colloquially is called vertebrates, but we prefer the name Craniota and its derivative craniotes (Lambertz, 2016a). Similarly with hexapods, traditionally called insects, which based on our current understanding are only a subordinated taxon of the Hexapoda. However, you will find the term insects along with hexapods within this book, because most research has been done on 'proper' insects. Having said all of this, we occasionally also deviate from a strictly phylogenetic framework and adopt a more grade-like than clade-like approach if summarizing certain general phenomena, observable in, for instance, 'reptiles' rather than speaking of non-avian and non-mammalian amniotes. The same applies to the usage 'invertebrates' as opposed to non-craniote animals.

So that's it. Let's see where this trip takes us.

CHAPTER 2

A very brief history of respiratory biology

A better understanding of what life is and how living organisms function has always been of crucial importance to humans, but 'biology' as a scientific discipline is quite young, the term being coined around 1800. Similarly, 'respiratory biology' as a discrete branch of biology is much younger and even today the term is not commonly used. However, the study of respiration-relevant processes and phenomena does look back on a long history. In the course of this book we will often give the names of scientists together with their dates of birth and death when their crucial findings are mentioned for the first time. The purpose of the present brief chapter is to place many of the major advances in respiratory biology and the names of key scientists in historical perspective and to point out their importance to the development of what we now call respiratory biology. Clearly this chapter—just like the rest of the book—reflects to some extent the authors' own interests, and we do not intend a comprehensive and unbiased approach. For this the reader is encouraged to consult other excellent works such as those of Fishman and Richards (1982), Otis (1986), Schmidt-Nielsen (1995), West (1996, 2011, 2012, 2015, 2016), Gunga (2009), and Fitzgerald and Cherniack (2012) and the references found there.

The clinical importance of lungs and breathing is recorded in papyrus documents dating back thousands of years, and Aristotle (384–322 BC) postulated vital power in the redness of blood. But the first suggestion that a vital substance is actually in the air taken up in the lungs and transported by blood vessels to the heart and from there to the rest of the body was made by Galen (*c.*129–200 or 216), gladiator surgeon and private physician to the Roman emperor Marcus Aurelius (121–180). The knowledge about structure and function of the human body in general and respiration in particular was codified in the works of the authorities Aristotle and Galen, and new investigation was strongly discouraged in the Christian world, until it surfaced during the Italian Renaissance. In Padua, the Spanish-born surgeon and theologian Miguel Serveto (1509 or 1511–1553), co-founder of Unitarianism, was the first to postulate pulmonary circulation (interestingly, published within his theological *opus magnum* of 1553, for which he was burned at the stake!). The first demonstration of capillaries in general, however, was provided much earlier by none other than Leonardo da Vinci (1452–1519) using wax injections of human cadavers, but it was never published. Among the first to describe the comparative anatomy of the craniote respiratory system was Claude Perrault (1613–1688). In 1669, he published illustrations of some 90 dissections, including chameleons with their aberrant lungs. G. Joseph Duverney (1648–1730) continued this work after Perrault's death.

Meanwhile, in the area of cardiorespiratory physiology, the prediction of blood circulation in 1628 by the English surgeon and comparative physiologist William Harvey (1578–1657) caused quite a stir, because it refuted both the Aristotelian concept that

Respiratory Biology of Animals: Evolutionary and Functional Morphology. Steven F. Perry, Markus Lambertz, Anke Schmitz, Oxford University Press (2019).
DOI: 10.1093/oso/9780199238460.001.0001

blood circulated due to its own vital power and Galen's belief that vitalized, nutrient-rich blood was consumed by body tissues and not returned to the heart. In addition, the idea that the heart was just a pump rather than a reflection of soul and mood just did not sit well. Harvey, a student of Hieronymus Fabricius ab Acquapendente (1533/37–1619), worked in Padua as well, but it was not until in 1661 in Bologna that Marcello Malpighi (1628–1694) actually observed blood entering and leaving a microvascular bed in the frog lung. Unaware of Malpighi's work, Antonie van Leeuwenhoek (1632–1723) made similar observations and another Dutch scientist, Jan Swammerdam (1637–1680), also made substantial pioneering contributions to pulmonary function.

This, together with biomechanical studies of breathing based on the mechanistic philosophical tradition dating back to René Descartes (1596–1650), made possible a purely biophysical approach to respiration and opened the door to mathematical modelling or simulation of gas exchange and breathing mechanics by subsequent generations of researchers. As the pieces were coming together regarding the structure and function of the animal body, including its respiratory faculty, the magic also was slowly disappearing from alchemy. It was known that something required for life was present in air, and for lack of a better word, it was called 'phlogiston'. Comparisons were made between the burning of a candle and respiration, both of which consumed 'phlogistonated' air and resulted in 'fixed' or 'dephlogistonated' air and water. Although the English theologian and chemist Joseph Priestley (1733–1804) is usually credited with the discovery of oxygen, the French chemist Antoine Laurent de Lavoisier (1743–1794) and the Swedish chemist Carl Wilhelm Scheele (1742–1786) may also share this honour.

Heating mercuric oxide had been long known to alchemists as a way of producing quicksilver, but the gas released had been ignored. Now that oxygen had been identified it could be used in experiments. The Russian chemist Dmitri Mendeleev (1834–1907) is usually credited with the development of the periodic table in 1869. However, numerical constants corresponding to atomic weights were known to the British eccentric chemist Henry Cavendish (1731–1810) a century earlier. By careful measurement of weights and volumes, Cavendish discovered that water is the oxide of hydrogen, the gas he is credited with discovering, although he expressed the reaction in terms of phlogiston. So by the end of the eighteenth century it was known that aerobic metabolism was chemically comparable to combustion. On this basis, in 1800 the Cornish chemist, Sir Humphrey Davy (1778–1829), in addition to conducting numerous other experiments that predated modern biochemistry, exposed arterial and venous blood to a vacuum and demonstrated that oxygen and carbon dioxide, respectively, were transported, but it still was a long way to our current understanding of the respiratory process.

In the footsteps of the Swiss Albrecht von Haller (1708–1777), the German Johann Friedrich Blumenbach (1752–1840) was among the leading anatomists and physiologists during the second half of the eighteenth century. In 1787, with the intention of explaining the different kinds of movements in animals, he postulated a *vita propria*, a special or inner life, for the movements of those organs and structures that he was otherwise unable to explain: not unlike Aristotle's vitalist views on the circulation of blood. *Vita propria* was assumed to account for ventilation of the lungs (i.e. spontaneous movement of the lungs was thought to move the ribs and not the reverse). It was actually one of Blumenbach's rather lesser-known students who first challenged this view: the English-born Robert Townson (1762–1827). He re-embarked on the path of Malpighi, Swammerdam, and others and conducted experiments on the respiratory movements in amphibians and turtles during the 1790s. He completely rejected the idea of the *vita propria* and instead declared movements of the throat and those of the lateral flanks, respectively, as responsible for ventilation in these animals (Figure 2.1). As we shall see later on, it was not until the second half of the twentieth century and the use of sophisticated electromyographic recordings that his pioneering work eventually became corroborated and accepted.

The nineteenth century could be considered the golden age of comparative anatomy and embryology. Beginning with the works of the French Georges Cuvier, the German Johann Friedrich Meckel the younger (1781–1833), and the German Martin Heinrich Rathke (1793–1860), accurate descriptions of the developing and mature respiratory organs of

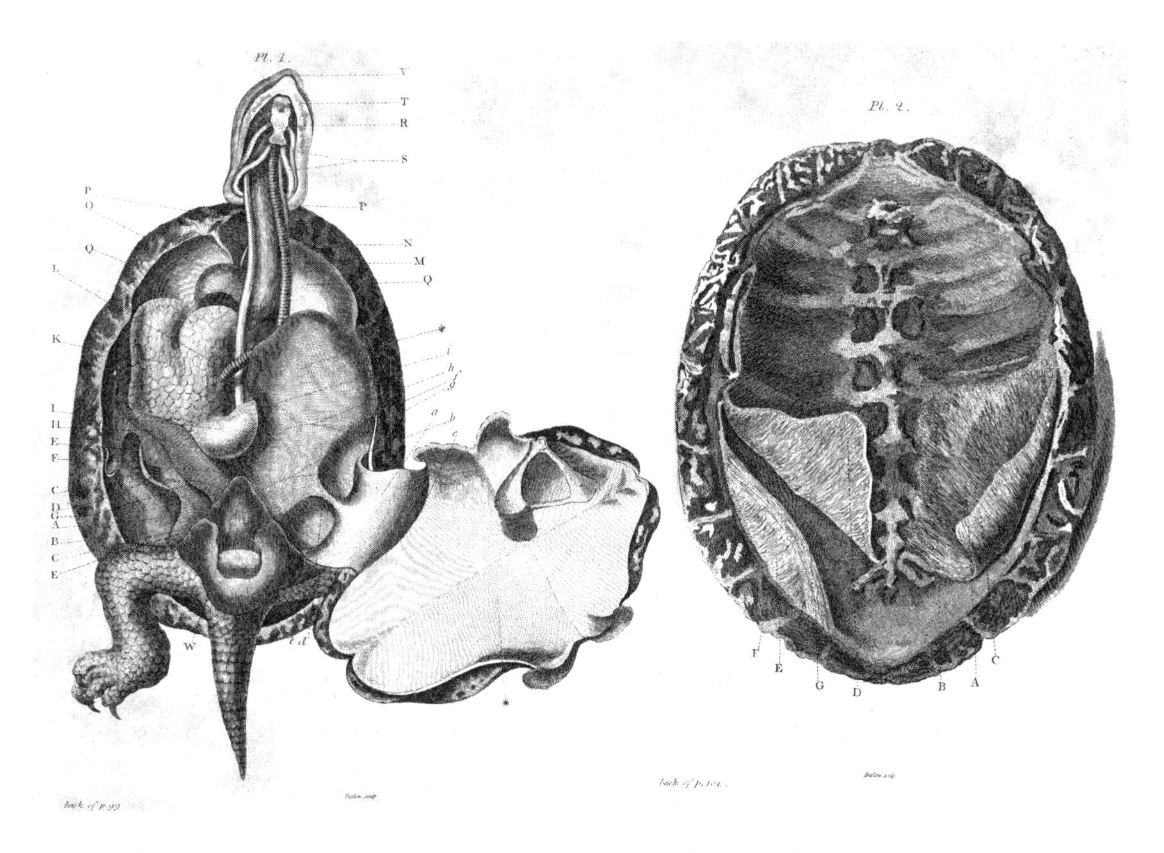

Figure 2.1 Reproduction of the two copper engravings that accompanied the work of Townson (1799) in which he first correctly described the principal mechanism of ventilation in turtles. Collection of M.L.

a large variety of animals accumulated. Numerous studies by others followed, of which particularly the seminal treatise on the mammalian bronchial tree by the Swiss Christoph Theodor Aeby (1835–1885) from 1880 deserves to be mentioned. Many of these old works remain—despite their age—indispensable sources of information for those interested in respiratory organ evolution. A huge problem for many current researchers, however, is that the majority of these studies were published in languages other than English, with rather cumbersome nineteenth-century German as the dominant one. Within this tradition fall also the slightly later embryological works on lungs and swim bladders by Fanny Moser (1872–1953), who eventually became a well-known parapsychologist, as well as those on the lungs and the organization of the coelomic cavity by the Swedish Ivar Broman (1868–1946).

But not only anatomical research progressed immensely during the nineteenth century, the same applies to respiratory physiology as well. One of the founding fathers of modern physiological research, Johannes Müller (1801–1858), for instance, discovered—while still a student (!)—the independent circulation of the mammalian fetus and its implications for respiration of the unborn life. This was carried out at the present authors' home institution, the University of Bonn, which remained an important centre of respiratory-related research throughout the century. Eduard Pflüger (1829–1910), whose name still graces one of the leading physiological journals, made substantial contributions in the field of gas exchange in Bonn. For instance, he took part in an academic dispute with Carl Ludwig (1816–1895) about how oxygen is transported from the lungs into the blood. While Ludwig was an advocate of active oxygen secretion, Pflüger favoured the passive process of diffusion as the sole agent for gas transport. Following up on the work of his former teacher Christian Bohr (1855–1911), who himself actually believed in oxygen secretion, it eventually was the Danish August Krogh (1874–1949), together with his wife Marie (1874–1943), who were able to show that diffusion is actually the driving force of gas exchange. Krogh was awarded the Nobel Prize in 1920, but also Bohr—in spite of being wrong about the normal mechanism of gas uptake—became immortal due to his work on the oxygen transport capabilities of haemoglobin, especially the sigmoid shape of its dissociation curve described in 1903. One of Pflüger's students, Nathan Zuntz (1847–1920), became one of the pioneers in high-altitude physiology, which took the problem of gas exchange into a more applied realm. And John Scott Haldane (1860–1936), a collaborator of Bohr, showed that at least under stress (such as high altitude), oxygen secretion actually can occur.

Research both on the anatomy and physiology of respiration continued throughout the twentieth century and especially the interplay of both disciplines became increasingly acknowledged. Fritz Rohrer (1888–1926) made substantial contributions to flow resistance and the pressure–volume relationships during ventilation in the mammalian lung. Together with the works of his Swiss colleagues Karl Wirz (1896–1978) and Kurt von Neergaard (1887–1947), they laid the groundwork for the mechanics of breathing as we know it today. Wallace Fenn (1893–1971), Hermann Rahn (1912–1990), Jeremiah Mead (1920–2009), and Pierre Dejours (1922–2009), to name only a few protagonists, further advanced the field of respiratory mechanics, gas exchange physiology, and the control of breathing, as well as the comparative aspects. Johannes Piiper (1924–2012), together with Peter Scheid (b. 1938), must be credited with establishing the currently used models of gas exchange, and Carl Gans (1923–2009), as one of the leading functional morphologists of the second half of the twentieth century, resolved many long-lasting controversial debates about breathing mechanisms in a variety of organisms, including those already initiated by Townson. George M. Hughes (1925–2011) also immensely advanced our understanding of how the respiratory system works, particularly that of fish.

Until now we have focused on past contributors, but a few additional names of the 'living history' who carried respiratory biology into the second half of the twentieth century should be mentioned. Hans-Rainer Duncker (b. 1933) and Anthony S. King (1922–2012), for instance, greatly increased our understanding of the anatomy of the avian lung and its intrapulmonary airflow, and the former also made important contributions to the functional significance of the organization and subdivision of the coelomic cavity for the ventilatory process in a large number of craniotes. Both also embarked on the path of quantitative functional anatomy: pulmonary morphometry. This latter field, however, was truly revolutionized by Ewald Weibel (1929–2019) through the first strict application of unbiased stereological approaches in 1963. Beginning with work on the mammalian and avian lung, these approaches were only rather recently applied to other organisms such as fish, reptiles, and a number of invertebrates. John B. West (b. 1928), in addition to his substantial contributions to high-altitude and comparative physiology, must be accredited for being one of the leading historians of our field.

Respiratory biology today continues to be a field of active research, with all of the different levels within anatomy, physiology, and molecular biology. But there are still a plethora of unresolved questions and phenomena out there looking for answers. We hope that reading this book will generate or increase your interest in respiratory biology, and that its history will continue.

CHAPTER 3

A primer on respiratory physiology

As you will notice, this chapter barely cites any references, which we did on purpose in order to maximize readability. We suggest that you consult standard physiological textbooks (e.g. Koeppen and Stanton 2017; Pape et al., 2018) and especially those devoted to comparative or even respiratory physiology (e.g. Hochachka and Somero, 2002; Randall et al., 2002; West and Luks, 2015) for a more in-depth overview of the topic and for access to relevant primary sources.

3.1 Properties of respiratory gases in air and water: the gas laws

The behaviour of respiratory gases (oxygen, carbon dioxide, and also nitrogen) has a great bearing on the respiratory faculties of organisms in the medium they live in. Of particular importance are the interactions of partial pressure, capacitance, solubility, temperature, density, and viscosity, defined and discussed in the following sections.

Life originated in the water and evolved on to dry land. Most major groups of land-based organisms also secondarily gave rise to aquatic forms. Since air and water are drastically different from one another and both usually differ from the internal milieu of animals, specialized mechanisms that favour gas exchange while at the same time protecting the animals from mechanical and osmotic challenges have evolved. In the course of this book we shall see how living organisms have taken advantage of these different properties in the evolution of their respiratory faculties. So let us first have a look at respiratory relevant gases and their inherent physical behaviour.

3.1.1 Pressure

Pressure is a concept we shall run into over and over again, so it is necessary to understand something about it straight away. Pressure is defined as the force applied perpendicular to a surface. The derived SI unit for pressure is the pascal (Pa, named after the French physicist Blaise Pascal (1623–1662) for his work on air pressure; 1 Pa = 1 N m^{-2}). In respiratory physiology we usually use kilopascal (kPa; 1 kPa = 1000 Pa). At sea level, the air pressure is about 101.3 kPa. Older units for pressure that still can be found in the literature are atmosphere (atm), bar, Torr (named after the Italian physicist Evangelista Torricelli (1608–1647), inventor of the mercury barometer), mmHg, or even pounds per square inch (psi): 1 atm = 101.3 kPa; 1 bar = 100 kPa; 1 Torr = 1 mmHg = 0.133 kPa; 1 psi = 6.9 kPa. Especially Torr and mmHg are frequently used in older physiological contexts and it should be noted that the air pressure at sea level equals 760 Torr or mmHg.

3.1.2 Boyle–Mariotte law

This most important gas law is named after its discoverers, the Irish physicist Robert Boyle (1627–1692) and his French colleague, theologist and physicist Edme Mariotte (*c.*1620–1684). It indicates that in a gas, the product of pressure (p) and volume (V) is constant provided that the gas is isolated from the

Respiratory Biology of Animals: Evolutionary and Functional Morphology. Steven F. Perry, Markus Lambertz, Anke Schmitz, Oxford University Press (2019). © Steven F. Perry, Markus Lambertz, & Anke Schmitz.
DOI: 10.1093/oso/9780199238460.001.0001

environment and the temperature remains constant. Thus, at constant temperature, the pressure and volume of a gas are inversely proportional (Eq. 3.1).

$$p \cdot V = \text{const. or, } p \sim 1/V \qquad \text{(Eq. 3.1)}$$

3.1.3 Gay-Lussac's law

According to Joseph Louis Gay-Lussac (1778–1850), a French chemist and physicist, either the pressure or the volume of a gas is directly proportional to the absolute temperature (T) if one or the other is held constant (Eq. 3.2).

$$V \sim T, \text{ if } p = \text{constant, and} \quad p \sim T, \text{ if } V = \text{constant} \qquad \text{(Eq. 3.2)}$$

3.1.4 Avogadro's law

Amedeo Avogadro (1778–1856) was an Italian savant and best known for his work on the molarity of substances. He also formulated Avogadro's law, which states that equal volumes of ideal gases at the same temperature and pressure contain equal numbers of molecules. As an example, equal volumes of molecular oxygen and nitrogen would contain the same number of molecules as long as they are at the same temperature and pressure and conform to ideal gas behaviour (see section 3.1.5). Unfortunately, an ideal gas is only a hypothetical entity, as it is made up of identical particles of zero volume, with no intermolecular forces and with molecules that are in constant random motion. Nevertheless, this principle comes very close to reality. One mole of any ideal gas contains $6.02214076 \cdot 10^{23}$ particles, which amounts to approximately 22.4 l at 0°C (273 K) and 101.3 kPa. In tribute to Avogadro, the number of particles is called Avogadro's number.

The combination of Boyle–Mariotte, Gay-Lussac's, and Avogadro's laws leads to the …

3.1.5 General equation of state for a gas: the ideal gas law

The general equation of state for an ideal gas tells us that, if the temperature and the volume of a gas remain constant, the pressure of the gas is directly proportional to the number of molecules of the gas. Accordingly, if temperature and pressure are kept constant, then the volume of the gas is directly proportional to the number of molecules (Eq. 3.3),

$$p \cdot V \sim n \cdot R \cdot T \sim M \cdot R \cdot T \qquad \text{(Eq. 3.3)}$$

where p is the pressure, V is the volume, n is the number of molecules of a gas or M the mass of a gas, T is the absolute temperature (in K, kelvin) and R is the universal gas constant (8.314 J K^{-1} mol^{-1}).

3.1.6 Dalton's law of partial pressures

The English chemist and physicist John Dalton (1766–1844) is best known as one of the founding fathers of modern atomic theory, but he also contributed fundamentally to the gas laws with his 'law of partial pressures'. This states that in a mixture of different gases, each gas contributes to the total pressure in direct proportion to the relative volume it occupies. Thus the total pressure (p) equals the sum of the individual partial pressures (P) (Eq. 3.4).

$$p_{total} = P_1 + P_2 + \ldots + P_n \qquad \text{(Eq. 3.4)}$$

The physiologically important gases for respiration are oxygen and carbon dioxide, and also nitrogen as a carrier gas. All these gases, together with water vapour and inert gases (mainly argon), are present in atmospheric air. Thus, the total air pressure (p_{air}) is composed of:

$$p_{air} = PN_2 + PO_2 + PCO_2 + P_{\text{inert gases}} \text{ (dry air)}$$

or

$$p_{air} = PN_2 + PO_2 + PCO_2 + P_{\text{inert gases}} + PH_2O \text{ (humidified air)}.$$

The fraction of gases is the same everywhere on Earth, whatever the altitude. But because the total air pressure changes with altitude or depth compared to sea level, the amount of oxygen per litre of gas differs when animals do not live at sea level. High-altitude specializations that compensate for this are found at the molecular, physiological, and anatomical levels (see Chapter 5).

3.1.7 Henry's law

The English chemist William Henry (1775–1836) did fundamental research on gases that are soluble in liquids. The best known of his discoveries is Henry's law (Eq. 3.5), which states that the content of a gas dissolved in a liquid is not equal to, but rather, proportional to the partial pressure of the gas in the gas phase. The proportionality constant is the Bunsen solubility coefficient (α), which varies with the type of gas, the temperature, and the type of liquid (named after the German chemist Robert Wilhelm Bunsen (1811–1899), best known for the invention of the burner that bears his name).

$$C_{gas} = \alpha \cdot P_{gas} \qquad \text{(Eq. 3.5)}$$

Henry's law combines with Dalton's law to yield the Henry–Dalton law. This states that in a mixture of gases, each single gas dissolves in a fluid independently of the other gases according to its partial pressure and solubility coefficient. Thus, although the partial pressure of oxygen in the air and in the water of a nearby pond are equal, the amount of oxygen contained in a litre of water is approximately 30 times less than in air due to the low solubility of oxygen in water.

The Bunsen solubility coefficient is inversely proportional to the temperature and also to the ionic strength of the solvent. This means that in cold, fresh water, α is greater than in warm sea water, and would further decrease with increasing salinity of the sea water. These physical constraints become quite relevant, for example, to inhabitants of tidal pools, which are filled with cold seawater at high tide and exposed to sunlight and dry air as the water recedes. In addition, α is much greater for carbon dioxide than for oxygen and nitrogen. The ratio of α_{CO_2}:α_{O_2} lies between 23 and 35 and is not constant because the decrease in α with increasing temperature is greater for oxygen than for carbon dioxide.

According to Henry's law, the oxygen content (C_{O_2}) of water can be calculated as follows: in a fresh water lake at sea level and at 20°C, 1 litre of water contains 6.38 ml of oxygen ($C_{O_2} = \alpha \cdot PO_2 = 20.73$ kPa $\cdot$ 0.308 ml l^{-1} kPa^{-1}) while at 0°C it would contain 9.97 ml of oxygen. On the other hand, in the ocean, 1 litre of water at 20°C would contain only 5.14 ml of oxygen (20.73 kPa $\cdot$ 0.248 ml l^{-1} kPa^{-1}), and 7.94 ml of oxygen at 0°C.

The calculation of carbon dioxide in water is more complex because of its reaction with water:

$$CO_2 + H_2O \rightleftarrows H_2CO_3 \rightleftarrows H^+ + HCO_3^- \rightleftarrows 2H^+ + CO_3^{2-}$$

So carbon dioxide will become hydrogencarbonate (bicarbonate) or even carbonate to a certain proportion when dissolved in water and will not simply stay as dissolved molecular CO_2.

Henry's law can be used not only in liquids but in modified form also for the medium air (Eq. 3.6). Instead of the solubility coefficient α, the capacity coefficient (β) is used:

$$C_{gas} = \beta \cdot P_{gas} \qquad \text{(Eq. 3.6)}$$

For gases-in-gas, β can be derived from the ideal gas law (Eq. 3.3). The gas content C_{gas} is M/V (in moles per litre) and thus $\beta = C_{gas}/P_{gas} = 1/(R \cdot T)$. So in air, the capacity coefficient for any ideal gas—including oxygen and carbon dioxide—is the same ($\beta_{gas} = \beta_{O_2} = \beta_{CO_2}$).

In fluids, however, β contains not only solubility constants but also includes chemical and physical reaction factors. As we saw previously for the solubility coefficient α, the capacity coefficient β is also much greater for carbon dioxide than for oxygen. This means that 1 litre of air would contain the same amounts of carbon dioxide and oxygen if the partial pressures of those two gases were the same, but that in water under the same conditions, much more carbon dioxide would be dissolved. In blood or haemolymph, chemical binding by respiratory proteins further increases the amount of oxygen and carbon dioxide that can be contained and transported.

3.1.8 Standard gas conditions

For respiratory biologists, it is important to know the conditions under which measurements were carried out in order to compare the results with the data of other experiments on the same or different species. Gas volumes are always reported for specified standard conditions and can be converted into other standard conditions. As the number of molecules per unit volume of a gas depends on pressure, temperature, and humidity, these conditions should always be stated together with the volume of a gas:

ATPS: ambient temperature and pressure, saturated with water vapour.
BTPS: body temperature, atmospheric pressure, saturated with water vapour
STPD: standard temperature pressure dry that means 0°C, 101.3 kPa, no water vapour pressure.

Although experiments are almost never actually conducted under STPD conditions, it is the one most often used for comparative purposes.

3.2 The basics of respiratory physiology

Now that we understand something about the properties of gases we shall be dealing with, let us move on to what is going on inside a living, breathing animal, namely the basics of respiratory physiology. To begin with, oxygen is outside the animal and the aerobically metabolizing cells are somewhere inside. So we have a problem. An organism consists of structures that are nested within one another, sort of like those Russian matryoshka dolls: organism > organ systems > organs > tissues > cells. To get from one organizational level to the other, respiratory gases need to cover appreciable distances and cross membranes. In many cases they cover long distances with the aid of convection, but ultimately diffusion is always involved. In order to understand what all this means and how it is achieved, we need at least some background in basic respiratory physiology.

3.2.1 Diffusion

Diffusion is the central process in respiration, whereby metabolically relevant gases move passively (i.e. without energy-consuming active transport) between an external medium and an internal transport medium in the gas exchanger, and are also released in the tissue and taken up by the cells via diffusion. This sounds like a pretty good deal, but since we all know that there is 'no such thing as a free lunch', let's take a closer look.

There is nothing magical about diffusion, but it is important that we get it right from the beginning. Let us imagine a closed container containing only gas X. As long as the temperature in this container is above absolute zero the gas molecules will move around and bang against each other in what is termed 'perfect elastic collisions'. In 1827, the Scottish botanist Robert Brown (1773–1858) reported seemingly random movement of pollen particles suspended in water. This apparently trivial observation, now called Brown's molecular motion or 'Brownian motion', was followed up on in 1905 by none other than Albert Einstein (1879–1955), who postulated and explained mathematically that the microscopically visible particles were themselves being moved by collision with water molecules. Einstein in fact used Brown's observation to explain the existence of atoms and molecules, and in 1926 the French physicist Jean-Baptiste Perrin (1870–1942) was awarded the Nobel Prize for his 1908 experiments supporting Einstein's interpretation of Brown's observation. Now, if we could label the individual molecules of gas X, we would find that they are moving around the container in such a way that the frequency and random direction of collisions is the same everywhere. In fact it is just this motion and these collisions that account for gas pressure, discussed in the first section of this chapter. The total pressure is caused by collision of all the gases involved, but if different gases are present, then we speak of partial pressures of these gases, reflecting the proportion of a particular gas in the mixture.

So let us introduce a quantity of gas Y into one end of the container. Due to the Brownian motion of all molecules in the container, after a certain time (which depends mainly on the temperature, molecular size, and total pressure), not only the X molecules but also the Y molecules will be evenly distributed. We say that Y got there by diffusion, which is really nothing more than Brownian motion where the starting situation was an unequal distribution. If the starting distribution had been equal, we would not be able to detect diffusion although molecular movement would have taken place.

In the case of a respiratory system, oxygen in the inspired air or water (high PO_2) is separated from the internal medium (low PO_2) by a complex membrane, in which oxygen is more or less soluble. This difference in PO_2 results in a 'driving pressure', implying one-way diffusion. This terminology is indeed unfortunate because it suggests that oxygen is being 'driven' across the membrane, when actually all that is happening is that the statistical probability of an oxygen molecule contacting the air/water side of the membrane is greater than the probability of one hitting the internal side. The greater the difference in partial pressure of oxygen on the two sides of the membrane is, the greater will be the amount that will make it across the membrane in a given time. Since diffusion through a respiratory membrane is very much slower than powered convection (ventilation and perfusion), diffusion is rate determining for the function of the system. A chain is only as strong as its weakest link.

In addition to the driving pressure one has to consider the morphology and biophysics of the membrane: how thick it is, how soluble the gas is in it, how large the gas molecule is, and how it might chemically react with other components of the membrane, to name only the most important variables. Many of these parameters are open to selective pressure and can yield different optimal trade-offs for different species. Once an organism has hit upon an evolutionarily successful solution, this condition tends to be maintained through natural selection. But we shall come back to this later.

The German cardiorespiratory physiologist Adolph Fick (1829–1901) characterized pulmonary diffusion mathematically in 'Fick's first law of diffusion' (Eq. 3.7):

$$\dot{M}_{gas} = d_{gas} \cdot \Delta C \cdot (S / L) \qquad \text{(Eq. 3.7)}$$

This means that the amount of a given gas (M_{gas}) transported per time ($\dot{M}_{gas}$)—normally expressed in moles or in volume (STPD) per minute—increases in direct proportion to the surface area (S) available for diffusion and the concentration gradient ΔC, and in inverse proportion to the thickness of diffusion barrier (L). The diffusion coefficient (d) given in Eq. 3.8 is expressed in terms of area per unit time ($cm^2\ s^{-1}$) and includes properties of the substance, the medium, and the influence of temperature on the diffusive process:

$$d = (R \cdot T) / (f \cdot N) \qquad \text{(Eq. 3.8)}$$

where R is the universal gas constant (see earlier), T the absolute temperature in kelvin, and f is viscous drag ($f = 6 \cdot \pi \cdot \eta \cdot r$, where r is the radius of substance molecules and η is the viscosity of the medium). N is just one of those magical constants that physiologists are always coming up with to make a morphologist's life difficult. It is called 'Loschmidt's number' (named after the Bohemian physicist and chemist Josef Loschmidt (1821–1895)) and it represents the number of molecules per unit volume of an ideal gas (see earlier) at standard temperature and pressure. Sometimes incorrectly referred to as Avogadro's number, N has the value of $2.6867811 \cdot 10^{25}\ m^{-3}$. The T and r factors do not make much difference under conditions compatible with life: 1 per cent for each kelvin, and the size difference between oxygen and carbon dioxide molecules is just a factor of 1.4 in favour of oxygen. But the viscosity is really important since water is about 60 times more viscous than air.

For our special application, respiration, Fick's law of diffusion must be slightly modified. In most physiological situations, the medium in which the respiratory gas diffuses is not the same in both compartments. For example, oxygen diffuses from air into blood in lungs or from water into blood in gills and these media have different capacity coefficients (β_{O_2}). Also, oxygen and carbon dioxide diffuse between 8000 and 9000 times faster in air than in water. This means that the concentration of gas alone cannot predict diffusion.

Partial pressure difference is really the 'driving' force for diffusion of gases and this process can even take place against a concentration gradient. For example, let us consider a trendy oil and vinegar container where the oil floats on top of the vinegar. There is a long, narrow, open spout on the vinegar side and a shorter, corked one on the side opening into the oil. The salesman told you the oil will never go rancid because it has no contact with oxygen in the air. When you pour out oil, air bubbles into the vinegar, not the oil. But he was not aware that he was dealing with a respiratory biologist! You know that β for oxygen is greater in oil than in water (vinegar is about 95 per cent water). Assuming that the oxygen concentration on both liquids is the same to begin with, oxygen will diffuse from the vinegar into the oil until the partial pressures in both liquids are the same, even when the oxygen content in the oil is greater (Figure 3.1). But now back to respiration.

3.2.1.1 Morphological diffusing capacity

Combining Eq. 3.6 and Eq. 3.7 for respiration results in:

$$\dot{M}_{gas} = d_{gas} \cdot \beta \cdot (S / L) \cdot \Delta P_{ga} S \qquad \text{(Eq. 3.9)}$$

We can now combine the two constants ($d_{gas} \cdot \beta$), with this new constant that is called Krogh's constant or the permeability coefficient (K_{gas}). The expression gets simplified to three terms:

$$\dot{M}_{gas} = K_{gas} \cdot (S / L) \cdot \Delta P_{gas} \qquad \text{(Eq. 3.9a)}$$

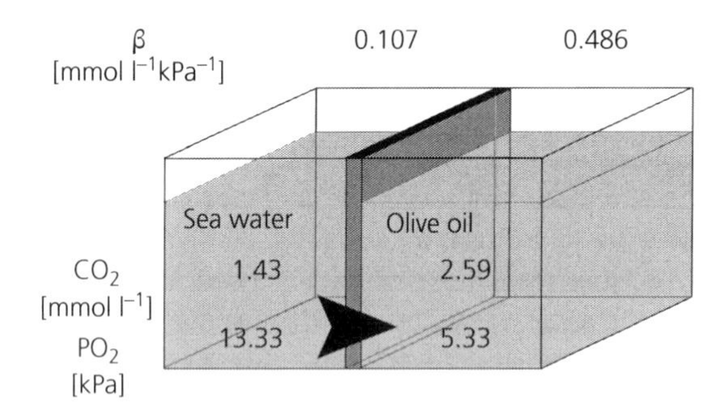

Figure 3.1 Partial pressures (PO_2) and content (CO_2) of oxygen in seawater and olive oil exposed to air at sea level. Note that the both values differ from atmospheric values due to the different capacity constants (βO_2) of all gases involved in the two liquids. Vinegar, mentioned in the text, would be similar to sea water. Original A.S.

Krogh's constant has the units $cm^2\ min^{-1}\ kPa^{-1}$, differs among media and gases, and is also temperature dependent (see Table 3.1 for examples). In general, K is four orders of magnitude greater in air than in fluids or tissues. Because oxygen is slightly smaller and lighter than carbon dioxide, it diffuses a little bit faster in air. But in tissues and fluids, due to the much greater solubility of carbon dioxide in water, the KCO_2 is some 15–40 times greater than the KO_2. In addition, respiratory proteins in body fluids enhance the diffusion of oxygen by binding this gas and effectively taking it out of solution, resulting in a high ΔPO_2. We refer to this phenomenon as diffusion facilitation. Diffusion of carbon dioxide is also facilitated due to the action of carbonic anhydrase, which catalyses the interconversion of carbon dioxide between the gaseous and ionic form, reacting with water and converting carbon dioxide to hydrogencarbonate (HCO_3^-) anions.

Taking another look at Eq. 3.9a we see that three components have quite different qualities: a complex physical constant (K_{gas}), an anatomical component (S/L), and a physiological component (ΔP), all of which contribute in different ways to determine the rate of gas diffusion in and out of an organism and within the tissue.

Table 3.1 Examples for Krogh's diffusion constant K at 20°C. After Altman and Dittmer (1971). Note that the value for chitin, the main component of arthropod cuticle, is an order of magnitude less than for other tissues

	KO_2 [$cm^2\ min^{-1}\ kPa^{-1}$]	KCO_2 [$cm^2\ min^{-1}\ kPa^{-1}$]
Air	$12.6 \cdot 10^{-2}$	$8.5 \cdot 10^{-2}$
Distilled water	$41.9 \cdot 10^{-8}$	$720.6 \cdot 10^{-8}$
Frog muscle	$13.8 \cdot 10^{-8}$	$522.5 \cdot 10^{-8}$
Chitin	$1.28 \cdot 10^{-8}$	N/a

We have already dealt with the physical constants. The anatomical part, often referred to as the anatomical diffusion factor (ADF), is determined microscopically, whereby the thickness (L) is determined as the harmonic mean thickness of the diffusion barrier (τ). The harmonic mean is defined as the reciprocal of the arithmetic mean of the reciprocal values. Since extremely long diffusion distances can occur, they would tend to bias arithmetic mean values in the direction of 'thickness', resulting in systematic overestimates of the diffusion barrier. Using the reciprocal values of diffusion distances is a more meaningful variant, since it means that extremely short distances would result in large numbers, thus biasing harmonic mean values in the direction of thin barriers, and this is exactly what the L factor in Eq. 3.9a is meant to express: thinness rather than thickness.

The new expression for gas transfer now becomes:

$$\dot{M}_{gas} = K_{gas} \cdot (S/\tau) \cdot \Delta P_{gas} \qquad \text{(Eq. 3.10)}$$

or, rearranging,

$$\dot{M}_{gas} / \Delta P_{gas} = K_{gas} \cdot (S/\tau) \qquad \text{(Eq. 3.10a)}$$

It is thus a cornerstone of respiratory biology and tells us how 'good' a gas exchange organ is: describing the 'diffusing capacity'. The diffusing capacity expresses the amount of gas that can be transferred per unit time and partial pressure difference. It is also a sort of 'Rosetta Stone' of respiratory biology, if you will, that should allow the interconversion of anatomy (right side of Eq. 3.10a) and physiology (left side of Eq. 3.10a). Theoretically anyway, but more about that later.

The factors K_{gas} and S/τ are often combined and, more specifically, since this definition contains parameters that must be measured using morphological (histological) techniques, called the morphological diffusing capacity (D_{morph}):

$$D_{morph} = K_{gas} \cdot (S/\tau) \qquad \text{(Eq. 3.10b)}$$

In respiratory organs and in the tissue, the boundaries over which gases have to diffuse often consist of consecutive layers made up of different materials.

In the mammalian lung, for example, oxygen has to pass the surfactant layer, pulmonary epithelium, basal laminae of the epithelium, and the endothelial lining of the blood vessel, the endothelial cell, blood plasma, and the membrane of the red blood cell to enter the erythrocyte. These barriers all have different attributes and are treated as multiple diffusive conductances in series.

Since the barriers are physically measured, we need the morphological diffusing capacity for each component, the reciprocals of which are added in order to determine the total diffusing capacity:

$$1 / D_{morph\ total} = 1 / D_{morph\ 1} + 1 / D_{morph\ 2} + \ldots 1 / D_{morph\ n} \quad \text{(Eq. 3.11)}$$

On the other hand, if we are dealing with respiratory systems that work in parallel to each other, such as the tracheal system and the book lungs of a spider, or the gills, skin, and lungs of a salamander, the diffusing capacity of each respiratory system is calculated separately and simply added up.

3.2.1.2 Physiological diffusing capacity

Until now we have been speaking of morphological diffusing capacity, but it is also possible to measure diffusing capacity physiologically. As already stated, the left-hand side of Eq. 3.10a also defines diffusing capacity, but consists of the physiologically measurable factors, and thus it called the physiological diffusing capacity ($D_{physiol}$):

$$D_{physiol} = \dot{M}_{gas} / \Delta P_{gas} \quad \text{(Eq. 3.10c)}$$

Theoretically, physiological and morphological diffusing capacities should result in identical values, but experimentally this is rarely the case, and the morphological value usually exceeds the physiological one. The reasons for this mismatch have yet to be completely understood and nearly identical values are obtained only in the simplest respiratory systems, such as the skin of lungfish.

Whereas the measurement of oxygen consumption ($\dot{M}O_2$) is routine in most physiological laboratories, the determination of ΔPO_2 is quite another matter. For instance, the tiny capillaries and air spaces in craniote lungs are virtually inaccessible. The PO_2 of the air is known and that of the arterial blood can be assumed to be at full saturation. So most physiologists must content themselves with the partial pressure difference between end-expired air and mixed venous blood (e.g. in the vena cava or sinus venosus of craniotes) to get an estimate of ΔPO_2. Even in the best-case scenario, though, we must assume that gas exchange along the length of capillaries does not take place in a linear fashion, but rather along a sigmoid curve reflecting the oxyhaemoglobin dissociation curve (see Chapter 4). The calculation is complex and requires knowledge of oxygen binding in the blood, haematocrit, and blood flow velocity. The Bohr integration used for estimating gas exchange in the skin, for example, is a technique mastered only by very few living respiratory physiologists.

3.2.2 Convection

A second important physiological process in respiration is the convective transport of gases. Convection facilitates diffusion by maintaining a high ΔP_{gas} at the gas exchanging surface by means of transporting respiratory gases in media (water or air) and in body fluids. These two forms of convective conductance are referred to as ventilation (movement of the respiratory medium) across the respiratory surfaces, and perfusion (movement of the transport medium: blood, haemolymph, or body fluids) within the body between the respiratory organ and the body tissues. Unless the animals rely on natural currents or ventilate as a by-product of their own locomotion, both processes require special energy consumption and can make up a considerable proportion of the total energy needs of the animal. On the other hand, convection allows animals to reach gigantic proportions and still provide their cells with oxygen.

Among craniotes, ventilation is used in lungs and gills. Air is moved in and out of the lungs and water is either propelled over respiratory surfaces of the gills, as in fish, or the gills are moved in the water, as it is the case in the external gills of caudate amphibians. Some researchers have used the term 'irrigation' to refer to ventilation with water, but we shall stick with 'ventilation', regardless of the medium. Perfusion (by a circulatory system) occurs in all craniotes and also in many invertebrates, especially those with respiratory proteins. Small animals or animals without respiratory systems (e.g. sponges and cnidarians) also lack a circulatory system. For

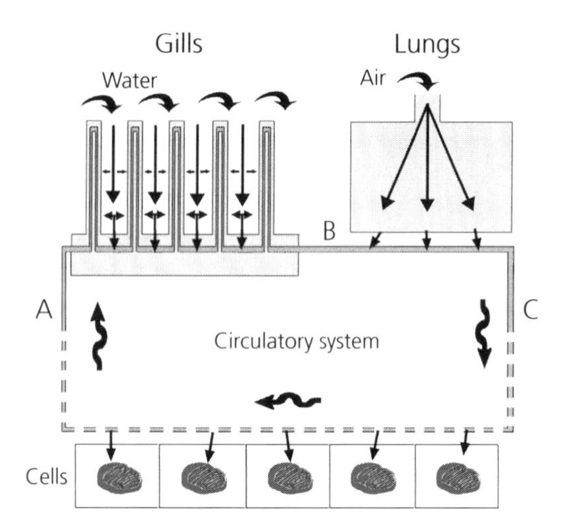

Figure 3.2 Schematic diagram of the pathway of oxygen in animals with a respiratory faculty, in which the circulatory system transports respiratory gases. In invertebrates the heart, in position C, receives haemolymph that has passed through the gas exchanges organs. In craniotes the heart in position A or (if gills are lacking) B receives deoxygenated blood. When true lungs are present, the oxygenated blood is returned to the heart via pulmonary veins and recirculated to the body cells. Except in birds and mammals an admixture of oxygenated and deoxygenated blood at the heart is possible. Original A.S./S.F.P.

animals that use convection of the external media and a circulatory system for the transport of respiratory gases, a schematic overview of the pathway of respiratory gases is given in Figure 3.2.

3.2.2.1 Fick's principle

If you were asked to design a respiratory system, you would certainly have as a major criterion the effective coupling of perfusion and ventilation. In addition, you would require an effective, feedback-controlled, anatomical and physiological relationship between the respiratory system and the metabolizing cells of the body. Based upon the assumption that these relationships exist, the German physiologist Adolf Fick in 1870 used the mathematical coupling of ventilation and perfusion as a non-invasive method for determining cardiac output in humans. The so-called Fick principle, which is not to be confused with Fick's first law of diffusion (Eq. 3.7), is based upon the assumption that the amount of oxygen taken up by the lungs must be the same as that used by the tissues. It also uses the fact that the output of the left and right ventricles in humans and other mammals must be equal due to the complete interatrial and interventricular septations in the heart. That is, in words:

Oxygen consumption = cardiac output · arteriovenous oxygen partial pressure difference,

or

Cardiac output = oxygen consumption / arteriovenous oxygen partial pressure difference.

3.2.3 Gas exchange models

Gas exchange models (Figure 3.3) are simplified mathematical versions of what is going on inside a gas exchange organ. They have the advantage over actual experiments in that one can easily change variables in mathematical formulas and predict how the animal should react to changes in external or internal conditions.

3.2.3.1 The open or infinite pool model

This model applies to cutaneous gas exchange of craniotes and invertebrates. The model assumes that inspired and expired air/water does not actually exist: the surrounding medium is considered for all practical purposes to contain an infinite supply of oxygen and maintains an approximately constant PO_2. The so-called mixed venous blood has a much lower PO_2 than the surrounding medium thus the ΔPO_2 is considerable.

3.2.3.2 Ventilated pool model

In this model, a finite volume (pool) of air is supplied to and removed from the closed respiratory organ. We find the ventilated pool model, for example, in the blindly ending lungs of mammals. Although mathematically similar to the infinite pool model, the ventilated pool relies on active ventilation to maintain an effective partial pressure difference between the medium and the blood in the gas exchange organ. In mammals, around 25 per cent of the inspired oxygen is usually extracted, but the percentage depends on the length of the non-ventilatory period. The maximum PO_2 of the blood is reached when it is in complete equilibrium with the gas contained within the gas exchanger. That is, it will be highest at end expiration, when much of the

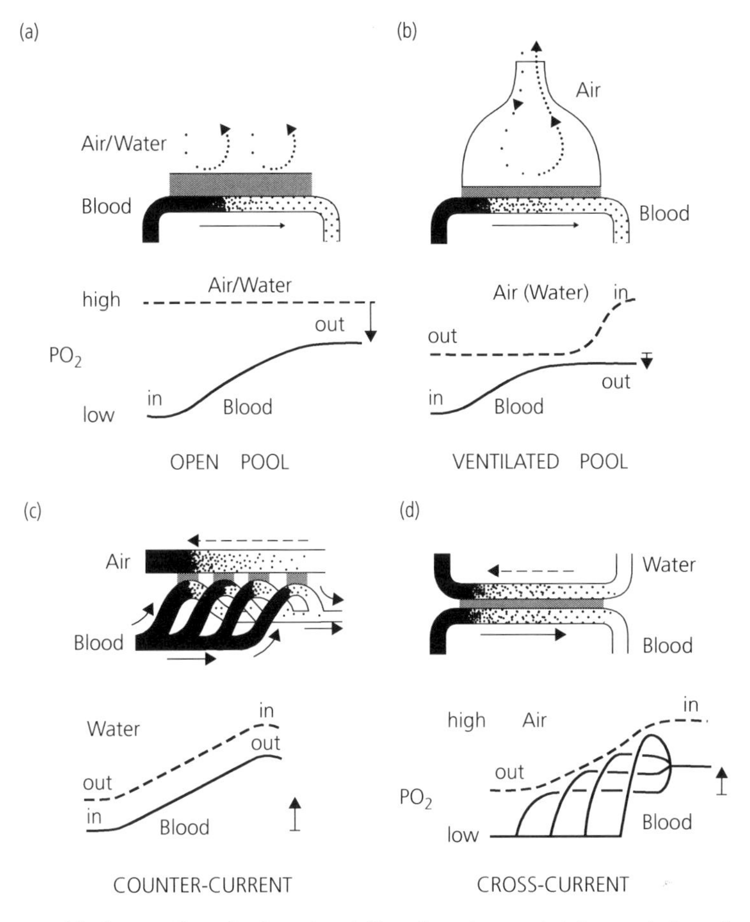

Figure 3.3 Gas exchange models. The upper illustration for each model is a schematic anatomical diagram; the lower illustration, a physiological model. For explanations of the models, see text. After Westheide and Rieger, 2015.

oxygen has already been removed from the pool. The ΔPO_2 behaves just the opposite, being greatest at end inspiration and virtually non-existent at end expiration.

3.2.3.3 Counter-current model

In the counter-current model, the flow of internal medium (haemolymph or blood) and external medium (air or water) are opposed to one another. For simplicity, we shall speak only of blood in this sections, although in invertebrates haemolymph is implied. That means that the first blood entering the system contacts the lowest PO_2 in the external medium, and that the last blood leaving the system has just been exposed to fresh medium with the highest PO_2. In practical terms: the PO_2 of fully oxygenated blood can approach that of the inspired medium and will be greater than that of the medium exiting the system.

Prerequisites for a counter-current system are directed flow both of internal and external mediums. In addition, the thinner the diffusion barriers are, the more efficient gas transfer will be. Although theoretically possible for air breathers, to our knowledge no terrestrial animal has actually evolved a true counter-current system. In the water, however, it is a different story. All fish groups—from hagfish to teleosts and the Australian lungfish—have counter-current systems, the most extreme development being found in high-performance teleosts.

The oxygen extraction from water in teleost fish can reach 80–90 per cent. Counter-current flow maintains a large ΔPO_2, which remains more or less constant across the gas exchange surface of the gill lamellae, and is most effective if the values for oxygen capacity per unit time are similar in blood and water passing the gills: that is, perfusion matches ventilation. Counter-current gas exchange also has been demonstrated in some crabs but with a lower efficiency than in fish. This is mainly due to the properties of the diffusion barrier and not to a less effective counter current.

3.2.3.4 Cross-current model

A situation in which mass blood flow is perpendicular to the medium flow constitutes the cross-current model, which so far has only been demonstrated in birds. In the avian lung, the orientation of the parabronchial tubes with their complex mantle of air capillaries with respect to the blood supply and drainage result in a cross-current physiological model. It should be emphasized, however, that this is a physiological model, not an anatomical one. Blood capillaries do not actually cross the parabronchi. Instead, blood that leaves the lung is a mixture of blood that passed different parts along the length of the parabronchial system. In other words, blood is exposed to different oxygen levels, depending on whether it contacts air from the beginning or the end of the parabronchial system. So the blood collected has different oxygen levels: that from nearest the entry of air into the parabronchi contributes the most oxygen because the ΔPO_2 is greatest there, and blood from the part near to the air exit has a lower PO_2. The efficiency (40–50 per cent) lies between the ventilated pool and the counter-current model; the expired air having about the same or even lower PO_2 than the fully oxygenated blood. The anatomical arrangement seen in birds is unique among extant animals.

3.2.3.5 Tracheal gas exchange in arthropods

There are two models for tracheal gas exchange (Figure 3.4): the terminal gas exchange and the lateral diffusion, or tracheal lung model. The terminal exchange model assumes that all the oxygen enters the spiracles and diffuses to the terminal tracheoles (longitudinal diffusion) where it is consumed by the mitochondria of the metabolizing tissue. The diffusing capacity in this process can be calculated with Eq. 3.12.

$$D_{term} = K_{air}O_2 \cdot (A/L) \qquad \text{(Eq. 3.12)}$$

where D_{term} is the terminal diffusing capacity, $K_{air}O_2$ is Krogh's diffusion constant of oxygen in air, A is

(a)

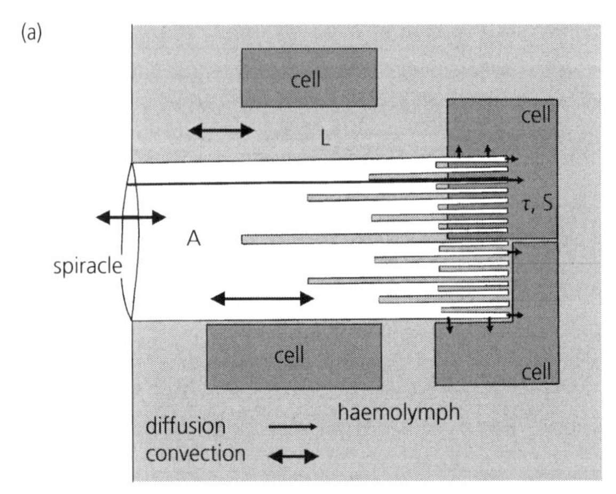

(b)

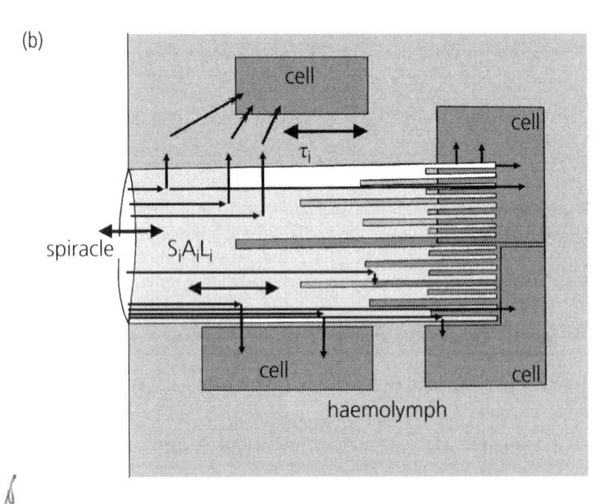

Figure 3.4 Terminal vs. lateral diffusion models for animals with tracheae. In the terminal diffusion model shown in part (a), it is assumed that the surface area (S) at the end of the branching tracheal system can be approximated by the cross-sectional area (A) of the spiracles, and the diffusion distance (L) is the mean length of the tracheae. This model ignores cells that are located long distances from the nearest tracheal terminus. The diffusion distance (τ) is assumed to be negligible. The more general lateral diffusion model in part (b) takes into account diffusion across the tracheal walls of incremental thickness (τ_i) and through tissue to the cells as well as terminal diffusion at the end of the tracheal system. It is assumed that the PO_2 within the tracheal system following inspiration is near atmospheric and that the surface area for diffusion is that of the tracheae. Original A.S.

the sum of the mean cross-sectional areas of all the tracheal orifices (spiracles), and L is the mean length of the tracheae from spiracle to the tracheoles. A is not constant from spiracle to tracheoles in most insects, but tends to increase as one moves down the tracheal system from spiracle to tracheole, so the spiracle is really the bottleneck. The diffusion distance within the tissue is considered to be negligible.

In the lateral diffusion model, oxygen is assumed to diffuse through the tracheal and tracheolar walls into the surrounding haemolymph. The lateral diffusing capacity component is assumed to be particularly great in arachnids and other groups with tracheal lungs, and can be expressed by Eq. 3.13.

$$D_{lat} = K_{trw}O_2 \cdot (S_{tr} / \tau_{tr}) \qquad \text{(Eq. 3.13)}$$

where D_{lat} is the lateral diffusing capacity, $K_{trw}O_2$ is Krogh's diffusion constant for oxygen in the tracheolar walls (cuticle and epidermis), S_{tr} is the surface of the tracheoles, and τ_{tr} is the harmonic mean barrier thickness of the tracheolar walls. This model assumes that gas exchange occurs over all the walls of all tracheae, not just at the terminal tracheoles. The lateral diffusion model is calculated for each point of the tracheal system using Eq. 3.13a.

$$D_{lat} = K_{trw}O_2 \cdot (S_{tri} / \tau_{tri}) \qquad \text{(Eq. 3.13a)}$$

where D_{lat} is the lateral diffusing capacity, $K_{trw}O_2$ is Krogh's diffusion constant of oxygen for the tracheal walls at the given point, S_{tri} is the surface of the tracheae at the given point, and τ_{tri} is the barrier thickness of the tracheal wall at the given point.

The terminal diffusion model holds true for most insects, especially for those with high metabolic rates. The lateral diffusion model on the other hand particularly applies to those arthropods that have tracheal lungs and respiratory proteins in their haemolymph.

3.2.4 Not only respiratory gases diffuse

One parting shot before we leave this topic: a major problem during terrestrialization is that both oxygen and carbon dioxide molecules are larger than water molecules and, like H_2O, are also polarized. This means that any membrane where respiratory gases can diffuse will lose water vapour at an even greater speed. This is not a problem in very humid environments where the ΔPH_2O is small, but for animals living in dry habitats, the openings to the respiratory organs must be protected, countersunk, or both. So it is not surprising that various mechanisms for closing or otherwise protecting the openings of the respiratory system have evolved repeatedly. These are further discussed in Chapter 7. Alternatively, in homeothermic craniotes, water can be condensed out and resorbed when warm, vapour-saturated expired air contacts cooled, moist surfaces in the upper airways.

3.2.5 The complicated story of carbon dioxide

As we shall see, although life originated in an aquatic environment, numerous major groups of multicellular animals also gave rise to terrestrial forms. On dry land, oxygen is abundantly available but carbon dioxide elimination becomes a problem. Perhaps not coincidentally, carbon dioxide, in the form of hydrogencarbonate, becomes a pivotal element of pH regulation in terrestrial organisms in such diverse lineages as snails, crabs, insects, and amniotes, and also serves as a key indicator of gas exchange in general. In addition to the biophysical differences between aquatic and terrestrial living, there are some limiting biochemical ones. Water-breathing animals regulate acid–base balance by modifying the excretion of hydrogencarbonate in expired water, whereas air-breathing animals regulate acid–base balance by modifying the elimination of carbon dioxide in expired air.

It is important to understand that pH is temperature dependent. This could be a problem for animals that live in habitats with wildly fluctuating temperatures, such as tidal pools or shorelines, if their enzymes can only function properly within a narrow pH range. In the 1970s, R.B. Reeves developed the 'imidazol alphastat' hypothesis, which essentially states that the dissociation of imidazol groups in proteins compensates for temperature-caused changes in neutral pH of water and guarantees proper allosteric function of enzymes (Reeves, 1972).

The pH of body fluids is very important to ensure the correct functioning of blood peptides and proteins (especially enzymes such as those involved in regulating blood pressure and blood clotting), antibodies, and respiratory proteins. Animals use homoeostatic mechanisms to maintain a constant blood pH despite temperature fluctuations or influences from respiration or food intake. A constant blood pH also helps to stabilize the pH of other extra- and intracellular body fluids. Animals possess natural pH buffers that compensate for the addition of relatively large amounts of an acid or base. Naturally occurring organic buffers are amino acids, peptides, and proteins—in particular albumin—and, in invertebrates, haemocyanin and hexamerines. The most important inorganic buffers are the hydrogencarbonate and phosphate systems.

In craniotes, the blood pH is slightly basic in amniotes. It is more basic in aquatic animals, reflecting the loss of carbon dioxide to the water. Regulatory mechanisms mainly have to cope with a continuous stream of metabolically produced carbon dioxide as well as transport products of ingested food and metabolic waste. Breathing also may cause a pH change for which the body has to compensate. Respiratory acidosis in amniotes is caused by a reduced carbon dioxide release, normally resulting from reduced ventilation. Carbon dioxide production exceeds the amount released and the pH falls. In contrast, hyperventilation causes an increase in carbon dioxide release relative to production and results in a rise in pH. This is called respiratory alkalosis.

Metabolic processes can also change the pH. Metabolic acidosis occurs, for example, when the H^+ concentration increases due to anaerobic lactate production during heavy exercise. Metabolic alkalosis might be caused by an electrolyte loss (e.g. as a result of vomiting or diarrhoea) or excessive protein ingestion (alkaline tide) and release of nitrate waste products to the blood.

3.2.6 Breathing mechanics

3.2.6.1 Surface tension and lung compliance

We all know that craniote lungs can be inflated and deflated, but the inflation does not occur without some resistance and the lungs themselves often, but not always, aid in their own deflation. The term 'resistance' has many different meanings and connotations. Fortunately, in the area of respiratory mechanics, the inverse concept is used instead: ease of inflation, or compliance.

A pivotal aspect in the mechanics of air breathing is the tendency of wetted, curved surfaces to collapse, explained by the Young–Laplace equation, often referred to by physiologists as the 'law of Laplace'. This states that the tendency of a bubble to collapse is directly proportional to the surface tension and inversely proportional to the radius of curvature. In other words, lungs with a low surface tension and large air spaces have a high compliance.

Compliance is measured in two different ways: as static compliance or dynamic compliance. In static compliance, the lungs are inflated and deflated slowly and stepwise, so that influence of airway resistance to airflow is negligible and the frictional resistance of the lungs and other internal organs is held constant. Static compliance is defined as the slope of the linear portion of the inflation curve of the volume–pressure diagram (Figure 3.5). In dynamic compliance, the lungs are inflated and deflated at frequencies and amplitudes similar to those seen in the living animal and/or as dictated by the experimental protocol. Dynamic compliance is more difficult to measure and to interpret, but better approximates what is going on in the animal. Experimental animals must be paralysed and artificially ventilated but in human experiments, the volunteer can be trained not to resist artificial ventilation. Dynamic compliance is defined as the slope of the elipsoid loops that result from the measurements.

In an experimental setup one can measure compliance of the lung and of the body wall separately. In many mammals, compliance of the lung and body wall are of similar magnitude, but in rodents, rabbits, and newborn babies, the body wall may be more compliant than the lungs. In reptiles and birds, however, the lungs or air sacs are usually at least an order of magnitude more compliant than the body wall.

How is compliance related to the work of breathing? The work of breathing for a single static-compliance inflation–deflation cycle is:

$$w = (½\ C) \cdot \Delta V^2 \qquad \text{(Eq. 3.14)}$$

where w is the work, C is the compliance, and ΔV is the respective change in volume.

For a spontaneously breathing animal, resistance cannot be ignored. Based on the pioneer work of Otis, Mead, and others, Crosfill and Widdicombe

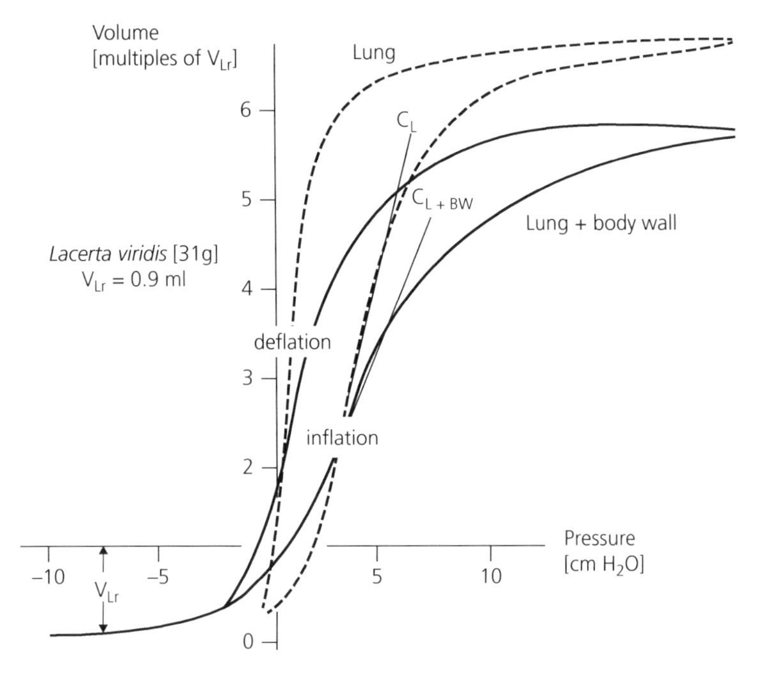

Figure 3.5 A typical static volume-pressure diagram using the European green lizard, *Lacerta viridis*, as an example. The volume unit used—the resting volume (V_{Lr}, residual volume following a maximal inflation manoeuvre)—is expedient when comparing animals of different sizes. Static compliance, C, is measured as the slope of the inflation curve along its straight portion, and is greater for the lung alone (dashed lines, C_L) than for the intact preparation of lung and body wall together (solid lines, C_{L+BW}). The isolated lung needs to be inflated to a greater volume than when it is in the intact preparation in order to reach the upper inflection point of the curve. The area of the space between the inflation and deflation curves (hysteresis) reflects the work lost in the manoeuvre. After Perry and Duncker, 1981

(1961) calculated the work of breathing for different breathing patterns in eight different mammalian species. It turns out that the breathing pattern that mammals habitually use is the one at which the work of breathing is minimal for a given hypothetical metabolic rate. In selected reptiles, experimental work came to a similar conclusion (Milsom and Vitalis, 1984; Vitalis and Milsom, 1986). Most reptiles are episodic breathers and these experiments explain the advantage that certain characteristic breathing patterns during a breathing episode could have.

3.2.6.2 Surfactant and the optimization of breathing mechanics in amniotes

Craniote lungs are derivatives of the gut, an organ system that has had to deal with gas bubbles (e.g. methane, hydrogen, and carbon dioxide) in a liquid medium long before any craniote ever 'thought of' breathing air. To effectively move intestinal contents by peristaltic waves, it is useful to separate the gas from the semi-solid material. But if digestive gas bubbles start out small, Laplace tells us that they cannot get bigger and foam would be maintained. The solution to the problem is to lower the surface tension of the bubbles, which can be achieved by a surfactant. Then as bubbles form, meet, and unite, the small ones with high surface tension will collapse into the larger ones. In fact, such surfactant-like particles secreted by the intestinal epithelium actually contain the major lipid components of pulmonary surfactant and also at least one surfactant protein (Eliakim et al., 1997). Surfactant is produced by all craniote lungs. The 'anti-glue function' of this chemical cocktail may explain how lungs lined with tiny surface area elaborations could have developed in the first place, and remain dynamically stable, inflatable structures without having to be stiff and calcified like the respiratory carapace of a terrestrial crab.

The surfactant produced and released by squamous pulmonary respiratory epithelial cells in lungfish and amphibians and by the special, surfactant-producing type 2 cells in amniotes is composed of protein and lipids. The proteins have multiple structures and functions, including importance in the immune response. For further information the reader is referred to Orgeig et al. (2016). The lipids, on the other hand, contain a small amount of neutral lipids such as cholesterol, which aid in spreading on the aqueous surface film of the lung, but the major component is phospholipids. These detergent-like molecules are characterized by a very low surface tension, and are made up of a polarized head and one or more long, non-polar aliphatic chains. The most common one in surfactant is dipalmitoylphosphatidylcholine.

The phospholipid molecules come to lie on the watery surface film of the lung, their polar heads extending down into the water and the aliphatic chains floating on the surface. This way they create a lipid surface on the water film and drastically reduce the surface tension. When the lung is inflated, the surface area increases but at first the lipid molecules resist movement as they find space to expand onto the new surface. Once the lipid chains have found their space, the lung inflates with increased compliance characteristic of the new surface coating. When the lipid molecules are maximally spread on the surface, further inflation is met by a decreased compliance, reflecting the surface tension of the water openings. When the lungs deflate, the aliphatic tails of the phospholipid molecules are at first scrunched together, resisting reduction in surface area until they find a way to follow the polar heads into the watery film. So at first there is a reduction in pressure with no decrease in surface area and volume, followed by a deflation curve that usually has the same or even greater compliance than for inflation. The deflation curve always lies above

the inflation curve in a standard volume–pressure diagram. In other words, the change in pressure required to increase lung volume by 1 ml is greater than that required to decrease lung volume by the same amount. The space between the inflation and deflation curves is called the hysteresis of the pressure–volume diagram and represents the energy loss or 'cost' of increasing and decreasing lung volume over a given pressure range.

Most reptiles have extremely flexible lungs. Compared with mammals, in which the compliance of the lung and of the body wall are similar, in lizards and snakes the lung is one to two orders of magnitude more compliant than the body wall. At first thought, this does not seem to make much sense but then we remember that surfactant is present inside the lungs but not outside them. The adhesive forces between the lungs and the surrounding viscera *outside* the lung are very much higher than the adhesive forces resisting inflation *inside* the lungs. Unlike mammalian lungs, which maintain a significant residual volume, at least parts of reptilian and amphibian lungs as well as avian air sacs habitually completely collapse and the surfaces stick together. The extreme flexibility of these respiratory structures together with their surfactant makes it possible to reinflate them using buccal or costal breathing alone, without any supplementary respiratory muscles or internal body-cavity septation. When we do see such accessory respiratory muscles we also usually find that these animals are adapted for strenuous and/or continuous activity, and that the lungs show dense packing of respiratory surfaces and relatively low compliance, as in tegu lizards. There we find a post-hepatic septum, to which some diaphragm-like function has been ascribed.

CHAPTER 4

Structure, function, and evolution of respiratory proteins

Respiratory proteins are complexes of proteins and metal ions. In haemoglobin (Hb) the metal is iron, in haemocyanin (Hc)—the most common invertebrate respiratory protein—it is copper. Each respiratory protein has its characteristic colour, which changes with the stage of oxygenation. For this reason they also have been referred to as respiratory pigments. Respiratory proteins are found in two places: in circulating body fluids, such as blood or haemolymph, or in body tissues.

In the former case—circulating proteins—they can be enclosed in cells, such as haemoglobin in craniote red blood cells (erythrocytes), or extracellular and dissolved in the haemolymph, such as haemocyanin in crustaceans and molluscs. These circulating proteins usually serve in taking up, transporting, and releasing respiratory gases: primarily oxygen, but also carbon dioxide. In some cases, they can also serve in gas storage.

In invertebrates, the most common circulating respiratory protein is haemocyanin. These molecules tend to be huge, at least one to two orders of magnitude larger than craniote haemoglobin. Circulating haemoglobin also exists in invertebrates but tends to form very large, complex molecules that consist of smaller subunits, each of which is capable of binding oxygen, meaning that the amount of protein and the number of binding sites is increased without increasing the number of molecules. One exception is the extracellular haemoglobin of certain insects (e.g. chironomid midges) which is less than half the size of craniote haemoglobin.

The latter case—tissue proteins—involves primarily proteins such as myoglobin, neuroglobins, and several others. They are contained in muscle and nervous tissue or, for example, in fat body cells and tracheal cells of insects, where they serve to facilitate removal of oxygen from circulating fluids and to store it in metabolically strategic locations.

4.1 Structure, function, and evolution of respiratory proteins, beginning with haemoglobin

This chapter deals with some pretty complex stuff that we treat in an evolutionary context, making it even more complex. Because of this, we will have to go through it in stages, first dealing with some characteristics of the structure and function of these macromolecules. Since haemoglobin is the best known of the respiratory proteins, we will use it as a case in point. We shall then turn to comparing the structure, function, and evolution of respiratory proteins: globins (in particular haemoglobin and myoglobin), haemocyanin, and haemerythrin.

The perfusing fluids, called blood in craniotes and haemolymph in invertebrates, have many functions, including transport of nutrients, hormones, excretory products, and components of the immune system. And they are important in pH buffering as

Respiratory Biology of Animals: Evolutionary and Functional Morphology. Steven F. Perry, Markus Lambertz, Anke Schmitz, Oxford University Press (2019).
DOI: 10.1093/oso/9780199238460.001.0001

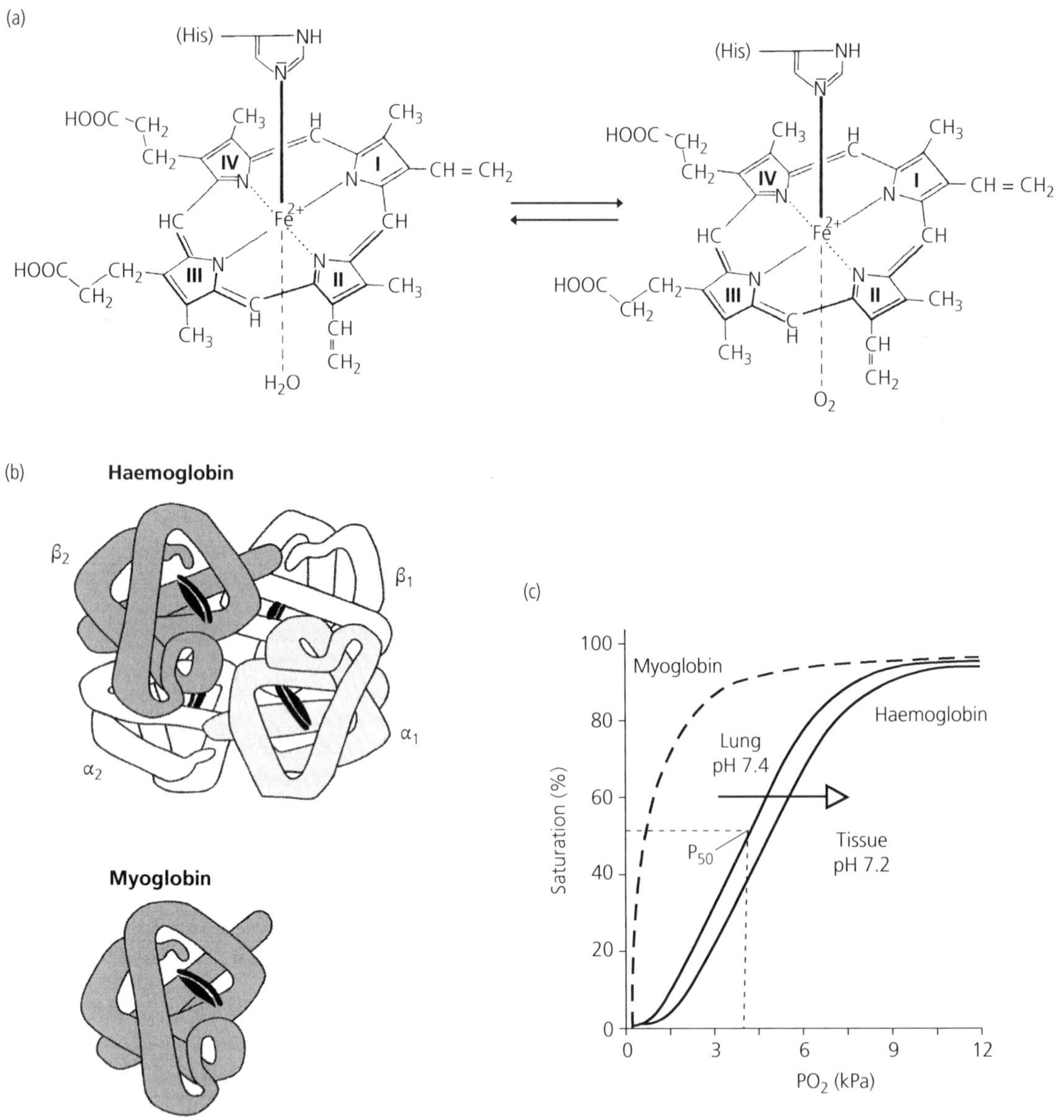

Figure 4.1 Structure and function of haemoglobin and myoglobin. Part (a) shows a haem porphyrin, its binding sites to the Fe^{2+} molecule and to a histidine residue of the α or β protein molecule (see part (b)) that surrounds it. When not binding oxygen, the Fe^{2+} binds water (see text). Part (b) is a sketch of the tertiary structure of a mammalian haemoglobin molecule, showing its four different component protein chains, each of which surrounds a porphyrin, indicated in black. Note the similarity in the tertiary structure of the β_2 molecule and myoglobin, below. Part (c) shows a typical oxygen equilibrium curve (OEC), commonly called a dissociation curve, indicating the PO_2 at half saturation, P_{50}. The shift of the OEC to the right (Bohr effect) is caused by allosteric changes in the tertiary structure of the haemoglobin molecule, resulting in a decreased oxygen affinity (increased P_{50}) at the lower pH in the tissue and release of oxygen. The S-shape (cooperativity) of the haemoglobin OEC caused by the interaction of the α and β proteins and is lacking in the OEC of a mammalian myoglobin, indicated by the dashed line. Modified from Eckert, 1998 and Penzlin, 2005.

well as in repair and healing following injury. But for a respiratory biologist, the uptake, transport, and release of respiratory gases are most important.

As pointed out in Chapter 3, the respiratory gases oxygen and carbon dioxide are quite different regarding their solubilities in aqueous transport media, oxygen being some 30 times less soluble in water than carbon dioxide. But, since the amounts of oxygen consumed and carbon dioxide released are similar, we have a problem: a system that can function efficiently for carbon dioxide will be woefully inadequate for oxygen, and if it works efficiently for oxygen it will be 'over-designed' for dealing with carbon dioxide, and too much may end up getting removed.

Happily, there are some metal ions that can level the playing field. They are relatively abundant in the environment and have the ability to bind and release oxygen. Two of these, iron and copper, are enveloped in protein molecules that can regulate the uptake and release of oxygen in animals. Iron in globins and also, parenthetically, the magnesium in chlorophyll of plants, is held in the middle of a flat molecule called a porphyrin: a 'haem porphyrin' when iron is involved (Figure 4.1a). The porphyrin molecule, in turn, is surrounded by protein or proteins (the 'globin' part of haemoglobin, myoglobin, and all other globins) (Figure 4.1b). But in haemerythrin and haemocyanin porphyrins are not present and the iron and copper ions, respectively, are bound directly to the proteins that surround them (see more detailed descriptions later in this chapter).

Respiratory proteins are characterized physiologically by their ability to reversibly bind oxygen and often also carbon dioxide, thereby temporarily increasing the effective solubility of blood or haemolymph for the transport of these gases. By binding gases, respiratory proteins take them out of solution and in this way also help to maintain a high partial pressure gradient. To function most efficiently, a respiratory transport protein (e.g. haemoglobin or haemocyanin) should bind the gas cooperatively (that is, to show a low affinity at very low ambient oxygen levels and an increasing affinity at higher levels until the molecule approaches saturation) and should be adjustable by effectors. Staying with haemoglobin, these include local pH or PCO_2, or genetically controlled physiological regulators such as 2,3-biphosphoglycerate (2,3-BPG; previously called 2,3 diphosphoglycerate or 2,3 DPG) that can affect oxygen affinity.

Before embarking on our discussion of the evolution of respiratory proteins in animals, we should like to direct your attention to something pointed out in Chapter 1. Once a key function has evolved, the structures around it can change while conserving the function. This appears to have been the case with the metallic porphyrins. Probably it is no coincidence that virtually identical porphyrin-containing molecules (cytochromes, tissue globins such as myoglobin, leghaemoglobin, neuroglobin, and cytoglobin) are found in many different organisms, ranging from bacteria to birds, but may serve quite different functions. One function that oxygen-binding molecules always have, however, is to maintain certain parts of cells or organs in a relatively oxygen-free state. To understand the importance of this function we only need to remember that life originated in a virtually oxygen-free environment and that the initial part of cellular metabolism, namely glycolysis, still requires this ancient condition.

A glance at the plethora of cytochromes—there are more than 1000 variants of cytochrome P450 alone—illustrates the wide variety of functions that haem proteins can assume (Anzenbacher and Anzenbacherova, 2001). In other words, the reverse of the previous statement also applies to haem proteins. While the general structure is conserved, the function can change.

4.2 How respiratory proteins work: haemoglobin as a case in point

As mentioned at the outset of this chapter, respiratory proteins have two major functions: transport and storage of respiratory gases, with an emphasis on oxygen. By binding oxygen in blood or haemolymph, respiratory proteins can increase the amount that can be carried in a liquid transport medium up to 50 times, and by taking oxygen out of solution, they accomplish this while at the same time maintaining a high partial pressure gradient. That is, as far as the circulating fluid is concerned, once the oxygen molecule is bound to the respiratory protein it no longer exerts a 'back pressure' to escape from the circulatory system, and is for all practical purposes 'invisible'. In the body tissues, the conditions are quite different. The surroundings are oxygen poor, carbon dioxide rich, and have a low pH. Oxygen is released, carbon dioxide is taken up, the pH of the blood is lowered, and the blood/haemolymph is returned in this condition to the respiratory organs. Obviously this cannot take place willy-nilly and every step is tightly controlled at all levels. In the sections that follow we shall learn more about how this control takes place at the level of the respiratory proteins and how these extraordinary substances are not only biophysically and biochemically but also physiologically malleable with respect to the specific needs of the animals. How is gas transport adapted to the lifestyle of the animal? And how can we explain the repeated occurrence of a respiratory protein such as haemoglobin when another, such as haemocyanin, is already present?

At least mammalian haemoglobin does possess some characteristics reminiscent of an enzyme and that might help you to understand how it works. One of these is that the globin molecules are allosteric: they change their tertiary structure when they engage in a reversible chemical reaction. In the case of haemoglobin, binding of oxygen increases its affinity for binding further oxygen molecules. In other words, it catalyses its own reaction with oxygen by entering a high-affinity state. But haemoglobin also undergoes allosteric transformation when it reacts with carbon dioxide and returns to a lower-affinity state.

The sigmoid shape of the oxygen equilibrium curve (OEC, see section 4.3) also resembles what you may be familiar with from enzyme kinetics. The affinity decreases at the upper end of the OEC because the substrate (oxygen) binding sites are exhausted as in the case of an enzyme.

4.3 The oxygen equilibrium curve

Oxygen binding and release, regardless of which protein we are dealing with, are best described by the OEC (Figure 4.1c), usually referred to simply as 'the dissociation curve', which plots the percentage of oxygen saturation (or oxygen content in mM l^{-1}) of the respiratory protein as a function of PO_2. In the case of haemoglobin, positive interaction (cooperativity) among subunits of the globins results in a characteristic sigmoid curve. This is because the four-subunit (tetrameric) craniote haemoglobins occur in two characteristic conformations: the low-affinity, T (tense) state, which is constrained by salt bridges and hydrogen bonds and is typical of the low pH, high PCO_2 blood leaving the tissues, and the high-affinity, R (relaxed) state that is found in blood leaving the gas exchange organs, characterized by higher pH and lower PCO_2. But how does the respiratory protein 'know' when and how to enter the T or R state? Is this an all-or-none phenomenon or are there intermediate stages?

4.4 The Hill coefficient

To better understand the T and R states of haemoglobin let us first take a look at the process of transition from one state to the other. Plotted against PO_2, the T-state and R-state curves for percentage saturation lie parallel to each other, the R-state curve lying above the T-state one. Intermediate stages are possible, but not stable. The slope of a straight line lying on a double logarithmic plot of the curve connecting the T state at about 10 per cent saturation with the R state at about 90 per cent saturation on the OEC and measured at the point where the line reaches the 50 per cent saturation level is known as the Hill cooperativity coefficient or simply the 'Hill coefficient' (n_H), named for the British biophysicist Archibald V. Hill (1886–1977), who received the Nobel Prize in 1922 for his work on heat production and mechanical work in muscles. The Hill coefficient is actually given by the slope of the double logarithmic plot: $\log_{[oxy]/[deoxy]}$ when plotted against $\log PO_2$ where the line passes through the 50 per cent saturation point. If the concentrations of oxygenated and deoxygenated haemoglobin are equal, n_H has the value of 1 and no cooperativity exists, as in myoglobin or in shark haemoglobin. For mammalian haemoglobins, n_H lies between 2.4 and 2.9, and some large invertebrate haemoglobins even have a n_H of 5, indicating strong cooperativity.

So the n_H tells us how the respiratory protein dynamically reacts to oxygen: at low PO_2, low affinity, and at higher PO_2, increasing affinity until saturation is approached. This amounts to a certain plasticity with respect to environmental conditions. Somewhere along the X axis, cooperativity will begin, the heterogeneous haemoglobin molecules will interact with each other and the ability of the respiratory protein to bind oxygen will increase. The apparent latency in the beginning of cooperativity helps define a window for oxygen binding, which is relevant for survival. But there is more. The position and distance along the x-axis between 10 per cent and 90 per cent saturation on the y-axis also help define the affinity, or the ability of haemoglobin to bind oxygen at low partial pressures, but it is not the only factor involved. Shark haemoglobin, for example, shows a very high affinity but no cooperativity at all.

Since it is envisioned that the haemoglobin molecule actually evolved several times independently from a more physiologically myoglobin-like one that showed no cooperativity, we must also assume that cooperativity is a derived state that has arisen through mutation and was maintained through selection. In other words, the fact that cooperativity exists at all and evolved independently in haemoglobin and haemocyanin is proof of its importance.

4.5 Affinity and P_{50}

Reduced iron will react with atmospheric oxygen. So if this reaction will take place anyway, what do we need the protein for? The answer to this question is not simple, but can probably be approached best if we think on a geological time scale. Respiratory proteins and life probably originated together and those combinations of iron, copper (or magnesium in chlorophyll) with protein and in many cases also porphyrins that proved compatible with (or beneficial to) the preservation of life were positively selected. As we shall see again in the discussion of the evolution of respiratory systems including the control of breathing, evolution often is concerned with the refinement of controlled inhibition of chemical reactions such as that of oxygen and metal ions or the all-or-none neuronal discharge, resulting in a controlled reaction that fits into the multidimensional complex we call life.

Depending on the metabolic rate of the animal, whether it lives in an oxygen-rich or oxygen-poor environment, and on whether it is breathing air or water, it may be advantageous if the respiratory protein reacts rapidly to low oxygen levels (high affinity) or more slowly (low affinity). Affinity can tell us how fast a respiratory protein can take up oxygen but it does not tell us where on the dissociation curve this takes place. That is, how high does the environmental PO_2 have to be before oxygen can be effectively taken up? This information is given by the P_{50}: the partial pressure at which the respiratory protein is half saturated. A high-affinity protein that has a low P_{50} lies to the left on the OEC plot. If a respiratory protein with the same n_H has a high P_{50} and lies far right on the OEC curve, it can release oxygen rapidly but take it up only at high ambient oxygen levels. A low-affinity respiratory protein, on the other hand, can take up oxygen over a broad range of ambient PO_2.

The PO_2 of the blood is lowest as it enters the gas exchange organ, where the respiratory protein binds oxygen, and rises to a species-specific saturation level in the gas exchanger. In the tissues on the other hand, the PO_2 is low where the respiratory protein releases oxygen. Since oxygen uptake is fastest when the ΔPO_2 across the gas exchange barrier is high and respiratory proteins help maintain a high ΔPO_2, it follows that respiratory proteins are important for efficient gas exchange. When oxygen is bound to the respiratory protein in the gas exchange organ, the ΔPO_2 remains constant until the protein approaches saturation. Conversely, when oxygen has been released from the protein in the tissues, a high ΔPO_2 will exist there as the mitochondria in the cells consume the oxygen. The OEC and the P_{50} can be influenced by all sorts of things, including the type of respiratory protein involved, blood pH, effectors that directly influence oxygen affinity, species-specific differences in respiratory protein structure, age and body size of the animal, temperature, and even the time of day and season of the year.

Species with a high metabolic rate tend to have low-affinity respiratory proteins, which in turn mean that the oxygen is readily released in the tissues, whereas those with low metabolic rates have higher affinity respiratory proteins, making it possible for them to extract oxygen in oxygen-poor environments. The haemoglobin of reptiles, for example, characteristically has a higher oxygen affinity than that of mammals. Within a given taxon of animals, large-bodied species (e.g. elephants) tend to have a lower P_{50} than do small-bodied ones (such as mice).

Mackerels (Scombridae), which are highly active fish with a high metabolic rate and live in oxygen-rich water, have a haemoglobin with low oxygen affinity compared with less active species and/or ones that live in low-oxygen environments such as catfish.

4.6 Effectors, modulators, and their consequences

4.6.1 Organic phosphates

Organic phosphates such as 2,3-BPG in most mammals, ATP and GTP in fish, and inositol pentaphosphate (IP_5) in birds are present in the erythrocytes and reduce the affinity of the respiratory protein. 2,3-BPG is nearly equimolar with haemoglobin in erythrocytes of mammals, its concentration being higher in animals with a generally high oxygen affinity (rats, dogs, and horses) and lower in those with a low-affinity haemoglobin (cats and some artiodactyls such as goats). 2,3-BPG concentration also increases during physiological adaptation to high altitude.

How does 2,3-BPG work? When ambient oxygen levels are low and breathing is stimulated you expire more carbon dioxide than normal: the hydrogencarbonate level in the blood falls and the pH rises. Under these conditions, called respiratory alkalosis, oxygen would not be readily released in the tissues, and the blood would return to the lungs still highly saturated. 2,3-BPG binds to the β subunits of the haemoglobin molecule, and reduces the cooperative interaction of the protein molecules and the oxygen affinity. In this way it compensates for the effect of low PCO_2 by lowering the oxygen affinity of the haemoglobin to what it would have had at lower pH values. As previously mentioned, myoglobin does not show cooperativity and it also does not respond to 2,3-BPG.

4.6.2 Water

Even water can have an allosteric effect on haemoglobin within intact red blood cells, decreasing the oxygen affinity of haemoglobin by binding water molecules upon the transition from the T to the R state (Weber and Fago, 2004). The hagfish *Myxine glutinosa*, a representative of the most basally branching lineage of extant craniotes, is an osmotic conformer whose erythrocytes are able to swell and to remain swollen under hypo-osmotic conditions. When water is bound, the water molecules appear to stabilize the low-affinity, oligomeric state of the haemoglobin. This effect resembles that of the haemoglobin in the marine clam *Anadara broughtonii* (Furuta and Kajita, 1983), but is opposite to what is observed in most other invertebrates (Müller et al., 2003). So in this regard, hagfish haemoglobin appears to be functionally transitional between typical craniote and invertebrate haemoglobins.

4.6.3 pH and the Bohr effect

The reduction in oxygen affinity caused by a lowered pH, usually the result of increased PCO_2, is called the Bohr effect, or Bohr shift, named after the Danish physiologist Christian Bohr (1855–1911). To give an example: decreasing the pH by 0.2 units can reduce the oxygen affinity by 20 per cent. In addition, carbon dioxide binds to the haemoglobin molecule (see Haldane effect, in section 4.6.7.) causing allosteric changes that directly reduce oxygen affinity and resulting in oxygen release. The carbon dioxide produced by respiration therefore helps release oxygen in the tissues, and in the respiratory organs it aids oxygen uptake when the carbon dioxide is released from the haemoglobin into the medium. The effect of the pH is not the same in all animals: for example, in mammals it correlates inversely with the body size of an animal, the effect being greater in small animals than large ones. Teleologically speaking, this only 'makes sense' since small animals also need rapid oxygen delivery to the tissues because of their high mass-specific metabolic rate.

4.6.4 Negative Bohr effect

Compared with haemoglobin, myoglobin is relatively insensitive to pH changes, but haemocyanin (e.g. in crustaceans) does exhibit a Bohr effect. In the chelicerate horseshoe crab, *Limulus polyphemus*, and in some gastropod snails and even in some fish that inhabit hypoxic waters, a reversed (also referred to as 'cathodic') Bohr effect has been reported: a decrease in pH actually results in an increase in oxygen affinity (Breepoel et al., 1980; Weber, 1997). This phenomenon may exacerbate oxygen uptake under hypoxic conditions when anaerobic metabolism results in a low pH of the circulating medium.

4.6.5 Root effect

The Root effect, a phenomenon that originated and evolved separately among teleost fishes, cephalopod molluscs, and crustaceans was first described in 1931 by R.W. Root. It is characterized by a strong reduction of the oxygen-carrying capacity of blood/haemolymph, caused by an extreme sensitivity to a decrease in pH. Oxygen binding is so reduced that 100 per cent saturation cannot be even theoretically attained.

In physoclist fish (those with no connection of the swim bladder to the pharynx), the Root effect is instrumental in the release of oxygen into the swim bladder following a controlled release of lactic acid through the local activity of the enzyme lactate dehydrogenase. Another example is the retina in the fish eye, the organ with the highest metabolic rate. Since the eye is very sensitive to pH changes (a pH of less than 6.4 causes blindness), it is absolutely necessary to avoid excessive hypoxia.

In the pseudobranch (see Chapter 11), oxygen is secreted into the efferent vessels that supply the retina. Then the blood entering the choroid plexus beneath the retina is acidified by lactate (released by local activity of lactate dehydrogenase) and by carbon dioxide from the retina's metabolism: the Root effect drives the oxygen out of the blood. So the pseudobranch finely tunes conditions in the retina without running the risk of pH there becoming too low (Bridges et al., 1998).

4.6.6 Structural changes

Actual changes in the amino acid composition of craniote haemoglobin can also play a role in the adaptive response to high-altitude hypoxia. Transport of oxygen by haemoglobin is a combination of the intrinsic oxygen affinity and its allosteric interaction with cellular effectors. Whereas short-term *acclimatization* to high altitude is dominated by allosteric interactions, long-term *adaptations* are genetically coded and involve changes in the structure of haemoglobin molecules. Such changes comprise substitutions of amino acid residues at the effector binding site, that is, where the haem contacts the subunits of the protein molecule and stabilizes low-affinity or high-affinity structures. Also, molecular heterogeneity can be present in the form of coexistence of different haemoglobins within a single individual, resulting in a broad spectrum of oxygen-binding properties and a correspondingly great adaptability of the organism (Weber, 2007).

Animals living constantly at high altitude, such as the lama and the bar-headed goose, often possess high-affinity haemoglobins compared with other species, but during development of an individual, a change in haemoglobin structure can also take place. A mammalian fetus, for example, must take up oxygen under physiologically adverse conditions. A unique heterogeneity of the globin proteins allows fetal blood to bind oxygen better than maternal blood. How this is accomplished at the molecular level is discussed under 'Craniote haemoglobin' (see section 4.7.1).

4.6.7 Carbon dioxide transport by haemoglobin and the Haldane effect

The blood transports the carbon dioxide, both in the plasma and in the blood cells (Figure 4.2). Carbon dioxide reacts with water in the plasma to form carbonic acid (H_2CO_3) that readily dissociates into hydrogencarbonate (HCO_3^-) and carbonate (CO_3^{2-}) ions. In addition, carbon dioxide reacts with the hydroxyl ions to form hydrogencarbonate. The proportion of the carbon dioxide present in the ionic form depends on several mutually interacting factors, which include pH, temperature, and ionic strength of the solution. In mammalian blood, hydrogencarbonate is the dominant ionic form. At pH 7.4, the ratio of carbonic acid to hydrogencarbonate ions is 1:20.

In the tissue, carbon dioxide enters and leaves the blood as molecular carbon dioxide, which has a much higher diffusion coefficient than the ionic form. But once in the blood, carbon dioxide becomes ionized, allowing more carbon dioxide to enter. Now the enzyme carbonic anhydrase enters the picture. It is one of the fastest enzymes known and catalyses the interconversion of carbon dioxide and hydrogencarbonate. It is present in blood vessel endothelium and in erythrocytes but not in blood plasma, so most of the carbon dioxide is transported in the red blood cells as hydrogencarbonate. In mammals, reptiles, and birds it also binds to the free

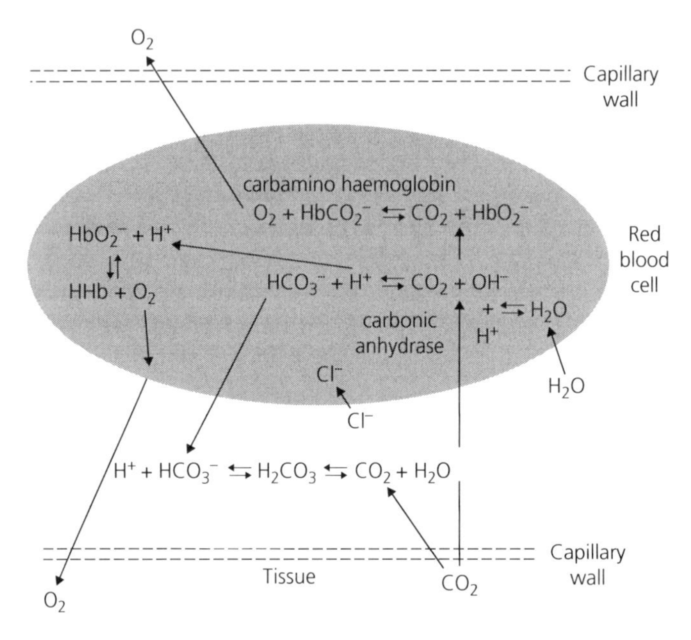

Figure 4.2 A schematic mammalian red blood cell (erythrocyte), showing some of the synergistic effects involved in oxygen release and carbon dioxide uptake and transport involved in the Haldane effect (see text). Modified after Eckert, 1998.

NH_2 groups of the globin (*not* to the binding site of oxygen) forming a negatively charged carbamate. This change from a positive to a negative charge favours the allosteric conformation of the low-affinity T state. In fish and amphibians this binding cannot happen because the amino groups of the globin are not free, but acetylated.

So far so good, but now things become complicated. Carbonate and hydrogencarbonate 'should' leave the erythrocyte in the lung because of the concentration difference relative to the plasma, but the membrane of the erythrocytes is only permeable to the anion HCO_3^- and not to the H^+ cation. This results not only in an ionic disequilibrium but also in a surplus of H^+ ions in the erythrocytes, which would cause a drastic fall in pH, were it not for an influx of chloride ions (Cl^-). This so-called chloride shift effectively neutralizes the acid (note that HCl is a strong acid, whereas H_2CO_3 is a weak one). The loss of chloride ions from the plasma is only temporary, since the whole process is reversed when carbon dioxide is released in the lung (Figure 4.2).

The allosteric effects of oxygenation of haemoglobin cause H^+ release, which in turn shifts the hydrogencarbonate buffer equilibrium within the erythrocytes towards carbon dioxide formation, facilitating carbon dioxide release in the lung. Deoxygenation, on the other hand, causes H^+ to bind to haemoglobin, reducing the change in pH in the surrounding tissue and facilitates carbon dioxide entry into the blood. This phenomenon together with allosteric facilitation of oxygen release associated with HCO_3^- and carbamate formation in the erythrocytes is collectively known as the 'Haldane effect', named for the Scottish pioneer in respiratory biology John Scott Haldane (1860–1936).

4.7 Evolution of haemoglobins and myoglobins

Haemoglobins are phylogenetically ancient molecules, as demonstrated by their widespread occurrence in archaea, eubacteria, protists, fungi, plants, and animals (Hardison, 1996; Blank and Burmester, 2012). The ancestral globin gene appears to have evolved at least 1800 million years ago (mya) and was already present when oxygen started to accumulate in the atmosphere. This suggests that one important property from the very beginning of life was to bind toxic oxygen, carbon monoxide, and nitric oxide. This binding function, which was exapted to this day, keeps the metabolic machinery of the cell in an almost oxygen-free state: a prerequisite for glycolysis to take place. While almost all craniotes possess tetrameric haemoglobin and monomeric myoglobin, non-craniote haemoglobins present a huge variety of structures and functions.

Evolution of respiratory proteins is difficult to explain, as haemoglobins and myoglobins seem to be distributed unsystematically among a number of major animal taxa, but they do not necessarily have respiratory function. This is due to the universally present porphyrin part of the molecule, which is found in plants and animals including cellular enzymes and coenzymes known as cytochromes. With only minor changes such a molecule can be changed into one with respiratory transport or storage function, or the reverse can happen. Haemoglobin and myoglobin evolved from tissue globins several times independently probably because there was the 'need' for a respiratory transport protein.

Haemoglobin also binds and releases nitric oxide (NO), which is now recognized to be pivotal for some key physiological processes such as vasodilation, neurotransmission, and immune defence. Myoglobin, which is much smaller than haemoglobin, also can cause diffusion facilitation: by binding oxygen, it reduces the PO_2 at the end of the diffusion pathway thereby increasing the partial pressure difference, the so-called driving pressure.

4.7.1 Craniote haemoglobins

4.7.1.1 Myxinoida

An extant representative of the earliest branching group of craniotes, the hagfish *Myxine glutinosa*, has haemoglobin consisting of three monomeric globins, which form heterodimers (three groups of two different protein pairs) and heterotetramers (three groups of four different proteins) when deoxygenated. This is so different from that of gnathostomes, that it most probably originated separately (Schwarze et al., 2014).

4.7.1.2 Petromyzontida

The blood haemoglobin of the lamprey *Petromyzon marinus* also differs radically from that of gnathostomes discussed in the following section. It has a molecular weight of only 17 kDa, like that of muscle haemoglobin, contains only one haem, and its amino acid composition bears similarities both to invertebrate (nematode and insect) haemoglobins and craniote ones that serve primarily in oxygen storage rather than transport. Its OEC is parabolic, like myoglobin or shark haemoglobin, but has a low oxygen affinity and a very large Bohr effect, making it as effective an oxygen transporter as most craniote blood haemoglobins (Wald and Riggs, 1951).

4.7.1.3 Gnathostomata

The haemoglobin of all jawed craniotes is probably of common origin. It is made up of two dimers, each of which consists of two subunits, an alpha (α) and a beta (β) chain, and has a total molecular weight of around 68 kDa. The dimers are designated $\alpha_1\beta_1$ and $\alpha_2\beta_2$, respectively. Each of the four globin chains contains a porphyrin, which, in turn, binds an iron molecule in its ferrous state (Fe^{2+}). This iron makes four links with the porphyrin molecule, one with each of the pyrrole nitrogens of the two subunits, one link with the histidine of the globin, and one link with the oxygen. Doing our sums, we come up with four oxygen molecules that can be bound reversibly to each haemoglobin.

The αβ dimers are connected to each other by salt bridges that are altered by oxygenation, resulting in conformational changes (cooperativity) that facilitate the binding of more oxygen, as discussed at the outset of this chapter. It is important to note that the iron molecules do not actually become oxidized, but rather that the oxygen is bound reversibly and the iron remains in its ferrous state. Oxidation to the ferric state would result in non-functional methaemoglobin (pronounced methaemoglobin). If this happens—which it fairly often does—the oxidation is reversed by the enzyme methaemoglobin reductase, which reinstates the ferrous state.

Another form of the haemoglobin is the carboxyhaemoglobin, which results from the binding of carbon monoxide to haemoglobin. Since the affinity of haemoglobin for carbon monoxide is some 200 times greater than for oxygen, it is very difficult to displace. Carbon monoxide poisoning is, of course, well known to anyone who watches cops and robbers programmes on TV, but even ambient levels of carbon monoxide in city traffic can impair brain function.

The change of a single amino acid in a globin molecule can dramatically alter the properties of the haemoglobin. One example is mammalian fetal haemoglobin. There, the β chain is replaced by a gamma (γ) chain. As opposed to β, γ is not capable of binding 2,3-BPG, which would normally reduce the oxygen affinity of haemoglobin, particularly under conditions of low ambient PO_2. This change results in the high oxygen affinity in fetal haemoglobin, which is necessary in order to extract oxygen from the placenta. Another example is sickle cell anaemia, in which the exchange of a single amino acid in the β chain causes hydrophobic association of haemoglobin molecules, which aggregate to bundles and cause the erythrocytes to change their form from a biconcave disc to a sickle-like shape. In humans, the homozygotic state for this change normally results in death during childhood. But in the heterozygotic state, the deformation of erythrocytes is only slight but sufficient to inhibit the proliferation of *Malaria tropica* in the erythrocytes, and infected red blood cells are cleared in the spleen.

Birds show some particularly interesting haemoglobin constellations. Some species residing at high altitude show haemoglobins with different amino acid sequences than those living at lower altitudes. In addition, some haemoglobins of amphibians, reptiles, and birds exhibit a super-cooperativity of oxygen binding, which depends on reversible deoxygenation-dependent tetramer–tetramer association to form an assemblage with a very low oxygen affinity. The modified oxygen-binding curve results in an increase in the amount of oxygen delivered to the tissues, which in birds could be especially valuable during high altitude migrations. We shall return to this topic in Chapter 5.

4.7.1.4 Craniote red blood cells: solution to the small-haemoglobin-molecule problem

Non-craniote chordates—tunicates (Tunicata) and amphioxus (Cephalochordata)—completely lack red blood cells, and although a large number of different globins are found they appear to have no respiratory function (Ebner et al., 2010). A circulating, intracellular vanadium-rich substance is present, but it is not known to transport respiratory gases.

Compared with haemocyanins and most annelid haemoglobins, craniote haemoglobin is a small molecule. From the very beginning, craniotes encapsulated the haemoglobin in specialized blood cells: the erythrocytes. If craniote haemoglobin were freely dissolved in a closed circulatory system, this would result in a plethora of problems ranging from osmotic stress to clogging the ultrafiltration apparatus of the kidneys as seen, for example, in pathological haemolysis, in which red blood cells rupture. The large size of haemocyanin molecules, particularly molluscan haemocyanin, all but eliminates these problems in these animals. Craniotes took another path. If free haemoglobin does end up in the circulatory system, damage can be avoided by binding it to a class of so-called scavenger proteins called haptoglobins. In general, this mechanism involves a primary molecular sequestering of the haemoglobin. This keeps it away from sensitive sites such as vascular endothelia. Later, the haemoglobin–haptoglobin complex is taken up by macrophages/monocytes and eventually ends up in the liver, where the porphyrin molecule with its bound oxygen is converted to carbon monoxide, bilirubin (a bile component), and iron, which is recycled for haemoglobin neogenesis.

In all craniotes except mammals, the erythrocytes are primarily nucleated. Lungless salamanders (Plethodontidae) and other amphibians have fragmenting red blood cells but only mammals have intact, nucleus-free erythrocytes in healthy animals. They are usually also biconcave, except in many ruminants, where they may have unusual shapes, ranging from polygonal in the elk (*Cervus elaphus*) to spherical in the mouse deer (*Tragulus kanchil*). In camels and their close relatives, they are fusiform like the nucleated erythrocytes of non-mammalian craniotes. That is why you sometimes see the false statement that camel erythrocytes are nucleated.

There are many 'just-so stories' that explain 'why' mammalian erythrocytes 'must' lack nuclei to support a high metabolic rate, but bird blood has the same oxygen-carrying capacity as in mammals in spite of nucleated avian erythrocytes (Hawkey et al., 1991) and birds can aerobically outperform any mammal. So we just don't know 'why' mammalian erythrocytes are enucleate.

4.7.1.5 Craniotes that lack respiratory proteins

Some teleost fishes, most prominently icefishes (Notothenioidei), living at temperatures around 0°C and reaching body lengths up to 70 cm, completely lack haemoglobin. Some icefish even lack myoglobin as well. Low body temperature increases the solubility of oxygen in the blood plasma but also increases the viscosity of blood and stresses the heart muscles. One compensatory strategy is to drastically reduce the number of erythrocytes or to eliminate them entirely, thereby reducing blood viscosity, and at the same time to increase blood and heart volume. This strategy is seen not only in icefish, but also independently in the leptocephalus larvae of eels (Anguillidae) and also the larvae of other species, in particular smelts (Galaxiidae), which appear to use transparency as a mechanism to avoid predation (Busse et al., 2006). In these fishes, the heart actually decreases in size as the animals acquire haemoglobin during metamorphosis of the larvae to the adult form.

4.7.2 Non-craniote haemoglobins

Non-craniote haemoglobins range from single-chained globins in protists to large multisubunit and multidomain molecules found in annelids, nematodes, crustaceans, and molluscs. Haemoglobins are found both in the tissue and as circulating haemoglobins, but haemoglobin-containing circulating cells (haemocytes) are rare and never attain the functional significance of craniote erythrocytes. It should be noted that a myoglobin-like intracellular haemoglobin is also present in the tissues of free-living flatworms (Plathelminthes) (van Holde, 1997; Mangum, 1998).

4.7.2.1 Annelida

The extracellular giant haemoglobin of some annelids has as many as 144 binding sites for oxygen. Annelid haemoglobins and myoglobins exist as circulating intracellular haemoglobin, non-circulating intracellular haemoglobin, myoglobin, and extracellular haemoglobin, all having most probably evolved from the same ancestral globin (Bailly et al., 2007).

Chlorocruorin is a low-affinity, green-coloured respiratory protein found in the coelomic fluid and in the blood of annelids. It has a molecular weight of 3400 kDa and its molecular structure is so similar to that of haemoglobin that it is now widely accepted as actually a special form of haemoglobin.

4.7.2.2 Mollusca

In molluscs, haemoglobin has been reported in all major taxa except the Cephalopoda and the Monoplacophora. Very often haemoglobin occurs in one species but is lacking in others of the same genus, and its occurrence does not necessarily correlate with the environmental conditions or with the physiology of a given species (Terwilliger and Terwilliger, 1985; Terwilliger, 1998). To make matters worse, there are four quite different amino acid sequences found in gastropod haemoglobins, but they are not always homologous, even among members of the same taxon. Since globins have been around so long and are present in virtually every life form, it has been suspected that numerous globin templates are present in the gastropod genome and are activated at random or perhaps were acquired independently. Closer examination, however, reveals that haemoglobins and myoglobins actually evolved according to the phylogeny of the species. The trick is that haemoglobin and myoglobin can evolve quite easily from a tissue globin, which did not have necessarily a respiratory function.

Invertebrate haemoglobins not only are used in transport and storage of oxygen, but also transport sulphide, recently revealed to have reactive oxygen species (ROS) activity (Hancock and Whiteman, 2016) and play a role in maintenance of acid–base balance, oxygen scavenging, and oxygen sensing. They occur in gills or in muscles and especially haemoglobins in haemolymph-bourne haematocytes may support metabolism at low ambient PO_2. In gill tissue, haemoglobins not only facilitate oxygen uptake by the organism, but also help support the metabolism of symbiotic bacteria, as in the deep-sea vent caenogastropod *Alviniconcha hessleri*.

Evolution of respiratory faculties often works in strange ways: here a case in point. Sure, molluscs have haemocyanin but there often seems to be a functional trade-off with haemoglobin. The freshwater pulmonate ramshorn snail *Biomphalaria glabrata* (Planorbidae), which inhabits oxygen-deficient water, actually evolved a new haemoglobin from its molluscan myoglobin. This haemoglobin, with more than ten globin domains and a molecular weight of more than twice that of craniote haemoglobin, seems to have replaced haemocyanin functionally.

In support of this idea is the fact that *Biomphalaria* has remnants of haemocyanin, which no longer are able to bind oxygen (Lieb et al., 2006). The selective advantage probably lies in its extremely high oxygen affinity, with a P_{50} of 0.8 kPa. By comparison, the pulmonate genus *Lymnaea* also inhabits stagnant waters, but has only a haemocyanin of much lower oxygen affinity. (How about a new Gary Larson cartoon of *Lymnaea* trying to bribe the haemoglobin gene off *Biomphalaria*?) Similarly, the deep-sea clam *Calyptogena kaikoi*, which also lives under hypoxic conditions, has extracellular haemoglobin, but uses this myoglobin-like protein for oxygen storage rather than for circulating it (Mangum, 1986; Lieb et al., 2006).

4.7.2.3 Crustacea

Crustacean haemoglobins are always extracellular and are found in Branchiopoda, Ostracoda, Copepoda, rhizocephalan Cirripedia, and one group of amphipodan Malacostraca. Haemoglobin is best known in the branchiopods *Daphnia*, *Artemia*, and *Triops*. In the water flea *Daphnia*, haemoglobin has a molecular weight about seven times that of craniote haemoglobin and consists of 16 subunits, each containing two haemes. Gene expression is inversely related to the ambient PO_2, resulting in intracellular concentration between 1 and 17 g l^{-1}. The colour of *Daphnia* changes accordingly from colourless to deep red, reaching a maximum after about 12 days of hypoxia. But that is not all: the subunit composition also changes, resulting in a lower P_{50} at an unchanged n_H of the OEC (Kobayashi et al., 1988; Paul and Pirow, 1997). Unlike *Daphnia* and the brine shrimp *Artemia*, the hypoxic response of haemoglobin in the tadpole shrimp *Triops* is not reversible (Guadagnoli et al., 2005).

4.7.2.4 Hexapoda: Insecta

Until 1999, only insects from hypoxic environments were known to have intracellular or extracellular haemoglobin, which supports oxygen uptake or storage under extreme conditions. But since then we know that intracellular haemoglobin is part of the standard equipment in insects.

Intracellular haemoglobins were first demonstrated in fat body and tracheal cells of some holometabolic insects: Diptera (*Drosophila*, *Anopheles*, *Aedes*, *Glossina*), Hymenoptera (*Apis*), Hemiptera (*Acyrthosiphon*, *Aphis*), Coleoptera (*Dascillus*, *Tribolium*), and Lepidoptera (*Bombyx*) (Burmester and Hankeln, 2007). In *Drosophila*, the haemoglobin is a high-affinity protein that probably serves in oxygen storage or in protection from hyperoxia, the protective function being the more important one: it intercepts the oxygen between tracheal system and cells or degrades and detoxifies ROS and NO in the cells.

An interesting illustration of the functional versatility of insect haemoglobin is shown by case studies of its roles in two evolutionarily diverse lineages: the dipteran horse botfly *Gasterophilus intestinalis* and the backswimmer bug *Anisops bueona*. In *G. intestinalis*, where the larvae live as endoparasites in the gastrointestinal tract of horses, donkeys, and mules, the haemoglobin has the function of craniote myoglobin. In young larvae it is mainly found in the fat body, muscles, and epidermis, whereas in older larvae it is concentrated in specialized tracheal cells at the posterior end of the body, where it facilitates oxygen uptake from air bubbles swallowed by the host during feeding. Adult and larval backswimmers, on the other hand, swim and forage under water and breathe via a physical gill (see Chapter 6). High concentrations of haemoglobin (about 20 mM), found mainly in special tracheal organs, are low-affinity proteins (P_{50} = 3.7 kPa) with high cooperativity. Their main function is to extend the length of the underwater excursion through storage of oxygen, which is released into the bubble surrounding the tracheal organ.

However, probably the best and longest known insect haemoglobin is that of chironomid midge larvae, *Chironomus plumosus*. This haemoglobin in the ovaries and eggs of adults and in the haemolymph of larvae and pupae is presumably derived from intracellular haemoglobins. It is produced mainly by the fat body cells and then transported to the haemolymph. It is a high-affinity protein (P_{50} = 0.07–0.21 kPa) with a low molecular weight (31 kDa). Such a low-molecular-weight haemoglobin would have catastrophic osmotic effects in the renal ultrafiltration apparatus of a craniote, and would be sequestered and destroyed. Since the Malpighian tubules of insects do not use ultrafiltration, such a small extracellular respiratory protein is possible. The protein transports oxygen at low PO_2—the normal habitat of these midges. Haemoglobin increases the oxygen-carrying capacity of the haemolymph and functions as a short-term oxygen store, significantly increasing the survival time of haemoglobin-containing midges compared with haemoglobin-free ones. Many more examples could be listed but all show the same tendency: haemoglobin in invertebrates has positive effects on elevated performance under normoxia and on normal performance at low PO_2. For further examples, see the reviews by Terwilliger (1998) and Weber and Vinogradov (2001).

4.7.3 Myoglobin

In craniotes, myoglobin is a single-chain, porphyrin-containing molecule in which the protein is similar to the β chain of haemoglobin. It has a much higher oxygen affinity than haemoglobin and unlike the latter, myoglobin demonstrates a hyperbolic—rather than sigmoid—OEC. In craniotes, it is not present in red blood cells, but rather in tissue, in particular in muscle, where it is important in oxygen storage. In addition, myoglobin also serves as a sink for NO and regulates oxygen influx into the skeletal and heart muscle cells and oxygen consumption by mitochondrial cytochrome oxidase (Wittenberg and Wittenberg, 2003).

Although myoglobin is one of the best studied tissue globins, many of its potentially numerous functions remain to be clarified. While its respiratory role in oxygen storage and facilitated diffusion in diving craniotes is not challenged, it is still unclear whether in other tissues it only facilitates the oxygen diffusion from the cell surface to the mitochondria or whether it also serves in short-term oxygen storage (Wittenberg and Wittenberg, 2003). In the African lungfish *Protopterus annectens*, at least seven distinct myoglobin genes have been found, suggesting numerous functions. The presence and structure of tissue myoglobins also recently has been used to infer aquatic ancestries of such varied groups as echidnas, moles, hyraxes, and elephants (Mirceta et al., 2013).

Myoglobin is common in molluscs, being found in several separate lineages: Polyplacophora, Scaphopoda, Bivalvia, and in the Gastropoda, where it can reach particularly high concentrations. Whereas the craniote myoglobin is monomeric (17 kDa), molluscan myoglobin can be mono- or dimeric and can even demonstrate cooperativity. This can prove advantageous in facilitating diffusion from the respiratory protein to the tissues in species that are virtually sessile and live at an environmental PO_2 as low as 3 kPa.

4.8 Other globins

In addition to haemoglobin and myoglobin, a host of haem proteins, including neuroglobins, cytoglobins, globin E, globin X, globin Y, and androglobin, are found in craniote tissue. Androglobin, neuroglobins, and globin X appear to have been present in the metazoan common ancestor of all animals, and the genome of the coelacanth (*Latimeria chalumnae*) and also turtles (Schwarze et al., 2015) contain copies of all eight different globins (Burmester and Hankeln, 2014).

4.8.1 Haemerythrin

Haemerythrin (Figure 4.3a) differs from other iron-containing proteins in that it does not contain a porphyrin molecule. It is found in such diverse groups as Brachiopoda, Annelida, Sipunculida, and Priapulida. This polypeptide forms four antiparallel α helices in which two Fe^{2+} molecules are embedded. It shows no cooperativity and, like myoglobin, its OEC is hyperbolic. Also the oxygen binding mechanism of haemerythrin is unique (Figure 4.3b): binding of oxygen causes the oxidation of Fe^{2+} to Fe^{3+}. In deoxyhaemerythrin, the iron atoms are bridged by a hydroxyl group. An electron transfer to the irons and a transfer of the bridging proton to the oxygen occur during binding. Unlike haemoglobin, haemerythrin does not bind carbon monoxide.

4.9 Haemocyanin

Haemocyanins (Figure 4.3c) are large, copper-containing, allosteric respiratory proteins of arthropods and molluscs. Two Cu^{+} ions in two copper-binding sites serve as oxygen carriers. Haemocyanins occur freely dissolved in the haemolymph and are not found intracellularly. Oxygen binding is mediated by the pair of copper atoms that are coordinated by six histidine residues, three per copper atom

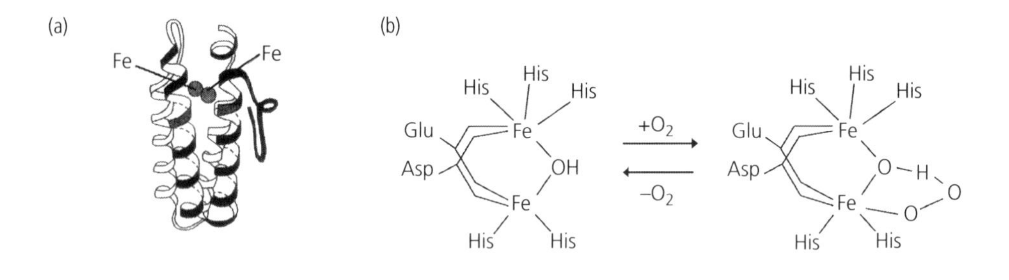

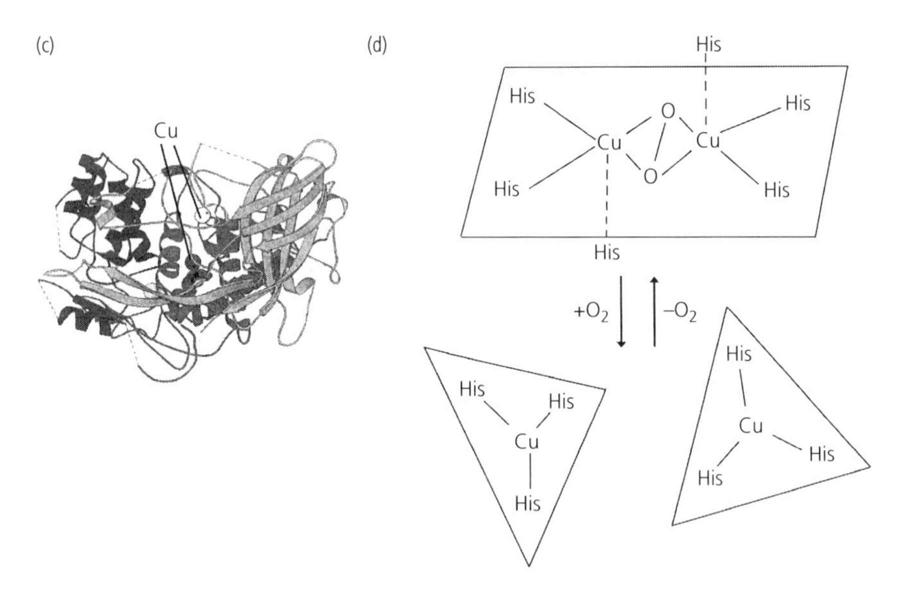

Figure 4.3 (a), Haemerythrin. Sketch of tertiary structure showing location of oxygen-binding sites. part (b) shows the mechanism of oxygen binding in haemerythrin. Part (c) is a sketch of the tertiary structure of the arthropod haemocyanin of the horseshoe crab, *Limulus*, showing three domains (dark, medium and light shading) and the location of oxygen-binding copper (Cu) moities in the middle one (medium shading). Part (d) illustrates how two Cu moities combine to bind one oxygen molecule, removing it from ($-O_2$) or releasing it to ($+O_2$) the surroundings. Asp, Glu and His, indicate critical aspartate, glutamate and histidine residues in the respiratory proteins, respectively. Diverse sources.

(Figure 4.3d). Haemocyanin also transports carbon dioxide according to the same general principles as haemoglobin, but its binding capacity is lower.

The haemocyanin is composed of several subunits, which results in a cooperative oxygen binding and a hyperbolic OEC. The haemocyanins of molluscs and arthropods are very similar in the type of oxygen binding but differ significantly in sequence and structure and are, at first glance, so different that a common origin is not discernible (van Holde and Miller, 1995; van Holde et al., 2001). On closer analysis, both arthropod and molluscan haemocyanins appear to be derived from enzymes with phenoloxidase activity, possibly dating back to a remote ancestral lineage conceivably predating the origin of these two taxa (van Holde et al., 2001; Burmester, 2002).

4.9.1 Arthropod haemocyanin

Arthropod haemocyanins and phenoloxidases, together with crustacean pseudo-haemocyanin and insect hexamerins (the latter two being unable to bind oxygen), belong to a superfamily of proteins. Cu^{+} ions are present in haemocyanins and phenoloxidases but lacking in hexamerins and in pseudo-haemocyanins.

Haemocyanins evolved from phenoloxidases (tyrosinases and catecholoxidases) somewhere in the stem-line of the Arthropoda and it is probable that the haemocyanin superfamily even emerged among stem Bilateria (Immesberger and Burmester, 2004). Typical arthropod haemocyanin consists of hexamers or oligo-hexamers with subunits, each of which comprises 70–80 kDa (Markl, 1986; van Holde and Miller, 1995; Burmester, 2002). They may combine to multimers of up to 8×6 subunits, depending on the taxon or physiological conditions. The size of a subunit is about the same as that of a haemoglobin molecule (75 kDa as opposed to 68 kDa for haemoglobin) (Markl, 1986; Markl and Decker, 1992; van Holde and Miller, 1995). Multimers result in macromolecules that are about 50 times as large as mammalian haemoglobin. Each subunit contains one binding site for oxygen: each having two Cu^{2+} ions, which together can bind one oxygen. As in haemoglobin, reversible oxygen binding is modulated, enhanced, and adapted by allosteric interactions between subunits (see later). In addition, temperature and pH are important effectors.

Arthropod haemocyanins occur in Chelicerata, in some Crustacea (Malacostraca, Remipedia, Ostracoda, Branchiura), in Onychophora, Chilopoda, Diplopoda, and not to forget Hexapoda (Jaenicke et al., 1999; Kusche et al., 2002; Burmester, 2015). Phenoloxidases are Cu^{2+}-containing enzymes of the melanin pathway, which itself is involved in immune response, cuticle formation, wound healing, and free radical binding, and are composed of hexamers made up of subunits with a molecular mass of about 71 kDa (Decker and Terwilliger, 2000; Burmester, 2001; Jaenicke and Decker, 2003). Since some chelicerates lack phenoloxidases, it seems that the haemocyanin itself serves this function. Haemocyanins also show similarities with the enzyme tyrosinase and catecholaminoxidase (Decker and Tuczek, 2000). So it is possible that haemocyanin was first present as an intracellular protein that reversibly bound oxygen and protected the cells from oxygen toxicity, and later evolved as an extracellular oxygen transporter. But haemocyanin is also involved in other processes such as osmoregulation, buffering protein storage, cuticle synthesis, and in the formation of antimicrobial peptides (Destoumieux-Garzon et al., 2001; Lee et al., 2003).

In arthropods, haemocyanin makes up about 80 per cent of all proteins. But haemocyanin raises the oxygen content of the haemolymph by not more than a factor of four. Between 5 and 7 ml O_2 l^{-1} is physically dissolved in haemolymph, and haemocyanin raises this value up to 28 ml l^{-1}. This compares with about 200 ml l^{-1} for human blood.

Subunit evolution took place independently in chelicerates, crustaceans, myriapods, and insects (Burmester, 2001). Cheliceran and crustacean haemocyanins appear to have separated some 600 mya, and the apomorphic subunits of the cheliceran haemocyanin evolved between 550 and 450 mya. Among crustaceans, haemocyanins were thought to be restricted to the Malacostraca, and are best known in the Decapoda. The malacostracan haemocyanin separated from the rest only about 200 mya and pseudo-haemocyanins evolved in decapod crustaceans about 215 mya. Recently haemocyanin was found in other crustacean taxa as well: Remipedia, Ostracoda, and Branchiura (Ertas et al., 2009; Marxen et al., 2014; Pinnow et al., 2016). The structure of ostracod and branchiuran haemocyanin indicates ancient divergence of these two groups (Pinnow et al., 2016). Also, insect haemocyanins and hexamerins appear to have evolved independently around 440–430 mya (Burmester, 2001). Two haemocyanin subunits exist (Hc1 and Hc2) which appear to have developed before hexapods originated (Ertas et al., 2009).

There is the tendency for the protein aggregates to be small in crustaceans and large in myriapods and chelicerates. The plesiomorphic structure within non-decapod malacostracans is assumed to be a 1×6-mer. Decapods can have 1×6 or 2×6 blocks, as seen in spiny lobsters (*Panulirus*) and lobsters (*Homarus*), respectively, and thalassinid shrimps have upped this to a 4×6-mer structure. In general, malacostracan haemocyanins show a high variability. Three types of subunits occur and some subunits might be present only under special physiological conditions or in some developmental stages (Markl, 1986). In contrast, the structure and the subunit composition of the haemocyanins of chelicerates and of myriapods are conserved and presumably have been retained for more than 500 million years (Markl et al., 1986; Burmester, 2002; Rehm et al., 2012).

Among chelicerates, sea spiders have a simple hexameric haemocyanin (1×6-mer). But the horseshoe crab *Limulus polyphemus*, a representative of the most basally branching extant chelicerate radiation, has a 8×6-mer haemocyanin, while scorpions, whip scorpions, whip spiders, and orthognath spiders (e.g. Araneidae, Linyphiidae, and Theridiidae) all possess a highly conserved 4×6-mer haemocyanin that consists of at least seven distinct subunit types (termed a–g). Some 470 mya, when Xiphosura and the remaining chelicerates diverged, they seem to have possessed a 4×6-mer. Independent gene duplication events gave rise to the other distinct subunits in each of the 8×6-mers of Xiphosura (Rehm et al., 2012). Almost half of the known species of arachnids (including Agelenidae, Lycosidae, Salticidae, Thomisidae, and Pisuaridae) possess a 2×6-mer.

Within the chelicerates, the evolution of the haemocyanin and its loss appears to correlate with the configuration of respiratory organs. One example within the Araneae (web-making spiders) is the genus *Dysdera* (Rehm et al., 2012). It breathes mainly with its well-developed tracheal system while the book lungs are reduced and it appears to lack haemocyanin. Other chelicerates that do not have book lungs as main respiratory organs have lost haemocyanin as well. They breathe with tracheae or exclusively cutaneously: Solifugae, Opiliones, Pseudoscorpiones, and Acari. In Opiliones, haemocyanin was first described (Kempter et al., 1985; Markl et al., 1986), but later it was found that the putative haemocyanin probably is a vitellogenin-like protein (Rehm et al., 2012).

4.9.1.1 Onychophora

Onychophora (velvet worms)—albeit only being closely related to arthropods but not belonging to them—are listed here for convenience. They have haemocyanin, but since they lack specialized respiratory organs and have only bush-like tracheal lungs that extend from the stigmata into the haemolymph, the evolution of their haemocyanin appears to have been coupled primarily to the evolution of an efficient circulatory system rather than to that of respiratory organs (Kusche et al., 2002).

4.9.1.2 Myriapoda

When haemocyanins occur in Chilopoda (centipedes) and Progoneata (millipedes) they are structurally similar. Within the centipede *Scutigera*, and the millipede *Spirostreptus* for example, 1×6 to 6×6-mers occur. In other genera, both among centipedes and millipedes, haemocyanin can be lacking. The haemocyanins, when present, are homologous and at least three subunit types developed before the separation of centipedes and millipedes some 420 mya (Kusche et al., 2003). The 4×6-mer haemocyanin of *Scutigera* may in part compensate for the mismatch between the simple structure of the tracheal system, which consists of dorsal tracheal lungs, and the high oxygen demands of these fast-running animals. The haemocyanin of *Scutigera* is a low-affinity protein with high cooperativity (Mangum, 1986; Jaenicke et al., 1999).

In millipedes, however, the tracheal system is well developed. Since it reaches the organs and tissues, like insect tracheae, the animals should not need haemocyanin for oxygen transport. Accordingly, the haemocyanin of *Spirostreptus* is a high-affinity protein with low cooperativity, and might actually represent an adaptation for oxygen storage in the oxygen-poor, subterranean biotope where the animals spend the day.

4.9.1.3 Hexapoda

Respiratory proteins were long thought to be neither present nor necessary in insects, since the tracheal system supplies oxygen to the tissues directly. But recently haemocyanin has been confirmed in many hexapod taxa, including Collembola, Archaeognatha, Zygentoma, Dermaptera, Plecoptera, Orthoptera, Mantophasmatodea, Phasmatodea, Mantodea, Blattodea, and Isoptera, and its presence might even be plesiomorphic for the group.

The first indication that haemocyanin is present in insects was its discovery as an embryonic haemolymph protein, expressed in haemocytes of grasshopper embryos, where it is present in relatively high concentrations. This protein has even preserved residues involved in oxygen binding, oligomerization, and allosteric regulation of oxygen transport proteins in embryos, but it is lacking in later developmental stages and in adults (Sanchez et al., 1998). In adult hexapods, haemocyanin is present in entognaths and in some hemimetabolic insects, but has not yet been demonstrated in winged basally branching groups such as Odonata (dragonflies) and Ephemeroptera (mayflies), or in holometabolic insects. Presumably it was lost during the evolution of the later branching groups (Burmester, 2015).

4.9.1.4 Hexamerins are not oxygen transporters

Hexamerins are copper-free, haemocyanin-related haemolymph proteins. They are structurally diverse and widespread in insects, where they carry out practically every function of haemocyanin except oxygen transport (Burmester, 1999; Burmester and Hankeln, 1999, 2007). They are especially highly concentrated in larval stages and during metamorphosis. Specifically, hexamerins function in storage and transport of hormones, in the humoral defence, and cuticle formation, but no longer play a role in the transport of respiratory gases (Hagner-Holler et al., 2004).

Hexamerins appear to have evolved from crustacean-like haemocyanins, providing further evidence for the suggested close relationship between insects and malacostracan crustaceans (see also Chapter 10). According to this scenario, this haemolymph-based, initially respiratory protein was conserved. During the terrestrialization of insects and the evolution of their tracheal system, conservation of the oxygen-binding function became counterproductive.

4.9.2 Molluscan haemocyanin

In molluscs, haemocyanin occurs in Polyplacophora, most Gastropoda, and Cephalopoda, but is lacking in Scaphopoda and in most Bivalvia (Bergmann et al., 2007; Markl, 2013). In general, the molecules are huge. The molecule is organized in dodecamers or dodecamer aggregates of polypeptide chains, each aggregate consisting of ten subunits. Each subunit is 350–450 kDa in size and the molecule is therefore much larger than arthropod haemocyanin. Each subunit contains seven to eight domains, or functional units, arranged like pearls on a string, each with seven to eight oxygen binding sites. So the oxygen-carrying capacity per molecule of molluscan haemocyanin is much greater than that of its arthropod counterpart.

In cephalopods, the molecule is a 1×10 block of 4500 kDa, but in snails, the molecule is twice as large: a 2×10 block. It has 20 subunits with eight oxygen binding sites each, resulting in 160 binding sites per molecule compared with about one-third that number in arthropod haemocyanin and a paltry four sites in craniote haemoglobin!

4.9.3 Physiological properties of arthropod and molluscan haemocyanins

Like haemoglobin, haemocyanin can show a normal Bohr shift. In cephalopods it is particularly pronounced, but in some chelicerates (e.g. *Limulus*) and in some gastropods, a reverse Bohr shift is observed: a decrease in pH causes an *increase* in oxygen affinity (see section 4.6.4).

In arthropods, some ions, monoamines, urate, and lactate, which occur under hypoxic conditions and during anaerobic metabolism, can alter the oxygen affinity of haemocyanin. Just two examples illustrate how tightly the oxygen affinity of haemocyanin can be coupled with environmental conditions: only a slight decrease in the PO_2 of the respiratory medium causes the increase of uric acid, since the enzyme that degrades urate requires oxygen. Urate, in turn, causes an increase in the oxygen affinity of the haemocyanin. Also lactate, which is produced whenever animals use anaerobic pathways, increases the oxygen affinity of the haemocyanin by influencing the cooperativity of the haemocyanin molecule. At the same time, lactate decreases the pH of the haemolymph, shifting the OEC to the right (Bohr effect) and thus facilitating the release of oxygen in the tissues. But lactate also weakens the tendency to release oxygen, meaning that the haemocyanin still can be fully saturated as long as the environmental oxygen concentration remains stable. In terrestrial crustaceans the lactate effect is weak while it is important in aquatic species (Bridges, 2001).

Now, metabolically produced urate and lactate have medium- to long-term effects on the oxygen affinity of haemocyanin, and after more than 7 days of hypoxia are often followed up by an increase in the concentration of the respiratory protein, convergent with high-altitude polycythaemia in craniotes. But a rapid, short-term increase in haemocyanin oxygen affinity in crustaceans is brought about right at the onset of exercise-induced hypoxia, by release of monoamines such as dopamine from the pericardial organs. This is much quicker than waiting for anaerobic metabolism to set in and release lactate.

In addition, arthropod haemocyanin is influenced by organic ions, by salinity changes, and divalent electrolytes such as Mg^{2+} and Ca^{2+}, which normally increase the affinity, while Cl^- more often decreases it. The influence of environmental salinity depends on the osmoregulatory ability of a species and can modulate affinity and subunit composition (Bridges, 2001).

As is the case for craniote haemoglobin, an increase in temperature decreases the affinity in arthropod and molluscan haemocyanin. Indeed, in cephalopods, temperature and pH-dependent changes appear to be the only modulators of oxygen affinity.

CHAPTER 5

Coping with extremes

In a Panglossian best of all possible worlds, animals would have to deal only with ideal conditions. But reality says that what is ideal for one could also be ideal for another, or seen from the other side of the looking glass, there is room for species that can handle less than optimal conditions and even more space for those that can survive extreme conditions. The most frequently encountered extreme condition with regard to respiration is hypoxia. We distinguish between external (environmental) and internal hypoxia, the latter being caused by large oxygen demands such as exercise. But extreme conditions also include hyperoxia, which can happen when aquatic animals are surrounded by photosynthetic plants on sunny days, or exposure to very high or very low temperatures, dehydration, or the like. And the list goes on.

The first part of this chapter deals with hypoxia, first the more general principles and mechanisms then the way in which specific animal groups actually deal with the problem. In the second part, we shall look at other extreme conditions and the way these are dealt with.

5.1 Hypoxia

The discovery of high-altitude hypoxia is interwoven with the history of aviation, extreme mountain sports, and underwater exploration. Throughout this chapter it is important to keep in mind that the bottom line of respiratory biology is to keep the cells—regardless of their very different requirements—constantly supplied with oxygen at a partial pressure between 1 and 3 kPa, which keeps oxidative metabolism supplied but is not so high as to shut it off. If this requirement cannot be met, the most common default solution is anaerobic metabolism which, as we shall see, is not really such a good idea because it quickly uses up energy reserves for a small ATP production and ends up with problematic waste products. However, it is still better than the alternative: death. Then again, various forms of dormancy have evolved that circumvent the disadvantages of purely anaerobic metabolism. But we are getting ahead of ourselves.

For practical purposes we define 'normoxia' to be a PO_2 of 20.73 kPa at sea-level atmospheric pressure of 101 kPa. The fraction of oxygen even on the top of Mount Everest is the same as at sea level (around 20.9 per cent) but since the total atmospheric pressure is so low, the PO_2 is only about one-third of the sea-level value and not sufficient to support human life for an extended time.

According to this definition, normoxic conditions are only valid for a small group of animals that actually live at sea level and most others are often or habitually exposed to some degree of hypoxia: in terrestrial environments at altitude and in aquatic habitats at depth or at night, when plants, plankton, and microorganisms deplete oxygen content. So for many animals, slight to moderate hypoxia is normal and these levels elicit no compensatory reactions. For species that have to cope with hypoxia sporadically, we speak of *acclimatization*: mechanisms that are reversible when the animals return to their normoxic conditions.

Respiratory Biology of Animals: Evolutionary and Functional Morphology. Steven F. Perry, Markus Lambertz, Anke Schmitz, Oxford University Press (2019).
DOI: 10.1093/oso/9780199238460.001.0001

However, tissue oxygen level is not only influenced by the ambient oxygen content. Exercise or particularly high-demand tissues such as muscles and the nervous system rapidly deplete oxygen reserves. But also feeding, illness, or gravidity/pregnancy might cause internal hypoxic conditions in affected organs or in the entire organism. While exercise and feeding require only short-term mechanisms, illness or pregnancy calls for longer-term *adaptation*.

An alternative definition of 'hypoxia' could be a species-specific one: external oxygen levels below the habitual state for a given species. Even extremely hypoxic environments might be normal for animals that habitually live at high altitude or in burrows. In these species, long-term adaptations have become incorporated into the genome. These animals have, so to speak, evolved a derived set point around which a new Gaussian curve of metabolic, behavioural, and physiological acclimatization centres.

In general, our definition of (non-species-specific) hypoxia is found more frequently in aquatic environments than in terrestrial ones, simply because of the relatively low solubility of oxygen in water (see Chapter 3) and its low diffusion velocity. In fresh-water habitats, temperature and altitude are also main contributing factors. For example, in a lake at 5300 m in the Andes, 1 litre of water contains only 4 ml of oxygen, as opposed to about 7 ml at sea level. Also, high temperatures reduce oxygen solubility and during the winter the oxygen content of ponds and lakes is reduced by ice, which prevents replenishment by diffusion from the atmosphere. In addition, many fresh-water bodies can partially or completely dry out, and animals living in tidal pools are also confronted with cyclic evaporation, overheating, or undercooling. In the oceans, there are also zones with extremely low oxygen content. Off the Pacific coast of Mexico, for example, at depths between 100 and 900 m, the oxygen content is less than 0.1 ml l^{-1}. Below this zone the oxygen content rises to around 2 ml l^{-1} at 4000 m depth ('normoxic' would be around 6 ml l^{-1}). The main reason for these very low oxygen contents is the high biomass of animals living there for at least part of the day. This phenomenon is not uncommon and is found, for example, in the wake of large fish schools or zoo-plankton swarms. Another cause of aquatic hypoxia is eutrophication by agriculture and waste disposal, and the ensuing over-population with bacteria and other microorganisms, zooplankton, and phytoplankton. Among aquatic invertebrates, the most tolerant to hypoxia/anoxia are nematodes, flatworms, bivalve molluscs, and annelids: among the craniotes, cyprinid fish and turtles.

Although terrestrial environments have a higher oxygen content (in spite of equal partial pressures) than aquatic systems at the same altitude, there are a variety of habitats and conditions under which even terrestrial animals may also experience hypoxia. The most obvious is high altitude. For many animals, altitudes of 1500–3000 m with 180–140 ml of oxygen per litre of air, are easily compensated for by acclimatization. At 5300 m the oxygen content is only half as much as at sea level, and at 10,000 m it is less than 50 ml l^{-1} as opposed to about 210 ml l^{-1} at sea level. Among craniotes, only birds are able to cope with such conditions. But many invertebrates also live at high altitude, such as jumping spiders living in the Alps and at 6000 m in Tibet, and their adaptations have not been studied at all. At 19,000 m, air pressure and water vapour pressure are the same so it is impossible to breathe even using pure oxygen, since the lungs would fill with water vapour and the blood within the lungs would boil.

Also, hypoxic microhabitats for invertebrates and some burrowing craniotes commonly occur in soil or sediment habitats under leaf litter or in caves, due to limited oxygen accessibility and to oxygen consumption by microorganisms, fungi, and plant roots. Due to their small size and limited motility, many invertebrates also endure long periods of flooding, and some specialists even live in dung, rotting carcasses, craniote stomachs, or in sealed food storage containers in which oxygen content often is close to zero. Also conflagrations can consume so much oxygen that animals on the ground suffocate.

Diving is undoubtedly the most challenging life-style for air-breathing craniotes. Since most terrestrial animals are incapable of aquatic gas exchange, diving can be considered synonymous with anoxia. For this reason, we shall deal with diving in some detail later in this chapter.

Since hypoxia is a common stress factor for virtually all animal lineages, it should not come as a surprise that the genetic basis for coping with it is ubiquitous. Let us now take a look at these mechanisms.

5.1.1 General genetic and metabolic mechanisms dealing with hypoxia

5.1.1.1 Hypoxia-inducible factors (HIFs)

For at least two thousand million years, eukaryotes have been dependent on oxygen for survival. Adjusting to the harmful effects of high concentrations of stable oxygen and unstable, highly ROS such as O_2^-, hydroxyl radicals (HO•), and hydrogen peroxide (H_2O_2) was important during the origin and early evolution of life. Once aerobic metabolism became established, feed-forward mechanisms at the lower end of the oxygen spectrum, which allowed aerobic organisms to deal with environments that were critically low in or lacking oxygen, also became necessary. Accordingly, we find ancient pathways that signal hypoxic conditions and set regulatory processes in action that alleviate it or accommodate to it. In animals, many of these processes are mediated by HIFs (for reviews, see Semenza, 2004; Hoogewijs et al., 2007; Weidemann and Johnson, 2008).

It should be noted here that hypoxia is a stress factor and there are also a large number of stress proteins, originally called heat shock proteins (HSP) which will be discussed later and the production of which is upregulated by stress factors including hypoxia. HIF, however, is a separate entity not related to HSP.

HIF (Figure 5.1) is a sort of master switch that is capable of activating a host of responses, specifically to reduced oxygen at the cellular level, whereby the cell is not actually able to distinguish between external and internal sources of hypoxia. HIF is a DNA-binding transcription factor composed of two basic helix–loop–helix (bHLH) proteins of the so-called PAS family. The PAS domain gets its name from the proteins in which it was first discovered: PER (period circadian protein), ARNT (aryl hydrocarbon receptor, nuclear translocator protein), and SIM (single-minded protein). Many PAS proteins detect their signal by way of an associated cofactor such as haem, but this is not the case for HIF.

The two subunits that make up the HIF heterodimer are hypoxia inducible, whereby one of them is also found under both normoxic and hypoxic conditions (that is, it is 'constitutive'), whereas the other, 'non-constitutive' one is rapidly degraded in normoxia. The non-constitutive subunit exists in

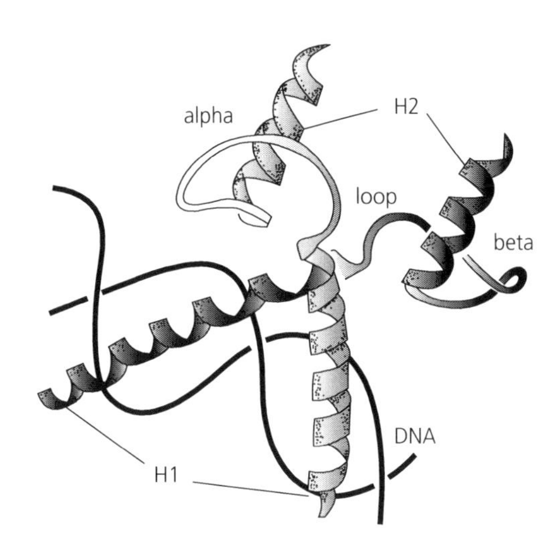

Figure 5.1 Tertiary structure of the Alpha (non-constitutive, lightly shaded) and Beta (constitutive, darker shaded) subunits of a hypoxia inducible factor (HIF) transcription factor heterodimer, showing the manner in which the helix-loop-helix molecules straddle DNA molecules, where they induce transcription of genes that code for enzymes involved in cellular-to-organismic level response to hypoxia. Only the H1 helix subunits are capable of interaction with DNA, at approximate locations on the HIF molecules shown in the drawing. The constitutive subunit alone cannot induce transcription and the non-constitutive subunit is destroyed in the presence of oxygen (see text). Modified from diverse sources.

three principal isoforms, 1, 2, and 3, which have non-redundant functions (for review, see Weidemann and Johnson, 2008). All three subunits, however, are capable of forming a dimer with the constitutive protein, resulting in what is called HIF-1, -2, or -3 (see Hoogewijs et al., 2007).

Oxygen-dependent regulation of HIF is two-pronged. HIF-1, for example, has important regulatory sites in its C-terminal transactivation domain (C-TAD) and also in the oxygen-dependent degradation domain (ODD). The latter contains two LXXLAP motifs (L = leucine, A = alanine, P = proline, X = any other amino acid). Under normoxia, hydroxylation of the proline residues (Pro 402 and 564 in human HIF-1, for example) destabilizes the protein and the molecule is quickly destroyed by a nuclear protein scavenging proteasome. A second high oxygen-dependent hydroxylation takes place on an asparagine residue (Asn 803 in human HIF-1) in the C-TAD domain, ultimately resulting in prohibiting transcriptional activity of the target genes (hypoxia-response elements (HREs), see later in this section) in the nucleus (see Hoogewijs et al., 2007). It should be pointed out that the substrate for prolyl hydroxylation is oxygen itself. So those enzymes are actually oxygen sensors at the molecular level (Semenza, 2004). Under hypoxia, the HIF non-constitutive subunit enters the nucleus where it binds to the constitutive subunit. Once formed, the dimer complex attaches to HREs of numerous genes. This initiates transcription, which leads to subcellular, cellular, organismic, and behavioural responses to hypoxia, including glycolysis, erythropoiesis, angiogenesis, and apoptosis (Semenza, 2004). In total, more than 100 genes are directly affected by HIF (Weidemann and Johnson, 2008).

HRE sequences in a wide range of oxygen-mediated control mechanisms for glycolysis in plants and animals have been conserved, and these mechanisms can be traced back to Archaean organisms (Webster, 2003). The same applies to HIF in metazoans. Among craniotes, there is an overall conservation of molecular structure, but with some environmental specificity seen in fish (Law et al., 2006). And when distantly related groups are compared—for example, humans and limpets (Weihe et al., 2009)—one sees that key amino acids such as the two oxygen-sensitive proline residues and the one asparagine mentioned previously are conserved. Although the medical relevance of HIF has encouraged the use of craniote—in particular mammalian—models, features of hypoxic signalling by HIF, such as the well-conserved HIF–prolyl hydroxylase pathway (Semenza, 2004), were actually disclosed through the study of invertebrate model systems (Hoogewijs et al., 2007).

5.1.1.2 ATP-producing metabolic pathways

Although already mentioned in previous chapters, we shall now take a somewhat closer look at those aspects of anaerobic and aerobic metabolic pathways relevant to survival of animals. Anaerobic metabolic pathways very often use HIF-controlled alternatives for a fast and effective possibility of ATP production in the short term. The substrate hydrogen atoms cannot be transferred to molecular oxygen and reduce it to metabolic H_2O because under anoxic conditions oxygen is—by definition—lacking. Instead, other organic compounds known as fermentation products are formed, the best known of which is lactate. This pathway is frequently used during internal hypoxia, also termed 'functional anaerobiosis', for example, during exercise.

During functional anaerobiosis, the muscles first receive ATP of internal phosphagen stores: in craniotes this is creatine phosphate, and in many invertebrates, phosphoarginine or phosphotauromycin. When this activated phosphate source is exhausted, the organism produces ATP by glycolysis. Glycolysis takes place in the cytosol of the cells. It is important to realize that first of all, glycolysis *must* take place, or all of metabolism comes to a screeching halt and, secondly, as we have repeatedly stated, glycolysis requires a low oxygen cytosol, simulating the conditions under which life first evolved. Thus seen, as pointed out in Chapter 1, one important function of mitochondria is to act as an oxygen sink.

In glycolysis, the six-carbon sugar glucose is split into two 3-carbon pyruvate molecules. Per mol of glucose, 2 mol of ATP are required but 4 mol of ATP (substrate level phosphorylation) and 2 NADH are produced. The further fate of the pyruvate depends on the presence or absence of oxygen: if oxygen is available, pyruvate is decarboxylated and incorporated into an acetyl coenzyme A molecule, whereby per mol of pyruvate, 1 mol of NADH and 1 mol of carbon dioxide result. The acetyl coenzyme A is fed into the citric acid cycle, which takes place in the mitochondrial matrix. This process occurs, for example, during the recovery phase after exhaustive exercise. If oxygen is not available, the pyruvate is converted to lactic acid, which then must be removed from the cells. For each molecule of glucose, 2–3 ATP molecules are produced which is very low compared to the aerobically produced 30–38 ATP molecules produced by oxidative phosphorylation on the mitochondrial membranes, but it is available much faster. In most animals L-lactate is produced but arachnids make D-lactate (Prestwich, 1983b). All enzymes of the glycolytic pathway are under the control of HIF, and are rapidly upregulated under hypoxic conditions.

Lactate can enter a pathway called gluconeogenesis. This not only gets rid of lactic acid but also produces glucose, which then enters the carbohydrate metabolism so to speak through the back door: via the citric acid cycle and also indirectly from protein metabolism. Gluconeogenesis is sort of glycolysis in reverse. The first step is catalysed by the enzyme lactate dehydrogenase. Gluconeogenesis requires 6 ATP molecules per glucose molecule, and since glycolysis only produces 4 mol of ATP but requires 2 mol of ATP in the process, gluconeogenesis coupled with glycolysis is a losing proposition. But it is a good way to get rid of excess lactate, and the pyruvate thus formed can be used in aerobic metabolism when oxygen becomes available.

Glycogen is a polymerized form of glucose. It occurs in granules and is particularly abundant in skeletal muscle and the liver of craniotes. Muscle glycogen can be rapidly split into glucose, and liver glycogen is used to maintain blood glucose levels. But the pathway for glucose production from glycogen uses glucose 6-phosphate as an intermediary and the phosphate has to come from somewhere. Nevertheless, in spite of its depletion of phosphate stores, glycogen serves well as a direct source of quick energy and is much faster than fat or proteins, although the latter are more advantageous as long-term energy reserves.

Some animals can rely on anaerobic metabolism even for very long periods for survival under external hypoxic conditions. Invertebrates are often confronted with long-lasting environmental hypoxia, and use glycolytic metabolic pathways other than those seen in craniotes. Like craniotes, the isopod *Stenasellus virei* responds to long-term hypoxia (PO_2 <0.03 kPa) with classical anaerobic metabolism and accumulates besides lactate also alanine, which is not unusual for a crustacean. In molluscs we also see other end products of glycolysis such as oxalacetate (which is used in the citrate cycle) and opines. Also in molluscs, such end products of anaerobic protein and carbohydrate (not just glycogen) catabolism such as alanine, succinate, and the volatile propionic and acetic acid occur (Hochachka et al., 1973). During active foraging under hypoxic conditions, many invertebrates—like craniotes—use their glycogen stores (Hervant et al., 1997), but leeches (Annelida: Clitellata) have only small glycogen reserves and instead metabolize malate to propionic or lactic acid (Zebe et al., 1981). In anaerobic tidal flats and deep sea hydrothermal vents, oxidized sulphur sources can actually be utilized instead of oxygen by some bivalve molluscs for mitochondrial ATP production (Parrino et al., 2000).

5.1.1.3 Oxygen conformers and oxygen regulators

As stated at the outset, we speak of reversible acclimatization and permanent adaptation to hypoxia, but we can also speak of oxygen conformers versus oxygen regulators (Figure 5.2). Conformers let their oxygen uptake decrease along with the

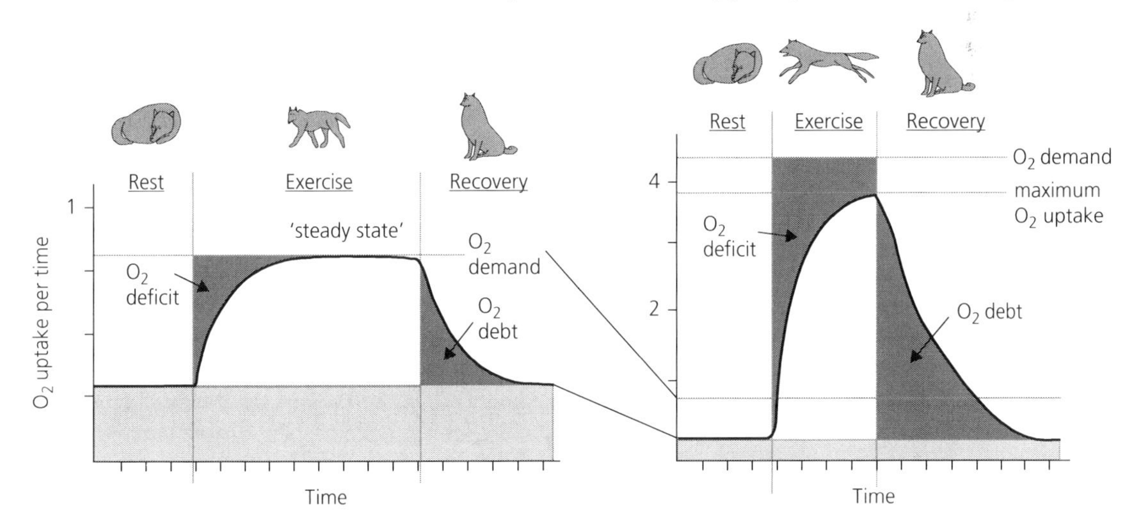

Figure 5.2 Responses to hypoxia. The response of a typical oxyregulator in response to internal hypoxia caused by mild (left-hand diagram) and strenuous (right-hand diagram) exercise is shown. An oxygen debt occurs in mild exercise only until a steady-state can be reached. During strenuous exercise, when aerobic steady-state cannot be achieved, the metabolic rate continues to be supported by anaerobic metabolism. In both cases, the accumulated waste products of anaerobic metabolism must be eliminated by energy-consuming metabolic pathways: the oxygen debt must be repaid. After Klinke and Silbernagl, 1996

decrease in oxygen content of the respiratory medium by reducing their metabolic rate, often in a very finely tuned manner (regulated conformism). On the other hand, in oxygen regulators (like most mammals and birds, for example), the oxygen consumption rate remains constant in spite of decreasing PO_2. But even regulators can become conformers, the PO_2 at which they shift to conforming being defined as the critical PO_2, or $PO_{2\,CRIT}$. The $PO_{2\,CRIT}$ is species-specific. For strict conformers it is usually just below normoxia; for extreme regulators it may be near zero.

5.1.1.4 Behavioural strategies

If HIF can activate more than 100 different genes in response to hypoxia then it should be obvious to the casual observer that numerous possibilities exist. But what is a 100-member orchestra without a conductor? Exactly how these responses are coordinated is the subject of much ongoing research, and is known in some detail only for a few species.

One way for animals to solve the hypoxia problem is to activate a behavioural suite that results in the animal leaving the hypoxic region. This is a nice trick if you can do it, but many animals are sessile, live in burrows, or are so small that they cannot escape. Others, for example, those living in the deep ocean, live part of the time in a hypoxic zone where the food is, but periodically migrate vertically to zones with higher oxygen. Such a vertical migration on a small scale is also typical of water fleas (*Daphnia*), as discussed later (see section 5.1.2.2). Amphibious animals (e.g. many crabs, isopods, and amphibians) can simply switch from water breathing to air breathing and back again as the conditions change (see also Chapters 7 and 8).

The activation of such strategies is three-staged: first, the animal has to detect hypoxia at the cellular level and activate the behavioural search programme. This is one role of HIF. Second, the animal must be able to sense where to go to improve the situation. This is the role of peripheral oxygen sensors, as discussed in Chapters 8 and 13. And finally, HIF must shut off the search programme when the species-specific normoxic condition has been restored.

Another behavioural strategy is to just tolerate low oxygen. As previously mentioned, the physiological role in a respiratory faculty is to maintain the PO_2 in the tissues at around 1–3 kPa. This hypothesis is supported by an unassuming group of tiny crustaceans, called seed shrimps (Ostracoda). These animals lack any regulatory mechanisms for adapting to changes in oxygen of their surroundings. Instead, they inhabit sediment where the oxygen is about 3–5 kPa, so one can assume that the PO_2 under the carapace of an ostracode would be lower: maximally 3 kPa (Corbari et al., 2004; Corbari et al., 2005).

Also, low activity levels or an actively reduced metabolic rate (hypometabolism) during periods of extended hypoxia might be considered to be part of a behavioural strategy, which is turned on by conformers early on, and by regulators, later. Both invertebrates and craniotes are capable of entering a hypometabolic state (discussed in the next paragraph) in which, for example, heart rate, respiratory frequency, and metabolic rate are reduced to very low values. Depending on whether it occurs during cold, winter periods or hot and dry periods, this behaviour is designated as hibernation or aestivation, respectively. Also, shorter coma-like states—called torpor—with reduced metabolism can occur. Species with extremely high metabolic rates such as hummingbirds, bats, and bumblebees commonly fall into torpor overnight.

Hypometabolism occurs in all animal groups not only as a response to hypoxia but also to extreme temperatures or as a response to reduced food availability as part of a host of reactions initiated by activity of 'stress protein' HSP genes (see 'Other metabolism-related extremes', section 5.4.). An advantage of hypometabolism as opposed to anaerobic metabolism is that it circumvents the acid–base consequences of anaerobiosis. Accumulation of lactate results in metabolic acidosis, which can be balanced in the short term by hyperventilation, but in the longer term this excess acid must be buffered, excreted, or converted to something else in an energy-consuming process.

Another behavioural strategy that will be further discussed in Chapter 6, particularly for survival of air-breathing species underwater, is, like a SCUBA diver, to take your oxygen with you. The water spider *Argyroneta aquatica*, and numerous aquatic insects do this.

Oxygen, however, does not need to be transported in gaseous form. Oxygen can be transported and stored to some extent in body fluids and haemolymph/blood plasma, but this is accomplished much more effectively on oxygen-binding proteins such as haemoglobin and myoglobin. This will be discussed later (see section 5.2.3), but one example from the beginning of the metazoan animal lineage that shows just how old these strategies are and is worth mentioning in the present context are Ctenophora (comb jellies) and Cnidaria, where the medusae (if present) can load their virtually acellular mesogloea with oxygen when in oxygen-rich waters and release it passively in hypoxic surroundings (Thuesen et al., 2005a; Thuesen et al., 2005b).

5.1.2 Group-specific examples of dealing with hypoxia

The strategies for dealing with hypoxia may be limited by the group-specific faculties that evolved in the normoxic state. In addition, they are dictated to some extent by animals' abilities to move to areas of higher oxygen levels and also by other conditions found there, which might negatively influence individual survival chances: for example, danger of predation, exposure to ultraviolet light, temperature extremes or desiccation. In general, groups in which haemoglobin is found are better capable of regulating, and maintaining, a high metabolic rate under hypoxic conditions. Although both haemocyanin and haemoglobin usually increase in concentration under hypoxia, the molecular structure of haemoglobin and myoglobin can also easily be changed and these molecules are found not only in the blood/haemolymph but also in the tissue (see Chapter 4). In the following sections we shall see some of the ways in which representatives of various animal groups deal with hypoxia. The list is not meant to be exhaustive, but rather to illustrate the wide variety of strategies that exist.

5.1.2.1 Annelida

Many annelids possess haemoglobin and are highly hypoxia tolerant. The aquatic sludge worm, *Tubifex tubifex* (Oligochaeta), is one of these. In spite of a markedly reduced metabolic rate in highly eutrophic waters, the sludge worm maintains elevated anaerobic ATP production and rapidly restores its energy stores during post-hypoxic recovery (Datry et al., 2003).

The familiar polychaete marine littoral sand worm *Arenicola marina* goes anaerobic below a PO_2 of 6.2 kPa (Wohlgemuth et al., 2000), but another polychaete uses another strategy: *Methanoaricia dendrobranchiata* has modified its parapodia to function as gills. With a respiratory surface area of 10–12 $cm^2\ g^{-1}$ and a water–blood diffusion distance of 3 µm it has a diffusing capacity similar to that of a goldfish gill. Its microhabitat can approach anoxia, with sulphide up to mmol levels, but with its high-affinity haemoglobin (P_{50} = 0.028 kPa) and a Bohr effect that is pronounced at high oxygen saturations, it can remain aerobic down to about 0.9 kPa oxygen. When fully saturated, the bound oxygen is sufficient for 30 minutes of aerobic metabolism (Hourdez et al., 2001; Hourdez et al., 2002).

5.1.2.2 Crustacea

With some 50,000 known species and an enormous variety of body shapes and sizes as well as habitat preference, crustaceans are quite comparable with craniotes. Here we discuss just a few of the numerous mechanisms that have evolved in relationship to acclimatization and adaptation to hypoxia among crustaceans.

In decapod crustaceans, the short-term response is comparable with that seen in craniotes: hyperventilation up to ten times the normoxic value coupled with increased cardiac output (either by increased stroke volume, heartbeat frequency, or both) and often selective perfusion of the gills at the expense of digestive organs. The respiratory alkalosis caused by hyperventilation increases the oxygen affinity of haemocyanin (negative Bohr effect). This increase in affinity makes the respiratory system more efficient, thus conserving energy.

Again convergent with craniotes, chronic hypoxia results in an increase in the concentration of respiratory proteins. This applies to *Daphnia* and to brine shrimp (*Artemia*), two branchiopod crustaceans that have haemoglobin (Paul and Pirow, 1997), as well as to the fresh-water giant prawn *Macrobrachium rosenbergii* and the white shrimp *Penaeus vannamei*, which lacks haemoglobin but shows the same response with haemocyanin (Cheng et al., 2003). In addition, the molecular structure of *Daphnia* haemoglobin is altered by a differential gene expression under hypoxia, increasing its oxygen affinity and allowing cardiac output to fall to nearly normoxic levels.

Crustaceans not only show acclimatization that is convergent with that of craniotes, but also can show a similar degree of adaptation. For instance, the giant red mysid, *Gnathophausia ingens*, discussed later in Chapter 9 because of its exceptional respiratory faculty: it is a pelagic shrimp that for all practical purposes 'thinks it is a fish'. The respiratory faculty is in fact comparable with that of pelagic fish, with gills having a surface area of 9–14 cm^2 g^{-1} body mass and a diffusion barrier of only 1.5–2.5 μm. It is a metabolic regulator that inhabits waters where the oxygen concentration may fall to 0.5 ml l^{-1}, but these extreme regulators maintain 'normoxic' metabolic rates down to 1 ml l^{-1} aerobically. Physiologically speaking, the haemocyanin has high cooperativity, a large Bohr effect, a high oxygen affinity (P_{50} = 0.19 kPa), and the circulatory capacity is high with up to 225 ml kg^{-1} min^{-1}. Combined with extreme ventilatory rates of up to 8 l kg^{-1} min^{-1}, this results in removal of up to 90 per cent of the oxygen from the inspired water.

As will be discussed in Chapters 7 and 10, the terrestrialization of crustaceans runs the whole gamut from marginally adapted to completely terrestrial. Also virtually all combinations of acclimatization and adaptation exist, so it is hard to distinguish where one leaves off and the other begins. Some aquatic crustaceans shift to air breathing under hypoxic conditions. The common yabby (*Cherax destructor*), an Australian crayfish, for example, changes to air breathing at a PO_2 below 2.7 kPa. The ensuing respiratory acidosis during air breathing results in an increased oxygen affinity of the haemocyanin caused by the Bohr shift (Morris and Callaghan, 1998), but this is not a universal response of crustaceans to hypoxia. The purple shore crab *Hemigrapsus nudus*, for example, is already well adapted and doesn't really seem to care if the water gets hypoxic. Its metabolic acidosis adequately increases the capacity of the haemocyanin to remain well oxygenated (Morris et al., 1996) by the mechanism discussed in Chapter 4.

Small terrestrial crustaceans may drown if they cannot flee when their habitat is flooded. The amphipod *Talitrus pacificus*, for example, regulates oxygen consumption down to only a PO_2 of 10 kPa (Mendes and Ulian, 1987) but will die if kept anoxic for 40–45 minutes. Some terrestrial crabs use their stiff gill lamellae and sometimes the carapace lining for air breathing, but the gills are often so reduced that they are no longer suitable for water breathing (Innes and Taylor, 1986). The physiologically advantageous conditions provided by air breathing mean that haemocyanin is fully adequate in normoxia (Morris, 1991) and exercise has little effect because of a small Bohr shift and reduced lactate (Greenaway et al., 1988; Morris et al., 1988; Adamczewska and Morris, 1994; Morris et al., 1996). The robber crab *Birgus latro* (Coenobitidae) drowns if kept immersed, in even normoxic water. On land, it compensates for hypoxia in a way convergent to terrestrial craniotes. Down to 12 kPa it is an oxygen conformer, showing little change in respiratory or heartbeat frequency, but below that, ventilation increases and oxygen extraction rises to three times the normoxic values (Cameron and Mecklenburg, 1973).

Other terrestrial crabs show more bizarre ways of circumventing their anatomical and physiological constraints. A relative of *B. latro*, namely terrestrial hermit crabs of the genus *Coenobita*, do just the opposite of the water spider (which, as discussed in Chapter 6, takes air under water with it): they fill the stolen snail shell with water, and carry it onto dry land (McMahon and Burggren, 1979). When exposed to hypoxia down to 3 kPa, ventilatory frequency increases up to 60 times the normoxic value while heartbeat hardly changes, but ventilation remains far more sensitive to oxygen than to carbon dioxide/pH, indicating that *Coenobita* never really left the water (McMahon and Burggren, 1979). And, of course there is the obligate air-breather *Cardisoma hirtipes*, which does just the opposite. When it enters the water, it traps air in its branchial chamber. In addition, it experiences facultative hypometabolism during immersion, decreasing the gas exchange rate to 20 per cent. Respiratory alkalosis (not acidosis!) occurs, while haemolymph glucose concentration and cardiac output decrease. The haemocyanin has a high oxygen affinity, but low pH sensitivity, which facilitates oxygen uptake from hypoxic environments, such as the branchial chamber (Adamczewska and Morris, 1996; Dela-Cruz and Morris, 1997; Adamczewska and Morris, 2000). But unlike craniotes, *C. hirtipes* does not store oxygen in respiratory proteins and it must either leave the water after about 20 minutes of immersion to replenish its oxygen supply or it becomes progressively anaerobic.

The Christmas Island red crab *Gecarcoidea natalis*, on the other hand, uses L-lactate as a modulator to *reduce* haemocyanin oxygen affinity (reverse lactate effect). High affinity is necessary in water, but not in air. This species has a high oxygen demand during exercise and the unloading of haemocyanin acts with the help of L-lactate to provide oxygen to the muscles during running (Adamczewska and Morris, 1998; Morris, 2002). These are just a few of the fascinating acclimatization and adaptation possibilities exploited by crustaceans. They show in an impressive way how the interaction of multiple variables within the cardiorespiratory faculty pushes the limits of the crustacean bauplan.

5.1.2.3 Myriapoda

Very little is known of the mechanisms by which centipedes and millipedes deal with hypoxia. Some diplopod millipedes possess high-affinity haemocyanins, which may represent an adaptation to their subterranean and potentially hypoxic habitats (Jaenicke et al., 1999; Kusche et al., 2003). Some exhibit discontinuous gas exchange (see Chapters 7 and 8), and therefore their tissues may experience periodic hypoxia (Klok et al., 2002). In addition, burrowing diplopods may be threatened by temporary floodings during which they either would have to come to the surface and risk predation or tolerate submersion and the associated hypoxia: problems they have in common, for instance, with earthworms (Dwarakanath et al., 1977). Most diplopods are able to regulate respiration in declining oxygen tensions, some even down to a PO_2 of 5 kPa. But others are oxygen conformers and their oxygen consumption rate decreases when the PO_2 falls below 15 kPa (Stewart and Woodring, 1973; Penteado and Hebling-Beraldo, 1991).

5.1.2.4 Arachnida

With certain notable exceptions, spiders tend to be sit-and-wait predators and high levels of activity are associated with anaerobic metabolism. Their mechanisms of hypoxia tolerance are still a matter of speculation. Although rarely exposed to hypoxia at sea level, some spiders, harvestmen, and pseudoscorpions may live at high altitude (Somme, 1989; Thaler, 2003). Pseudoscorpions also live under the bark of trees or under leaf litter, some spiders are cave dwellers, and sun spiders (Solifugae) may be exposed to hypoxic or hypercapnic environments in deep, sealed, underground burrows (Lighton, 1998).

Spiders show an interesting solution to the problem of local internal hypoxia caused by the presence of regions of relatively high metabolic rate. While the plesiomorphic gas exchanger in spiders is the book lung (see Chapter 7), which could also be described as an air gill, some groups such as jumping spiders and hackled orb weavers have a well-developed tracheal system that targets specific regions that have a high metabolic rate, such as the nervous system or the leg pairs used in web preening, respectively. Whereas the principle of developmental plasticity providing greater vascularization of organs with high metabolic rate is familiar to most of us, solving this problem by evolving a separate respiratory system is unusual.

Arachnids cannot ventilate their respiratory organs (Paul et al., 1987) but they can regulate opening and closing of spiracles, some species even showing discontinuous breathing (see Chapters 7 and 8). Among the potential responses to short-term hypoxia are increases in spiracular opening, increases in perfusion, leaving the hypoxic environment, reduction of aerobic metabolism, or switching to anaerobic metabolism. Scorpions and spiders rely greatly on anaerobiosis during activity, with, as previously mentioned, D-lactate (as opposed to L-lactate in other organisms) as end product in spiders (Prestwich, 1983b). These groups would therefore be potentially well suited to deal with at least short-term external hypoxic conditions.

5.1.2.5 Hexapoda

Regarding their ability to deal with hypoxia, hexapods are able to recover from severe hypoxic or even anoxic exposures that would kill most other animals. Some have even survived live examination in an electron microscope, which may be worse than being released on the moon.

Anoxia

Terrestrial insects respond to complete anoxia with loss of body coordination and posture, suppression of neuronal action potentials, transient convulsions, and loss of brain function (Wegener, 1987; Krishnan et al., 1997), and the metabolic rate falls to between 3 and 14 per cent of normoxic values (Hochachka et al., 1993; Wegener, 1993; Hoback et al., 2000), responses they have in common with most craniotes. However, insects possess one decisive advantage that allows them to recover from hours or even days of exposure to complete anoxia (Wegener, 1993; Hoback and Stanley, 2001). They can supply the tissues during reoxygenation by passive diffusion through the tracheal system alone, a mechanism that works even in the complete absence of ATP, which would be needed to power respiratory and circulatory muscles of other organisms. But ATP is not necessarily totally absent: even under complete anoxia, insects support some ATP production by normal anaerobic pathways, converting glycogen to lactate and alanine (Hochachka et al., 1993; Wegener, 1993; Hoback et al., 2000). But that is not all. Larvae of tiger beetles (Cicindelidae) inhabit areas that are often flooded for days or weeks. They enter a quiescent state without lactate accumulation and are thus able to survive anoxia for longer periods than any other terrestrial insect larvae (Hoback et al., 1998).

But how do insects regulate this anoxia tolerance? In the fruit fly, *Drosophila melanogaster*, anoxia tolerance requires RNA editing and the production of anoxia-specific proteins (Farahani and Haddad, 2003). A second finding is that accumulation of the sugar trehalose appears to be important for preventing protein aggregation during anoxia in flies (Chen et al., 2002). Overexpression of the trehalose-6-phosphate synthetase gene increases anoxia tolerance, elevates trehalose, and reduces anoxia-dependent protein aggregation, while deletion of the gene has opposite effects.

Hypoxia

Most insects are hypoxia tolerant to a PO_2 down to 2–3 kPa, which is equivalent to an altitude of 14,000–16,000 m, depending on the temperature one assumes, and they can usually sustain resting aerobic metabolic rates down to an ambient PO_2 of 5 kPa, or an altitude exceeding the height of Mount Everest. But the critical PO_2 required to support metabolic rates that are consistent with flight are greater: 10 kPa in fruit flies and 8 kPa in honey bees, equivalent to altitudes of about 6000–7000 m, respectively.

The first acute responses to short-term hypoxia are behavioural changes and also physiological mechanisms of the respiratory system. Those mechanisms include opening spiracles and reducing tracheolar fluid levels (see later in this section). Initially, a drop in atmospheric oxygen causes a drop in PO_2 of a similar magnitude inside the tracheal system. Spiracular opening results in an increase in convective gas transport and in gas exchange (see Chapter 7). Falling PO_2 also causes transient increases in abdominal pumping rates (Miller, 1966; Arieli and Lehrer, 1988; Greenlee and Harrison, 1998). At least in the locust grasshopper, *Schistocerca migratoria* (Lewis et al., 1973; Hustert, 1975), this increased ventilation exacerbates carbon dioxide emission while metabolic rate remains constant or falls, leading to a drop in internal PCO_2 levels and a rise in pH, which inhibits ventilation as in terrestrial craniotes (Miller, 1966; Harrison, 1989; Greenlee and Harrison, 1998; Greenlee and Harrison, 2004). This inhibitory hypocapnia explains why ventilation rates measured even minutes after exposure to hypoxia are often still below normoxic rates unless the PO_2 is below 5 kPa (Arieli and Lehrer, 1988; Greenlee and Harrison, 1998).

The removal of fluid from the terminal tracheoles during hypoxia represents a further potential for control of oxygen delivery, analogous to the local regulation of capillary perfusion in craniotes (Wigglesworth, 1930, 1983). The mechanisms responsible for removal of fluid from tracheoles are unknown, though there is some evidence that increases in cellular osmotic pressures, perhaps due to accumulation of metabolites, cause fluid transfer (Wigglesworth, 1930, 1983).

When all other short-term strategies fail, hypometabolism and anaerobic metabolic pathways are used. But when insects are exposed to hypoxia for long periods of time (hours to days), morphological changes in the respiratory system are induced that are analogous to hypoxia effects on capillary sprouting in craniotes. Rearing grasshoppers, bugs, and beetles in hypoxia increases the diameter of major tracheal trunks and the number of tracheoles, and also increases tracheal branching (Locke, 1958a, 1958b; Wigglesworth, 1983; Loudon, 1989; Henry and Harrison, 2004). Many hexapod species are naturally exposed to hypoxic conditions with a PO_2 below 1 kPa. As long as they do not live permanently under hypoxic conditions, they will try to escape or will survive by switching to low metabolic rates or anaerobic metabolism until conditions improve. But the soil-dwelling springtail *Folsomia candida* (Collembola) has another solution: it can breathe in water to a limited extent by forming a gas bubble around the entire body thanks to its non-wettable cuticle. The gas bubble works both as an oxygen store and as a physical gill, so that *F. candida* can survive several days in air-saturated waters and can even tolerate up to 18 hours of complete anoxia anaerobically (Zinkler and Rüssbeck, 1986). But temporary survival in water is not so uncommon among insects. Species with aquatic larvae, such as mosquitos, midges, and dragonflies, are well known, and are discussed in more detail in Chapters 6 and 9. But also some adult terrestrial beetles escape to water to avoid predation and survive there relying on reduced metabolism, cuticular gas exchange, or plastron respiration (Hoback and Stanley, 2001).

Species that normally live in hypoxic habitats show behavioural adaptations to these conditions. For example, the scarabaeoid beetles *Aphodius rufipes* and *Sphaeridium scarabaeoides* live in dung pats for feeding and reproduction, where they maintain normal locomotion and respiration for about 30 minutes in conditions of down to a PO_2 of 0.7 kPa, below which they reduce activity (Holter, 1994; Holter and Spangenberg, 1997). But they also seem to be able to sense oxygen and move to oxygen-richer areas of the dung pats.

In fact, if one looks carefully, practically every insect group has developed some way either of maintaining aerobic conditions or surviving periods of hypoxia or anoxia, whether generated externally or internally (Keister and Buck, 1961; Gäde, 1984; Krolikowski and Harrison, 1996). Some of these methods are quite bizarre, such as termites that ventilate their colonies using wing beats, or the mangrove ant, *Camponotus anderseni*, that keeps water from flooding the nest by soldier ants blocking the entrance with their heads (Nielsen and Christian, 2007).

Acclimatization

As stated previously, the tracheal system in insects responds to hypoxia during development. Terminal branching processes of the tracheoles have been especially well investigated in *Drosophila*, where their growth is regulated by the oxygen demand (Jarecki et al., 1999). In *D. melanogaster*, tracheolar branching is induced by release of fibroblast growth factor from oxygen-starved tissues under the control of the regulatory gene *branch* (Jarecki et al., 1999). Tracheae express HIF-1 and the proteins Similar and Tango, which function as HIF-α and HIF-β homologues, respectively, are expressed at higher oxygen levels there than in other tissues (Ma and Haddad, 1999; Lavista-Llanos et al., 2002). This suggests that tracheae have some mechanisms for sensing and responding to hypoxia (Loudon, 1988; Burggren, 1992; Greenberg and Ar, 1996).

Maximum body size is positively correlated with environmental PO_2: fruit flies raised at 10 kPa are smaller than controls (Frazier et al., 2001). But this phenomenon is not unique to insects—aquatic amphipods (Chapelle and Peck, 1999), snails (McClain and Rex, 2001), and nematodes (Soetaert et al., 2002) are smaller at high altitude (Peck and Chapelle, 2003) than at sea level.

5.1.2.6 Actinopterygii

Behavioural mechanisms of ray-finned fishes and hypoxia

A behaviour commonly observed among fish that inhabit warm, oxygen-poor water is surface skimming. Due to evaporation, the temperature at the surface of standing bodies of water can be appreciably cooler than just a few millimetres deeper. This means that this water can also contain more oxygen than lower in the water column not only because it is in contact with the air but also just because it is cooler. During the warm-water season, the South American pacú (*Colossoma macropomum*) develops a huge lower lip that aids in surface skimming (Val et al., 1998; Wood et al., 1998). Even sharks have reportedly been observed to force surface water and air over their gills, although this behaviour also could be related to cleaning the gills.

Hypoxia is known to result in behavioural changes that have a negative effect on escape from predators (Domenici et al., 2007). One of these is surface breathing, as discussed earlier, which exposes fish to avian and mammalian predators. Another is chaotic and panic reaction and interference with normal schooling behaviour. Large schools of fish can completely deplete oxygen, even in cold, well-oxygenated water. This means that the fish at the end of the school experience hypoxia or even anoxia, which would have a negative survival effect as long as the school moves in a straight line: the ones at the end get less oxygen, swim slower, lose visual acuity, and are more exposed to predation. So is not surprising that schooling behaviour is seldom follow-the-leader, but the leader changes when the school changes direction.

Anatomical adaptations

Crucian carp (*Carassius carassius*) maintained in 'normoxic' water have gills that lack protruding lamellae, the primary site of oxygen uptake in fish (see Chapter 11). Instead, the lamellae are embedded in a cell mass. This implies that these fish are permanently adapted to a low-oxygen environment, and, when exposed to normoxic water, they have a reduced respiratory surface area. But when the fish are kept in hypoxic water, a large reduction in this cell mass occurs, making the lamellae protrude as they do in most other fish, increasing the respiratory surface area by nearly 7.5-fold. Carp with protruding lamellae have a higher capacity for oxygen uptake at low oxygen levels than fish with embedded lamellae, but water and ion fluxes also increase, meaning greater osmoregulatory costs (Sollid et al., 2003). This change in gill structure with respect to ambient oxygen is not restricted to the Crucian carp, nearly 40 years ago it was already known that the hypoxia-intolerant rainbow trout (*Oncorhynchus mykiss*) can increase the diffusing capacity of its gills in response to hypoxia (Soivio and Tuurala, 1981).

In addition to the release of modulators such as nucleoside triphosphates, which regulate the oxygen affinity of haemoglobin in the erythrocytes (for references see Val, 1993), catecholamines secreted by paracrine chromaffin cells following hypoxia have far-reaching effects, ranging from enhanced erythrocyte release from the spleen to swelling of red blood cells, elevation of their internal pH and increased permeability of the gill epithelium (reviewed by Randall and Perry, 1992).

A closer look at hypoxia-tolerant fish

Although fish do not actually 'dive', they can, in fact, drown if the respired water does not contain enough oxygen. However, neither fish nor aquatic, water-breathing invertebrates have a problem with caisson disease (decompression sickness, discussed in section 5.2.3) because they never have to deal with oxygen and nitrogen in gas form.

Survival in complete anoxia is possible for only very few craniote species. While most craniotes, for example, mice or trout, die in anoxic conditions after minutes, some animals manage anoxic phases without problems. The most hypoxia-sensitive organ is the brain, which consumes most of the oxygen when animals are at rest. When ATP supply is insufficient, the animal dies. Mammals are particularly vulnerable in this respect. But anoxia-resistant species manage to maintain species-specific sufficient ATP levels, even under anoxia. They exploit large carbohydrate stores, mainly in the liver, to increase the glycolytic ATP-production, but this is not the only trick they use.

Developing zebrafish (*Danio rerio*) lack gills and rely on cutaneous gas exchange. While only larvae can survive a complete lack of oxygen, pre-exposure of zebrafish to hypoxia during early stages increases survival during exposure to extreme hypoxia or even anoxia at later stages. Decreased energy demand during hypoxia and a lower perfusion pressure following hypoxia are likely the most important factors (Rees et al., 2001). Adult zebrafish that have not been previously exposed to low oxygen levels are not very hypoxia tolerant. Typical hypoxia- and even anoxia-tolerant fish such as the goldfish (*Carassius auratus*) and the Crucian carp (*C. carassius*), like the zebrafish, reduce activity and seek out colder water during low-oxygen phases, but they remain active. The brain also shows more activity than in some other hypoxia-tolerant species such as turtles (see later), but is somewhat less than under normoxic conditions, and some sensory functions such as vision and hearing are reduced. There is a moderate release of adenosine and GABA. Thus, ATP depression is modest and physical activity is maintained during anoxia. The Crucian carp possesses very large carbohydrate stores in the liver and can generate ATP via glycolysis. The only limiting factor in anoxia is the depletion of its carbohydrate stores, but the amount of lactate produced is also too large to be stored in the tissue: the fish remove it by converting it into ethanol (!) in the muscles. The lactate is converted into pyruvate which is turned into acetaldehyde and finally into ethanol. From the muscles, ethanol enters the blood, where the alcohol content may reach 10 mM l^{-1} (= 46 mg 100 ml^{-1}), whence it diffuses out over the gills (Nilsson and Lutz, 2004). Since the legal blood alcohol limits in England (80 mg 100 ml^{-1}) and Scotland (50 mg 100 ml^{-1}) are above 46 mg 100 ml^{-1}, the anoxic goldfish would still be allowed to drive a car in Great Britain. Only recently have the molecular mechanisms regulating this fascinating phenomenon been elucidated in an evolutionary context (Fagernes et al., 2017). It appears that a pyruvate decarboxylase, analogous to the one found in brewer's yeast, in addition to a specialized alcohol dehydrogenase, are pivotal to this process, and appear to be an autapomorphy of the genus *Carassius* within the carp lineage. Whole-genome duplication resulted in some—but not all—of this multienzyme complex giving rise to a pyruvate decarboxylase that releases ethanol.

5.1.2.7 Tetrapoda

Also the gills of amphibians that are exposed to prolonged periods of hypoxia are larger than in normoxic ones (Burggren and Mwalukoma, 1983). Amniotes living for long periods under hypoxic conditions show adaptation of the molecular structure of haemoglobin associated with a change in affinity and an increase of the number of red blood cells (haematocrit), and of the concentration of haemoglobin per cell. All these conditions are found in animals living permanently in hypoxic environments but also occur as acclimatization in those exposed for shorter time periods or periodically to hypoxia. Also, underwater exploration and space travel have in common that the adventurers are exposed to artificial atmospheres at pressures that can differ drastically from those in which most terrestrial craniotes evolved. Since the species best studied under these experimental conditions is our own, we shall return to it in section 5.2.1.

Long-term, anoxic diving in a turtle

Slider turtles, *Trachemys scripta*, use a combination of metabolic depression and anaerobic metabolism and can dive for several hours. During the first diving phase they use their oxygen stores, especially in the lungs. Later on, metabolism is reduced to about 40 per cent of normal values, the final oxygen stores are depleted, and there is a shift to anaerobic ATP production. During the third phase, which may last several hours, metabolism is reduced to about 20 per cent of normal values and the animals are totally reliant on anaerobic metabolism (for overview and references, see Jackson, 2013).

The same species overwinters with reduced activity and metabolism in ice-covered ponds. It just swims to the bottom, where the temperature remains constant at 4°C, stops breathing, and falls into a comatose-like state. Since more than 50 per cent of the ATP used for the brain is needed to maintain electric gradients over cell membranes, which then are used for electric activity and transport of neurotransmitters and metabolites, it behoves the turtle to reduce this energy consumption. Unlike the mammalian brain, which is extremely sensitive to even short periods of anoxia, central neurons of anoxia-tolerant craniotes are protected from the damaging effects of anoxic depolarization and excito-toxic death by ion channel and spike arrest. This so-called spike arrest in the turtle brain involves both a decrease in excitatory glutamate release, probably originating from stellate interneurons, which causes a shunting current and prevents further depolarization of the neurons to threshold (Jonz et al., 2016). Both mechanisms reduce the metabolic demand of the brain. Glycolysis delivers the necessary ATP and the lactate formed is buffered partly by releasing calcium from the shell. Calcium binds to lactate and Ca^{2+} is also free for other processes (e.g. blood coagulation, heartbeat, and muscle contraction) whereby the Ca^{2+} concentration in the blood rises up to 16 times the pre-hibernation level. By forming a lactate–calcium complex, both free lactate and free calcium concentrations are held at reasonable levels. Degradation of the lactate and release of calcium have to wait until spring. Until then, the turtle slumbers in its dazed state at the bottom of the pond.

5.2 High altitude and diving in mammals (especially humans) and birds

5.2.1 Acclimatization and adaptation to altitude in humans

Down to an ambient PO_2 of about 18 kPa, which occurs at altitudes of around 1500 m, most lowland inhabitants can live comfortably without previous

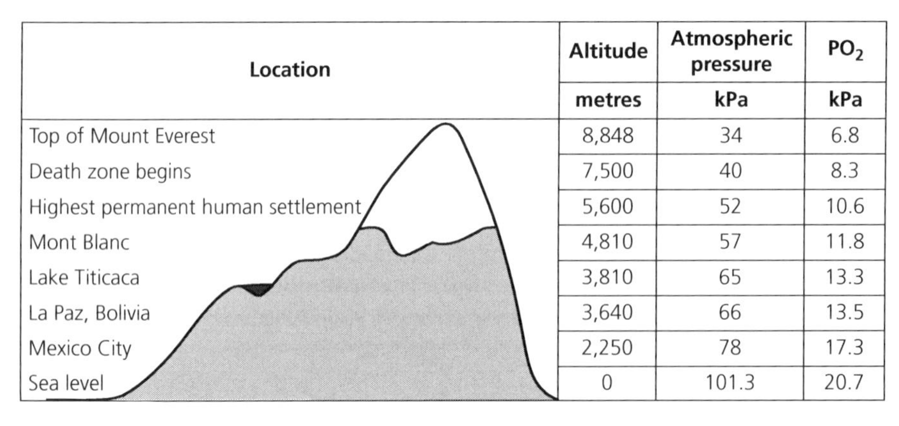

Location	Altitude	Atmospheric pressure	PO_2
	metres	kPa	kPa
Top of Mount Everest	8,848	34	6.8
Death zone begins	7,500	40	8.3
Highest permanent human settlement	5,600	52	10.6
Mont Blanc	4,810	57	11.8
Lake Titicaca	3,810	65	13.3
La Paz, Bolivia	3,640	66	13.5
Mexico City	2,250	78	17.3
Sea level	0	101.3	20.7

Figure 5.3 Although the PO_2 relative to the total atmospheric pressure (p_{atm}) remains the same at all altitudes in the troposphere due to the constant relative oxygen level, a reduction in p_{atm} results in an absolute drop in effective PO_2 with increasing altitude. The effective PO_2 at different altitudes relevant to human habitation is shown here. (Original, diverse sources)

acclimatization. In aircraft cabins, the pressure is most often a simulated altitude of slightly above 1500 m and causes no problems for healthy passengers. But about 140 million people live permanently at altitudes exceeding 2500 m and about 30 million, even above 3000 m. The highest permanent human settlements are at 5300 m in Peru and at 5600 m in Chile (Figure 5.3). Here partial pressure of oxygen is about half its sea-level value. Above 7000–7500 m the so-called death zone begins, in which long-term human habitation is not possible and a non-acclimatized person will die within minutes to hours. Very few, even well-acclimatized alpinists have survived the ascent of Mount Everest (8848 m) without additional oxygen. At that altitude, non-acclimatized persons will lose consciousness within 15 minutes.

And even in fully acclimatized alpinists, the mean arterial PO_2 when breathing ambient air is only 3.28 kPa, with a range of 2.55 to 3.93 kPa, compared with about 11 kPa at sea level. We are talking here about levels of *arterial* oxygen that are below the normal sea level *venous* value of about 4.6 kPa (Grocott et al., 2009). So just what happens that makes it possible for some people to survive at such altitudes?

Let us first talk about rapid, short-term acclimatization. In general, in lowlanders (people living at or below 1500 m) acute reactions and reversible acclimatization processes take 1–2 days at 3000 m and 2–3 weeks at 5000 m. Highlanders, on the other hand, have permanent adaptations which are irreversible and fixed in the genome. They also are capable of acute acclimatization when climbing to higher altitude, but their reactions are faster than in lowlanders. Both acclimatization and adaptation are multifaceted and involve adjustments in breathing frequency and tidal volume as well as in the blood and metabolism of the entire organism, in particular in the muscles and brain. These acclimatization processes do not occur singly, but in concert with one another in stages.

Hyperventilation, prompted by oxygen sensors in the carotid body, is the first acute reaction under hypoxia and is the most important feature of acclimatization up to 5000 m. At rest, non-acclimatized lowlanders begin hyperventilating at an alveolar PO_2 of 7 kPa, which would be reached at an altitude of 3000 m. At 4800 m, hyperventilation in a non-acclimatized person increases the alveolar PO_2 by about 1 kPa. acclimatized persons on the other hand start hyperventilating much earlier (i.e. at a higher alveolar PO_2), meaning that hyperventilation can increase the alveolar PO_2 considerably and that the effect is larger than in non-acclimatized people.

However, this hyperventilation also has its downside: the PCO_2 of the blood decreases by about 40 per cent. The respiratory alkalosis caused by this hyperventilation can result in reduced oxygen release in the tissues (small Bohr shift), and also a hypocapnic inhibition of breathing, with apnoeic phases that are dangerous especially during sleep. This problem usually lasts about 1 week, depending on the altitude, and is part of the long-term adaptation process. After this time, hydrogencarbonate release in the kidney restores the blood to pH values acceptable to the central PCO_2/pH receptors, and the peripheral oxygen sensors become more sensitive. Highlanders show a permanently blunted hyperventilation response and increase their alveolar PO_2 less than acclimatized lowlanders, but they also blunt the hypocapnic inhibition and rely on other adaptive processes (see later) for increasing oxygen uptake and delivery.

During hyperventilation, pulmonary vasoconstriction—mediated by intrapulmonary blood oxygen sensors—results in a more homogeneous perfusion of the lung and thus an improved ventilation/perfusion ratio. So the lungs are better off but the heart has a problem: in order to supply the lungs, the right ventricle has to do more work than it is accustomed to and is threatened by exhaustion or hypertrophy. Unlike skeletal muscle, cardiac muscle has very little glycogen reserve and a very limited capacity for anaerobic metabolism. Although the vasoconstriction in the lung is reduced and the resistance to blood flow is less than before acclimatization, lung function also reaches its limits in severe hypoxia. Blood passes quickly through the lung, resulting in reduced oxygen equilibration time. This, combined with a low ΔPO_2, means that the diffusive gas exchange is considerably reduced. In this respect, highlanders often have better lung function than even long-term adapted lowlanders can ever achieve.

The circulatory system increases cardiac output by increasing heart beat frequency (tachycardia) as a first, quick reaction to hypoxia. But the stroke volume of the heart does not change or may even go down. After a couple of days, heart rate is also reduced, and the final cardiac output might be even lower than under normoxic conditions. Peripheral circulation is reduced, while circulation to the heart, respiratory muscles, and brain is maintained.

Even more important than these acute reactions, however, is longer-term reactions mediated by vascular oxygen sensors. Under the influence of VEGF1 (vascular epithelial growth factor 1), which is controlled by HIF-1, the capillary density increases (angiogenesis), especially in well-vascularized organs. Thus the oxygen-carrying capacity is greater due to an increased blood volume. It may increase, for example, in Peruvian natives from 85 ml kg^{-1} at sea level to 110 ml kg^{-1} at 4500 m (Hochachka, 1998). But this also means more problems for the heart and lungs, since the lungs must accommodate the same volume as the systemic vessels with each heartbeat and there is no evidence of angiogenesis in the lungs. Increased pulmonary blood pressure could stretch the vessels in the lung to a certain extent, and also the air–blood barrier in the capillaries could be pulled to its minimum thickness, thereby increasing the diffusing capacity somewhat, but also risking haemorrhage. So lung anatomy sets a real, functional limit, which literally cannot be stretched.

The blood also shows medium to long-term effects, which start after 1–4 weeks at altitude. These include an increase in haemoglobin concentration per erythrocyte and also an increase in the number of erythrocytes: high-altitude polycythaemia. In lowlanders, both factors might increase by about 30 per cent at 4500 m and return to normal after returning to sea level. Although polycythaemia results in an increased carrying capacity of the blood for oxygen, it also has a disadvantage, since it increases the viscosity of the blood and leads to the risk of capillary blockage and thrombosis and/or haemorrhage.

Highlanders in general have high concentrations of erythrocytes and haemoglobin and also show a higher percentage of oxygen saturation than lowlanders. This is especially true for the Andeans, whereas Ethiopian highlanders do not differ from lowlanders (Beall, 2006). While the oxygen affinity in highlanders is only slightly increased, modulators and respiratory alkalosis influence the affinity of the haemoglobin in lowlanders. In addition, 2,3-BPG decreases the oxygen affinity (see Chapter 4) even below 4500 m, making it easier for the blood to release oxygen in the tissue. In the absence of 2,3-BPG, oxygen release would otherwise be a problem because respiratory alkalosis, caused by the hyperventilation, increases the oxygen affinity as hypoxia increases. Thus, 2,3-BPG allows oxygen to be released in the tissues, and through the pH-induced increased affinity, oxygen uptake in the lung is in concert with environmental conditions. So respiratory alkalosis has some advantage at extreme altitudes, but in long-term adaptation, renal compensation will eliminate it. For this reason it is recommended for high-altitude climbers to adapt at 5000–5500 m and then choose a short-term stay in higher altitudes to take advantage of the effect of respiratory alkalosis until renal compensation kicks in.

Metabolism during activity is adapted by increasing the anaerobic capability and by decreasing the aerobic demands of the muscles. The first acute response is upregulation of the glycolytic enzymes mediated by HIF-1. Medium- to long-term response, however, is downregulation of the glycolysis, reducing lactate levels. In highlanders, the glycolytic response is much less than in lowlanders, so the blood lactate level is also lower. $\dot{V}O_{2\ max}$ also decreases with altitude. At 4000 m in lowlanders it is 70–80 per cent of sea level values and at a simulated altitude of 8800 m only 20–25 per cent, and adaptation increases these values by only 5–10 per cent. These limits are set by lung function, reduced maximum heartbeat frequency, and the reduced perfusion in skeletal muscles, meaning that the skeletal muscles can only partly benefit from the hypoxic responses.

Chronic hypoxia even has a direct negative influence on the skeletal muscles of untrained lowlanders. Residing at altitudes above 5000 m, body mass and muscle mass can actually decrease by 5–10 per cent. Since the capillary volume remains unchanged and the muscle mass decreases, the increase in capillary density is only relative. In addition, mitochondrial volume density also decreases by about 20 per cent. Taking into account the reduction in muscle mass, the total loss of mitochondria is actually about 30 per cent and the oxidative capacity decreases by 25 per cent. The oxygen supply situation is then shifted and benefits the remaining mitochondria. Thus, the reduction in $\dot{V}O_{2\ max}$ is a combined effect of a reduction in muscle volume and in oxidative capacity (Hoppeler and Vogt, 2001).

In highlanders, the $\dot{V}O_{2\ max}$ is only 60 per cent of that of a lowlander, but acute hypoxia has a lower effect, reducing the value by maximally about 15 per cent. Also, oxidative capacity compared with lowlanders is 30 per cent lower but the muscle cross-sectional area is similar to lowlanders, whereas capillary density is smaller. This means that highlanders are at a distinct disadvantage when at sea level. In lowlanders, short-term training at altitude but living under normoxic conditions ('living low–training high') increases the oxidative capacity of skeletal muscles mainly by an increase in mitochondrial volume, in myoglobin concentration, and in capillary density. 'Living high–training low', on the other hand, increases mainly the haemoglobin concentration (Hoppeler and Vogt, 2001).

Summarizing, human responses to altitude begin with an acute increase in ventilation, lung perfusion, cardiac output, and glycolysis. Medium- to long-term acclimatization/adaptation include elevated ventilation and lung perfusion, normalization of cardiac output, and downregulation of glycolysis, while the oxygen transport capacity of the blood is improved by an increase of haemoglobin and erythrocyte concentrations. Adaptation of highlanders is characterized by a blunted ventilatory and lung perfusion increase, a downregulated glycolysis, and mainly an increased oxygen transport capacity of the blood together with a reduction in $\dot{V}O_{2\ max}$. In adapted lowlanders at altitudes above 7000 m, a very strong hyperventilation and a marked respiratory alkalosis occur, which is only incompletely compensated by the kidneys. An increased oxygen affinity, because of the respiratory alkalosis, helps to bind the small amount of oxygen in air and 2,3-BPG helps the blood to release oxygen in the tissue, but the $\dot{V}O_{2\ max}$ and anaerobic metabolism are greatly reduced. Finally, a continued weight loss (wasting) is symptomatic for the progressive deterioration experienced (West, 2006).

5.2.2 Other craniotes at altitude

Some birds are effectively adapted to hypoxia as they live in or pass through high mountain ranges. For example, Rüppel's griffon vulture (*Gyps rueppellii*) was observed flying at an altitude of 11,300 m, where it nearly collided with an aircraft, and the bar-headed goose (*Anser indicus*) breeds at 5600 m and flies over the Himalayas during its periodic migration. Special morphological adaptations could not be demonstrated but the avian lung, as such, is a highly efficient prerequisite for breathing at high altitude. *Anser indicus* shows a marked increase in tidal volume in response to hypoxia compared to low-altitude species (Scott and Milsom, 2007). Also, these two high-altitude species have haemoglobins with optimized oxygen-binding characteristics, achieved thanks to a modified structure of the molecule. Oxygen binding is more efficient compared with other birds and non-avian craniotes, and oxygen can be bound even at low PO_2. A potential super-cooperativity with an n_H value of more than 4 is attributable to the self-association of the haemoglobin with special subunits (see Chapter 4). The high-affinity haemoglobin of *A. indicus* shows a substitution of one amino acid which results in the loss of a hydrogen bond that otherwise stabilizes the low-affinity, deoxygenated T state. Another factor that influences oxygen affinity is the lack of a salt-bridge, which normally would stabilizes the T state, and create a large Bohr effect. *Gyps rueppellii* has four isohaemoglobins with different oxygen affinities due to amino acid substitutions that variously stabilize or destabilize the T and R states of the haemoglobin, depending on the demands, and ensure that oxygen loading and unloading is effective over a broad range of PO_2. In addition, the avian brain appears to continue functioning normally at levels of hyperventilation and the resulting respiratory alkalosis that would cause unconsciousness in humans (Shams and Scheid, 1989; Butler, 1991).

Compared with sister species from low elevations, high-altitude species tend to be larger and show a higher respiratory compliance. But since the highlanders also tend to choose a high tidal volume/low-frequency breathing pattern at rest and the work of breathing increases with the square of tidal volume, the higher compliance does not translate into decreased work of breathing (York et al., 2017). There may be some advantage in minimizing the friction in the body cavity associated with cyclic oscillation between inflation and deflation as well as a possible loss of gas exchange efficiency in reversing the airflow in the neopulmo. But this remains a hypothesis that needs testing.

In rats and mice, hypoxia exposure improves lung growth by an increase in lung volume, alveolar tissue volume, and surface area. But this process depends on the time period of development in which animals are exposed to hypoxia. While in the perinatal period and after birth, hypoxia even impairs lung development, lung development is accelerated by 13 per cent oxygen in the time period between days 14 and 40 immediately after birth. In guinea pigs and dogs raised under mild hypoxia, lungs result in higher diffusing capacities, caused by an increase in lung volume, in alveolar tissue volume, and in respiratory surface area, but also by a thinner diffusion barrier. But these changes are transitory: adults kept for longer periods at high altitude with comparable oxygen conditions do not show any effect in the respiratory system, and the measured effects are maturity-dependent increases of lung dimensions in response to hypoxia. Animals raised under hypoxia and returned to normoxia as adults are indistinguishable from those raised under normoxia (Burri and Weibel, 1971; Burri, 1974; Lechner and Banchero, 1980). Whether this holds only for mammals or applies generally remains to be tested. In a recent study, chicks of the domestic fowl bred under hypoxic conditions maintained an increased air capillary volume at least up to 10 days after hatching (Amaral-Silva et al., 2019).

In mammals other than humans, polycythaemia and increased erythrocyte haemoglobin concentration as part of the acclimatization response are also well documented. Some species such as the Tibetan antelope (*Pantholops hodgsoni*) have especially high erythrocyte numbers, and also show cardiac adaptations that allow them to further increase their heart rate when exposed to 12.5–14.6 per cent oxygen. These effects, however, are not seen in Tibetan sheep (*Ovis aries*) living at the same altitude (4300 m) (Rong et al., 2012).

Camelids of the high Andes have haemoglobin with increased oxygen affinity. There are six species of camelids: two lowland camels of Africa and Asia, and four highland species (the guanaco, the lama, the alpaca, and the vicuña) in the mountains of South America. All six species show a relatively low P_{50} of 2.26–2.93 kPa (in humans, for example, it is 3.46 kPa), but the high-altitude ones have higher affinity haemoglobins than the Old World species due to changes in the amino acids in the globin chains of the molecule, and also show a reduced 2,3-BPG interaction (Weber, 1995; Weber and Fago, 2004), not unlike fetal haemoglobin (see Chapter 4).

Of course, all of this is barely scratching the surface of this fascinating area of human and comparative adaptation and acclimatization to altitude. These are some of the most exciting chapters in respiratory biology and the reader is encouraged to check out accounts by John West, Sir Edmund Hillary (1919–2008), Hanns-Christian Gunga, Reinhold Messner, Jacques-Yves Cousteau (1910–1997), Hans Fricke, Peter Hochachka (1937–2002), Peter Ward, and others.

5.2.3 Diving craniotes

Diving has evolved in about half of the recognized major taxa of placental mammals, sometimes even several times within a single group such as rodents, while some groups such as seals, sea lions, sea cows, and whales have become exclusively aquatic predators or foragers. Whales and even hippopotamus give birth under water, which is quite a trick for an air breather. While on this topic, during birth, newborn mammals show similar responses in respiratory and circulatory system as diving mammals. For a while it was even trendy for mothers to give birth underwater. You may remember also from Chapter 4 that there is also some indication from the molecular structure of mammalian myoglobin that some terrestrial groups such as elephants may have harboured aquatic ancestors. And indeed, a recent study suggested that a potential Palaeozoic, pre-mammalian origin of the diaphragm may have been a key to an aquatic lifestyle (Lambertz et al. 2016).

Also, nearly half of the major taxa of neognath birds have evolved diving forms, some of them such as penguins and the recently extinct great auk giving up wings in favour of flippers. But unlike mammals, which are almost exclusively live bearing, all birds are oviparous and must come on land to lay eggs. In mammals, chemoreceptors near the glottis, nose, and mouth sense water and prevent inhalation during diving, while chemoreceptors of carotid and aortic bodies are ignored during diving. Ducks appear to do the same, going into breath-hold and bradycardia mode when water is applied to the beak (for references, see Butler and Jones, 1997). Looking at diving times and diving depths, it becomes clear that for routinely diving species special faculties must have evolved (Table 5.1).

And then there is the pressure. For every 10 m depth there is an increase in pressure of 1 atm. That means that at a depth of 400 m the pressure is about 40 atm. That is equivalent to a weight of about 400 kg on a man's chest, which would easily crush it. And yet sperm whales dive to a depth of 2 km! Even more threatening than being crushed is the fact that gases in the lungs are forced into solution in the tissues and blood, causing nitrogen narcosis (remember, air is 80 per cent nitrogen). The oxygen in the air gets consumed but the nitrogen stays in solution and is released as bubbles on resurfacing, causing decompression sickness (caisson disease). Think of the dents in the ceiling and stains on the

Table 5.1 Diving mammals, birds, and reptiles: routine diving times and depths, in (parentheses) maximum recorded values (after Kooyman and Ponganis, 1997). The northern sea elephant is the world record holder in diving times (up to 2 hours) and the sperm whale in diving depth (down to 2000 m)

Species	Habitual and (maximum) dive duration, minutes	Habitual and (maximum) dive depth, metres
Emperor penguin *(Aptenodytes forsteri)*	6 (22)	100 (500)
Leatherback turtle *(Dermochelys coriacea)*	11 (45)	200 (1000)
Weddell seal *(Leptonychotes weddellii)*	15 (82)	300 (700)
Northern sea elephant *(Mirounga angustirostris)*	25 (120)	400 (1600)
Sperm whale *(Physeter macrocephales)*	40 (75)	500 (2000)
Human (*Homo sapiens*)	3 (11 min 35 s)	10–20 (214)

carpet if you open a bottle of champagne quickly, but if you open it slowly and carefully the bubbles stay in the bottle. The best solution to avoid caisson disease is slow emergence or if this is not possible, immediate removal of the victim to a decompression chamber for re-pressurization and simulated slow emergence. Helium as a carrier gas for longer dives has some advantages, since it does not cause narcosis and diffuses much faster than nitrogen out of the tissues. Breathing pure oxygen is not an option, because of ROS formation and oxygen toxicity.

But the fact remains that whales, seals, and penguins can dive to tremendous depths and rapidly emerge without getting decompression sickness. One trick is just to go for short dives of less than half an hour. Then not enough nitrogen gets into the tissue to cause any harm. A glance at Table 5.1 reveals that this is just what marine craniotes actually do. For longer dives, seals, for instance, exhale on diving, thereby reducing the amount of nitrogen in the lungs.

Another option is to reduce the amount of gas that can enter the blood by exhaling or storing the air in wide, non-respiratory airways. Whales appear not to exhale, but, as in seals, the trachea and proximal bronchi are voluminous and when diving, most of the air is pressed out of the lung tissue and into the incompressible airways. Also the blood flow to the lung is reduced during diving so that only minimal amounts of nitrogen can enter the blood from the lungs. More adaptations of diving animals are discussed in the rest of this section.

But the ultimate trick is to store oxygen not in gas form, but bound to myoglobin in the muscles. As the muscles exercise anaerobically they release lactic acid, which lowers the local pH and causes oxygen to be released directly to the tissue and the blood, thereby repaying the oxygen debt in the muscles and providing oxygenated haemoglobin for the brain, eyes, and other vital organs. It is interesting to note that seals, sea lions, whales, and penguins independently evolved the same solutions to diving problems.

In diving mammals, birds, and sea snakes, bradycardia is part of a so-called diving reflex, which also involves selective vasoconstriction to some organs, which then must rely on anaerobic metabolic pathways, whereas others, such as the brain and eyes, receive oxygenated blood. It has even been suggested that highly derived divers such as whales do not experience diving bradycardia, but rather surfacing tachycardia, during which time the muscle myoglobin is loaded with oxygen.

In diving mammals, oxygen is stored not in the lungs but in blood and tissues, especially in the muscles (Table 5.2). The blood of two unrelated diving mammals, seals and whales, has independently achieved a high oxygen carrying capacity of 30–40 ml oxygen per 100 ml, compared with human blood, with 10–15 ml oxygen per 100 ml. To reach and make use of such high values, these species have high blood haemoglobin and muscle myoglobin levels, but the haemoglobin also has low affinity since loading with oxygen at sea level is not restricted, and release of oxygen in the tissues is best achieved with a low-affinity protein.

Anatomically speaking, unrelated diving species, such as whales and manatees, possess a double capillary net in the lung and therefore have a large pulmonary blood volume. This is helpful if the blood haemoglobin is used as an oxygen store, because the lungs must receive the same blood volume as the rest of the body with each heartbeat. Now taking a second look at diving bradycardia, if the cardiac output were to remain constant during the dives, the partially collapsed lungs would not be able to accommodate all this blood volume and the result would damage both the lungs and the heart. So diving bradycardia may be seen as an exaptation for diving, or a consequence of it. But as

Table 5.2 Amount of oxygen stored in different body parts during diving in mammals. After Kooyman and Ponganis (1997)

	O_2 (ml kg^{-1})	% in lungs	% in blood and other body fluids	% in muscles
Human	20	44	47	3
Seals, sea-elephants, whales	60–87	5	66	29

pointed out later, peripheral vasoconstriction also takes place so that the blood volume in the lungs and body can remain about equal.

So how do these animals manage to stay down so long? To begin with, diving mammals use a combination of energy-saving mechanisms, their locomotion is very economical. They do not need to support their body weight and most diving species have a spindle-shaped, ergonomic body form with short hairs or they are even hairless. Also, swimming involves long gliding phases. Secondly, the animals conserve energy by being well insulated, and they may even have counter-current heat exchange in the extremities, greatly reducing heat loss there. Diving bradycardia has already been discussed. In addition, due to peripheral vasoconstriction, oxygen stored in the lungs, blood, and muscle is supplied by the circulatory system only to the brain, sense organs and spinal cord, the heart, and to some extent to the swimming muscles. Other organs cover their ATP requirements anaerobically. Lactate is produced and for the most part stored in the organs and is returned to the bloodstream when animals surface.

The aerobic dive limit is defined as the maximum duration of diving before lactate increases, and is species specific. Diving mammals use anaerobic pathways when the diving time exceeds time periods that vary from individual to individual. The Weddell seal *Leptonychotes weddellii*, for example, starts anaerobic ATP production after about 10 minutes of diving, and blood lactate increases from 1 to about 4 μmol ml^{-1} during a 45-minute dive. The peak of lactate is reached only after re-emergence. They can dive for long periods without any restrictions by lactate, but free-ranging animals normally make shorter dives and stay well within their safety margins.

The highly refined combination of aerobic and anaerobic metabolism is really the key to long diving. Certainly part of the solution is oxygen storage in the blood, including the double capillary net of the lungs and in the systemic system with its large blood volume as well as in the myoglobin of the muscles, whereby the muscles themselves exercise largely aerobically. But it is not always that way. In seals, during development, skeletal muscles switch from aerobic metabolism to a more anaerobic one. Pups and juvenile seals are non-diving and live on land where they have to cope with cold temperatures and use their aerobic capacities and shivering for thermoregulation. During maturation, muscles change their metabolism to be more adapted to the low levels of aerobic metabolism associated with long dives (Kanatous et al., 2008). This observation gives us some insight into how these astounding and highly refined, convergent mechanisms may have evolved in such diverse groups as whales (highly derived cetartiodactyls), seals and sea lions (highly derived carnivores), and dugongs and manatees (more closely related to elephants), as well as in diving birds and reptiles.

In fact, mammals and birds are not the only diving experts among craniotes. Diving reptiles do not end up with some of the aforementioned problems, since the heart is capable of circumventing blood around the lungs (right–left shunt). Far from this incomplete intercardiac separation being a deficit that reptiles 'still' exhibit, this bauplan provides possibilities that mammals and birds do not have. Sea turtles are probably the true experts in this realm. In addition to the intrapulmonary shunt possibility, they even have a sphincter in their pulmonary artery, which can be assumed to play a role in reducing pulmonary circulation during deep and long dives, thereby minimizing gas uptake to the blood (García-Párraga et al., 2018).

Which eventually leads us to the inappropriate question, why haven't seagoing turtles, mammals, and birds evolved gills, since we know from the German embryologist Heinrich Rathke (Rathke, 1825a, 1825b, 1828) that the anlagen are present? Or, wouldn't it be possible for divers to be outfitted with artificial gills? Much to the chagrin of science fiction writers, this solution does not work in warm-blooded, high-performance amniotes, because too much body heat would be lost via the gill, and the hydrogencarbonate buffering system in the blood would break down. In fish blood, hydrogencarbonate is not an important element in blood buffering and the pH is almost one entire unit higher than ours. Although sea snakes and some turtles have achieved at least some modicum of aquatic gas exchange (see Chapter 11), this is not sufficient to cover their entire oxygen demand and would certainly not be much help for a mammal or bird.

5.3 Oxygen as a toxic substance: craniotes

Oxygen is poisonous. In humans, inhaling pure oxygen for longer than 24–48 hours irritates the lung epithelium, and inactivates surfactant, possibly leading to partial collapse of the lung. Longer exposure can result in tracheobronchitis and pulmonary interstitial fibrosis. In addition, since oxygen inhalation suppresses breathing, carbon dioxide tends to accumulate, leading to carbon dioxide-associated damage to the central nervous system (CNS), and eventually to convulsions and death. With increasing hypercapnia, the CNS also becomes less sensitive to CO_2/pH, and following oxygen exposure acute respiratory distress syndrome (ARDS) can result in cessation of breathing. Prolonged exposure to a PO_2 of 200 kPa (100 per cent oxygen delivered at 2 atm pressure) leads to convulsions and a PO_2 of 300 kPa can cause death within hours. This is especially relevant to divers. At 50 m depth, in salt water the pressure is 502 kPa. Unlike the mountain climber atop Mount Everest, who gets only one-third of the sea-level amount of oxygen, the diver getting compressed air at a PO_2 of 20 kPa at sea level would actually be getting five times as much oxygen with each breath as at sea level. So for divers working at that depth, the gas mixture must be adapted accordingly, for example, 16 per cent oxygen and 84 per cent of a mixture of nitrogen and some inert gas such as helium. Very deep divers (more than 100 m) may even use a mixture of helium and hydrogen as a carrier gas. Since the diver would be exposed to hypoxia at shallower depths, this mixture is used as a so-called bottom gas.

In experimental animals, hyperoxia of greater than 30 per cent causes an increase in body size during development in *Drosophila*, trout, and crocodiles, for example, while in tadpoles the external gills are underdeveloped. This experiment actually occurred on a global level 300 mya, in the late Carboniferous and early Permian, because of the evolution of huge photosynthetic vascular land plants under greenhouse-like conditions (Berner et al., 2007). The atmosphere contained up to 35–40 kPa oxygen at sea level. Dragonfly relatives with a wing span exceeding that of modern-day hawks roamed the planet, and massive, millipede-like arthropleurids reached up to 2.3 m, becoming candidates for the *Guinness Book of Records* as the largest terrestrial invertebrates ever (Graham et al., 1995; Dudley, 1998). These animals probably relied on diffusion alone for gas exchange, and increasing the oxygen content from 21 per cent to 35 per cent increased diffusion by about 67 per cent (Dudley, 1998), making long gas-in-gas diffusion distances and large animals meaningful. This hypothesis was experimentally tested using four beetle species that vary in body mass by three orders of magnitude. These studies show that large insects have a greater tracheal volume in relation to body mass than do smaller ones, but that the long distances involved in the tracheae to the legs appear to limit the size of the insects. Thinking now back to the Carboniferous, the prevalent hyperoxia could have facilitated the evolution of giant arthropods because the legs and associated musculature could reach large sizes without the tracheal system becoming limited by spatial constraints (Kaiser et al., 2007) or mechanical/weight constraints of the chitinous cuticle (Dudley, 1998).

This Palaeozoic hyperoxic acceleration of terrestrial animal evolution was not limited to invertebrates. During this period, amphibians the size of modern crocodiles were the dominant (semi-?) terrestrial tetrapods. The first exclusively terrestrial craniotes, the amniotes, although relatively small in size, appeared as well. Much later, sauropod dinosaurs evolved during the Mesozoic under conditions comparable to those today, but reached their monumental size during a second oxygen-rich period, the Cretaceous, with a PO_2 of up to 28 kPa (Gale et al., 2001). Insect gigantism was also again seen, with huge mayfly relatives with a 45 cm wingspan. Both hyperoxic periods were followed by phases of relative hypoxia correlating with the disappearance of some groups and survival of those that were able to exapt their modifications to conform to the new conditions.

5.3.1 Reactive oxygen species

But also even under normal conditions oxygen may become toxic, as metabolites of molecular oxygen known as ROS are generated. ROS can be important in control of cellular function, and they may act as second messengers in plants and animals (Perry and Oliveira, 2010; Hancock and Whiteman, 2016), but they can also result in cellular dysfunction or even in cell death. Living in an oxygenated environment has required the evolution of effective cellular mechanisms that detect and detoxify ROS, which exist as superoxide anions (O_2^-), HO•, and H_2O_2. Only the latter is stable over longer periods, while the other two are extremely unstable.

The balance between ROS production and antioxidant defences determines the degree of oxidative stress that can be tolerated. Cellular proteins, lipids, and DNA are particularly sensitive to oxidative stress and influence the lifespan of an organism: oxidant-damaged nuclear DNA accumulates in ageing cells and organisms. This seems to be true for all organisms, and in *Drosophila* the overexpression of the antioxidant superoxide dismutase, SOD (see below), actually can increase the lifespan of the flies (Sohal et al., 1995).

Most intracellular ROS production is derived from mitochondrial activity. Mitochondrial superoxide radicals occur primarily at two points in the electron transport chain: at complex I (NADH dehydrogenase) and at complex III (ubiquinone–cytochrome c reductase). *In vitro*, mitochondria convert only 1–2 per cent of the consumed oxygen molecules into superoxide anions; but *in vivo* the amount is even less. Nevertheless, in keeping with the potential harmfulness of ROS, protective mechanisms, which were probably already part and parcel of the origin of life, limit production and release of ROS (Perry and Oliveira, 2010). The enzymatic scavengers of the SOD system, consisting of catalase, glutathione, and glutathione peroxidase, speeds the conversion of O_2^- to H_2O_2, which is converted to H_2O and O_2 by catalase and glutathione peroxidase (Fridovich, 1998). Also the role of melanin as an ROS scavenger (Bustamante et al., 1993), not only in the skin but also in the liver of turtles and the peritoneum of several lizard species (Duncker, 1968), should not be overlooked. In addition, enzymatic peroxiredoxins and some non-enzymatic, low-molecular-mass molecules (ascorbate, pyruvate, flavonoids, carotenoids, and glutathione) support ROS scavenging. It is interesting that these antioxidant substances are produced in plants, which—as a result of photosynthesis—experience extremely high oxygen levels not only in the leaves but also in other tissues due to direct oxygen transport by so-called aerenchyma tissue.

These are but a few of the mechanisms that are known to protect animals from ROS. Some fish, for example, guppy (*Poecilia vellifera*), goldfish (*Carassius auratus*), and the marine clownfish (*Amphiprion percula*), rid themselves of H_2O_2 by diffusion over the gills. In insects, oxygen damage is avoided by breathing discontinuously (Hetz and Bradley, 2005). The termite *Zootermopsis nevadensis*, breathes continuously but responds to elevated oxygen concentration by decreasing the spiracular cross-sectional area. This means that the carbon dioxide concentration is increased but the tracheal system and the cells are protected from oxygen damage. One explanation for the presence of this phenomenon, explicitly in a termite, is protection of oxygen-sensitive intestinal symbionts (Lighton and Ottesen, 2005).

5.4 Other metabolism-related extremes

5.4.1 Cold

In general, influences of temperature on metabolic processes are expressed in terms of the so-called Q_{10}: the effect that a 10°C temperature change has on the metabolic rate. In most temperature-conforming organisms, an increase of 10°C increases the metabolic rate by a factor of 2–3: in jargon, a Q_{10} of 2–3. Surviving extreme temperatures, however, involves special adaptations, which can evolve because cold temperatures are usually seasonal phenomena that occur slowly, and can be predicted and compensated for by hormonally regulated feed-forward mechanisms.

5.4.1.1 Freezing solid and freezing tolerance

One such adaptative strategy is for the animal 'simply' to freeze solid. One aspect of surviving freezing is cryoprotection: a chemically mediated reduction in the size of ice crystals to a size that will not destroy cells. But for a whole animal to survive freezing requires more than just cryoprotection. The coordinated regulation of all aspects of metabolism is required to reorganize priorities for ATP use and to maintain long-term viability in the frozen state. At extremely cold temperatures, hypometabolism (see section **5.1.1.4**) is a key to survival. Ion channel arrest—as described previously for some fish and turtles—is one mechanism of metabolic depression.

The frog *Rana temporaria* is often used as model for hibernating ectotherms that survive periods of low temperatures and low food availability in cold, hypoxic ponds and lakes. When a pond freezes over, the PO_2 falls and animals that are trapped under the ice can have a problem maintaining their oxygen consumption. At the cellular level, either energy-producing pathways are increased in efficiency and/or ATP-consuming processes are reduced. One of these efficiency-enhancing processes is the reduction of ion channel density or channel leak activity, which, in turn, reduces cell membrane activity and lowers the energetic costs of maintaining transmembrane ion gradients. The related phenomenon of ion channel and spike arrest in the CNS of the slider turtle has already been discussed.

Reptiles can use crystallization avoidance (supercooling) or freeze tolerance strategies, whereby a high percentage of body water is converted to extracellular ice. Some species such as the European viviparous lizard (*Zootoca vivipara*) and painted turtles (*Chrysema picta*) tolerate more than 50 per cent of their body being frozen. When cryoprotectants are not used to any great extent, metabolic and enzymatic adaptations that provide anoxia tolerance and antioxidant defence are important. Genes responsive to freezing encode proteins involved in ion binding and enzymes for antioxidant defence (Pörtner et al., 2004).

Insects use freeze tolerance or avoidance. Long-term winter survival is enhanced by minimizing and prioritizing ATP use, an adaptation of ion pumping and suppression of mitochondrial activity. Survival of anoxia/hypoxia that is induced by freezing, however, appears to be aided by upregulation of HIF-1 to coordinate hypoxia protection. For further reading on the molecular mechanisms of freezing tolerance see Storey and Storey (2013).

5.4.1.2 Cold tolerance

Under conditions less extreme than freezing, the same mechanisms are involved as in surviving hypoxia and anoxia. Daily torpor and seasonal hibernation are the most common measures used by endotherms to reduce their energy expenditure while in the cold. At the beginning of hibernation, metabolic rate is suppressed along with the reduction of heart rate and ventilation. Body temperature falls gradually to approach ambient temperature. Hibernation is terminated by a sudden increase of body temperature accompanied by the return of the metabolic rate during which time, ironically, the animals often sleep. During hypometabolism, protein biosynthesis is largely suppressed, ATP production from glucose is reduced, and lipids are the main source of energy. Interestingly, hypometabolism is not only used for coping with cold temperatures but also occurs in tropical animals to deal with a temporary lack of energy intake, such as the previously mentioned overnight torpor in hummingbirds. Finally, long-term viability is also aided by the increase of a variety of so-called stress proteins (see section 5.4.2.1), which stabilize other proteins in cytoplasma, endoplasmic reticulum, and mitochondria. For a thorough review of the subject see Ruf et al. (2012).

5.4.2 Heat

Looking at the other side of the temperature range, we also see a wide variety of adaptive strategies, ranging from the molecular to the behavioural level.

5.4.2.1 Stress proteins (heat shock proteins)

First discovered in association with heat stress, it was soon revealed that upregulation of genes that code for so-called HSPs are present in all life forms from bacteria to birds and bees, and can also be initiated by exposure to many kinds of environmental stress conditions in addition to hyperthermia (De Maio, 1999). These include infection, strenuous exercise, toxins (including ultraviolet light and other ROS producers), starvation, hypoxia, water deprivation, and hypothermia. So the name was changed to 'stress proteins' but the designation of these substances remains HSP. A wide variety of HSP exist, designated by their molecular weights in kDa: for example, HSP10 and HSP90.

In general, HSP are involved in the folding and unfolding of other proteins, and several function as so-called intracellular chaperones (for references, see De Maio, 1999). They can stabilize unfolded proteins and aid in transporting other proteins across cell membranes. Thus it is not surprising that HSP are always present to some extent, and exposure to stress factors just upregulates their production rather than initiating it. They are also involved in removing old proteins and helping to activate new ones as part of the cellular stress response. Among the reactions assisted by HSP are those involving the function of NO in vascular muscle relaxation, including pulmonary vessels.

5.4.2.2 Some examples of what stress proteins can do

The desert ants *Cataglyphis* have the highest known critical thermal maximum of any insect group. They tolerate temperatures of 50–55°C. The discontinuous respiration of these ants is temperature dependent: the amount of carbon dioxide emitted per cycle remains constant, but in the ventilation phase the carbon dioxide output increases while the duration decreases, meaning that the animal has to breathe less hot air (Lighton and Wehner, 1993). Also the ventilation frequency is modulated with increasing temperature. These high-temperature tolerance mechanisms are coordinated by HSP.

In marine annelids, sipunculids, crustaceans, and molluscs, an oxygen-dependent thermal tolerance also exists, brought about by progressively inadequate oxygen supply toward either very warm or very cold extremes. In the end, decreasing extracellular oxygen levels lead to this temperature-induced anaerobiosis. But acclimatization is also possible. In the rough periwinkle snail, *Littorina saxatilis*, for example, heat-induced anaerobiosis is caused by insufficient oxygen supply at high temperatures, but cold acclimatization results in an increase in aerobic metabolic rates and a downward shift of the upper critical temperature in animals from the temperate North Sea. Animals of the same species from the sub-arctic White Sea, which has more extreme winters, show a much more limited metabolic plasticity in response to cold (Sokolova and Pörtner, 2003).

Temperature-induced metabolic changes in the cuttlefish *Sepia officinalis* are quite different from what is seen in other molluscs. Either high or low temperatures cause a shift of aerobic to anaerobic metabolism in the muscles. At high temperatures, the circulatory system apparently breaks down, decreasing the arterial PO_2 and putting the brakes on aerobic metabolism. But at low temperatures, ventilatory and circulatory failure is not responsible for the decreasing metabolic rates. Instead, haemocyanin shows high affinity at low temperatures, meaning that oxygen is transported at very low PO_2, limiting the amount of oxygen that can be released into the cells and causing a shift to anaerobic metabolism in the muscle (Melzner et al., 2007).

CHAPTER 6

Respiratory faculties of aquatic invertebrates

Not all animals have a dedicated respiratory faculty, and when they do, we see a huge variety regarding the location and structure of gas exchange organs even within a single phylogenetic group. In other words, there is no primordial and universal gill that has become modified through the aeons. Also, many invertebrates—in particular arthropods—are secondarily aquatic and have evolved taxon-specific adaptations allowing an aquatic life. The present chapter gives insight into this enormous variety of morphological and physiological respiratory adaptations to aquatic life. Parenthetically, we do not distinguish between fresh-water and salt-water (marine) habitats and refer to both as 'aquatic'.

A respiratory faculty, as discussed at the outset of this book, has at least two components: a recognizable gas exchange organ or region and some mechanism for distributing respiratory gases within the organism. In more complexly organized animals, a central neurological control unit joins the faculty.

In many cases, the gas exchange organ or region may have other functions such as mechanical protection in the case of cutaneous gas exchange, or locomotion, or feeding. In addition, the internal circulation may also be intimately involved in locomotion such as the use of the coelomic cavities as a hydroskeleton in annelids and some arthropods. As the respiratory faculty becomes more refined, it accumulates further attributes such as a ventilatory mechanism for the gas exchanger and discrete vessels, a heart or hearts, respiratory proteins that facilitate distribution of respiratory gases in the animal, and a central neurological control unit as mentioned previously.

In the present chapter we shall stick to describing the structure and function of the respiratory faculty as it occurs in various animal groups. The evolution of the faculty is the subject of Chapter 9. Since there is a great deal of convergence in respiratory faculties, it is sometimes more expedient and avoids repetition if we use a functional morphological approach and deviate from the systematic one, but you will be warned when this happens.

6.1 Porifera, Cnidaria, and Ctenophora: skin breathing

Since we are dealing specifically with the respiratory faculty and non-bilaterian taxa don't have respiratory organs, we shall quickly move on to more fertile ground. But before doing that let's have a brief look at the non-bilaterians anyway. Poriferans (sponges) are filtering organisms 'built around the water'. The flow of water through the organism, caused by beating of the choanocyte flagella, expedites gas diffusion virtually everywhere. In cnidarians (corals, jellyfish), a similar situation occurs, but here the mesogloea of certain medusae can actually passively store oxygen, which becomes liberated when the oxygen content of the water falls. The Portuguese man o' war *Physalia physalis*, a colonial

Respiratory Biology of Animals: Evolutionary and Functional Morphology. Steven F. Perry, Markus Lambertz, Anke Schmitz, Oxford University Press (2019). © Steven F. Perry, Markus Lambertz, & Anke Schmitz.
DOI: 10.1093/oso/9780199238460.001.0001

hydromedusa, has something akin to functional specializations, in which the large surface area of the prey-catching tentacles also serves in gas exchange. In place of a circulatory system, the anastomosing gastrovascular canal distributes respiratory gases. In ctenophorans (comb-jellies) the energy-consuming rows of the plate-like cilia that give the group its name can be supplied with oxygen from the mesogloea when they drift into less advantageous waters.

6.2 Unsegmented 'worms'

In spite of their greater mobility compared with sponges and cnidarians, Plathelminthes (flatworms) and Nemathelminthes, (roundworms) lack specialized respiratory organs. Instead, many species have sensing mechanisms that guide them to better oxygenated places and/or they have very well-developed anaerobic survival strategies, which are quite effective given the modest metabolic rates of these animals, and they have special relevance for endoparasitic species.

But repeatedly, individual components of a respiratory–circulatory complex (faculty) have appeared even among unsegmented 'worms'. The Nemertini (ribbon worms), for example, are unusual in having a functionally closed circulatory system. Haemolymph is circulated by contractile vessels and by the body wall muscles, and some species are several metres long. But specialized gills are lacking and the skin is the gas exchanger. Also a group of flatworms, the Rhabdocoela, may have haemoglobin in their mesenchyme, and some nematodes have haemoglobin in the fluid surrounding the internal organs. But nobody knows to what extent these haemoglobins have anything to do with oxygen transport, exchange, or storage.

6.3 'Lophophorates'

The group 'Lophophorata' is no longer recognized by most systematists but will be considered together because of a similar anatomical structure, the feathery lophophore, which is the main organ contacting the respiratory medium: sea water. Regarding the respiratory faculty, Phoronida, Bryozoa, and Brachiopoda, form a kind of evolutionary plateau. They rely on the lophophore coelom for distribution of respiratory gases, but lack a circulatory system and a gill that is dedicated to respiration. Instead, the lophophore combines respiration and suspension feeding, something common not only among invertebrates but also—as we shall see—in basally branching chordates.

Phoronids are little-known marine organisms, also called horseshoe worms because their often elongated bodies form a hairpin loop within a chitinous tube, with the anus at the front, separated from the mouth by the lophophore. Convergent with sipunculid worms (see later), the lophophore has its own coelom which can contain haemoglobin and is separate from the general body cavity. Bryozoans form colonies of many very small individuals, analogous to corals. Gas exchange takes place over the entire integument but especially on the lophophore, which is protruded from the cuticular exoskeleton into the water. Brachiopods superficially resemble bivalve molluscs, with a hinged calcareous shell, but unlike clams, they are usually suspended on a stalk, the shell is closed dorsoventrally, and the upper half of the hinge overlaps the lower half. These 1–6 cm long animals were dominant in the Devonian but only a few species exist today. Brachiopods rely on their lophophore for gas exchange but also have the rare non-haem respiratory protein haemerythrin, which is otherwise found only in annelids, sipunculids, and priapulids (penis worms). Since haemerythrin (see Chapter 4) lacks significant cooperative oxygen binding and a Bohr effect, it is probable that it serves in oxygen storage and delivery when the shell is closed rather than in transport (Wells and Dales, 1974).

6.4 Echinodermata

The group Echinodermata (the name means 'spiny skin') consists of five subgroups: Crinoidea (sea lilies and feather stars), Asteroidea (sea stars), Ophiuroidea (brittle stars), Echinoidea (sea urchins and sand dollars), and Holothuria (sea cucumbers). In keeping with their external skeleton and non-vascularized integument, echinoderms have specialized thin-walled body areas where gas exchange takes place. This is particularly evident in the sea urchins, but also sea stars and sea cucumbers don't provide much possibility for cutaneous gas exchange over the entire animal. Instead, as discussed in the following sections, we see various types of structures that serve in gas exchange. So we count the echinoderms among the first animals with some semblance of a respiratory faculty, although a circulatory system is lacking.

6.4.1 Crinoidea

In stalked sea lilies and their mobile cousins, the feather stars, the arms themselves offer the main site of gas exchange and are ideally suited for this, since unlike other echinoderms the oral side faces upward. Convergent with lophophorates, cilia provide water current to the feathery arms, which contain ambulacral grooves from which the tube feet extend. These delicate structures are involved in catching food particles and passing them along to the mouth, but also are assumed to be the main gas exchange organs.

6.4.2 Asteroidea

In sea stars, in addition to tube feet, the animals have filamentous structures called papulae near the mouth. Cilia on the outside of the epidermis cause a water current flowing over the papulae. The tube feet are ventilated by their own movement. Preventing the tube feet from extending in the red sea star, *Asterias rubens*, reduces the gas exchange by 60 per cent, suggesting that the papulae account for the rest.

6.4.3 Ophiuroidea

In the brittle stars, a well-developed skeleton stabilizes the arms but prevents gas exchange over the body surface, and rather than gliding with the help of tube feet, the animals use their arms to walk. Gas exchange is restricted to ten bursae: thin-walled invaginations of the oral surface of the central disc. The bursae open to the outside through small slits, and cilia produce a water current. Ventilation can be aided by 'push-ups:' raising and lowering the oral or aboral disc wall.

6.4.4 Echinoidea

Sea urchins also have a formidable exoskeleton. In the mouth region they have five sets of papilla-like 'gills', which are claimed to be the main respiratory organs. These structures are evaginations of the peristomial membrane and inside the animal also come in direct contact with the peripharyngeal coelom, which surrounds the jaws: the region of greatest metabolic activity. Ventilation is accomplished mainly by muscular pumping and during feeding movement. In addition, the sometimes centimetre-long ambulacral feet (podia) and the equally long, pincer-like pedicellaria are certainly self-supporting with respect to gas exchange. Cilia on the surface of the podia produce a water current that runs in a direction opposite to that of the fluid contained in the podia, creating the prerequisite for counter-current gas exchange between respiratory water and coelomic fluid. In sand dollars, the number of 'gills' is often reduced to four and the counter-current podia seem of greater importance in gas exchange.

6.4.5 Holothuria

Sea cucumbers are among the few animals to have water lungs. We shall run into water lungs again in the independently originating hindgut tracheal gas exchange organ of dragonfly larvae (see section 6.9.6.) and also in some turtles (see Chapter 11). In sea cucumbers, these are branched diverticula of the hindgut forming two respiratory trees with a large surface area next to the gut. Blocking the water lungs reduces the oxygen consumption but papillae surrounding the anterior end of the animal can account for up to 50 per cent of the total oxygen uptake. In addition, there is at least one pelagic, actively swimming species of sea cucumber, *Pelagothuria natatrix* (Miller and Pawson, 1989). The tentacles surrounding its mouth are much larger than in 'normal' sea cucumbers. Until there is some experimental data to prove us wrong, we can assume that they support gas exchange.

6.5 Annelida

The segmented worms, Annelida, reach the largest size of any worm-like invertebrates. There are two main divisions of this group: 'Polychaeta' and Clitellata, the former now widely believed to be polyphyletic. The latter group is made up of the Oligochaeta (earthworms, for example) and Hirudinea (leeches).

Another group now considered to be related to annelids are the Sipunculida (peanut worms). They breathe mainly with their oral tentacles, which are thin-walled and always in motion, and have their own coelomic cavity, separate from that of the body. Haemerythrin is found both in haemerythrocytes and free in the coelomic fluid, highly reminiscent of the polychaete *Magelona papillicornis* mentioned later.

6.5.1 'Polychaetes'

Polychaete annelids include free-ranging and sedentary forms. Most species breathe with multiple gills or have petal-like lophophores. In small-bodied species, the entire skin serves in gas exchange. All these gas exchange regions contain branches of the circulatory system, and the haemolymph or the coelomic fluid transport oxygen using respiratory proteins (see Chapter 4). Gills vary greatly in both location and structure, indicating that they probably have originated independently several times together with an increase in body mass and/or the species-specific oxygen requirement. Most often, the gills are parts of the stubby appendages called parapodia (Figure 6.1). These bilaterally paired structures are particularly well developed in marine species, which make up the majority of polychaetes. The parapodia consist of two lobes: the dorsal notopodium and the ventral neuropodium, whereby the notopodium is most often modified into a gill. But the body wall itself can also form gills. Annelid gills

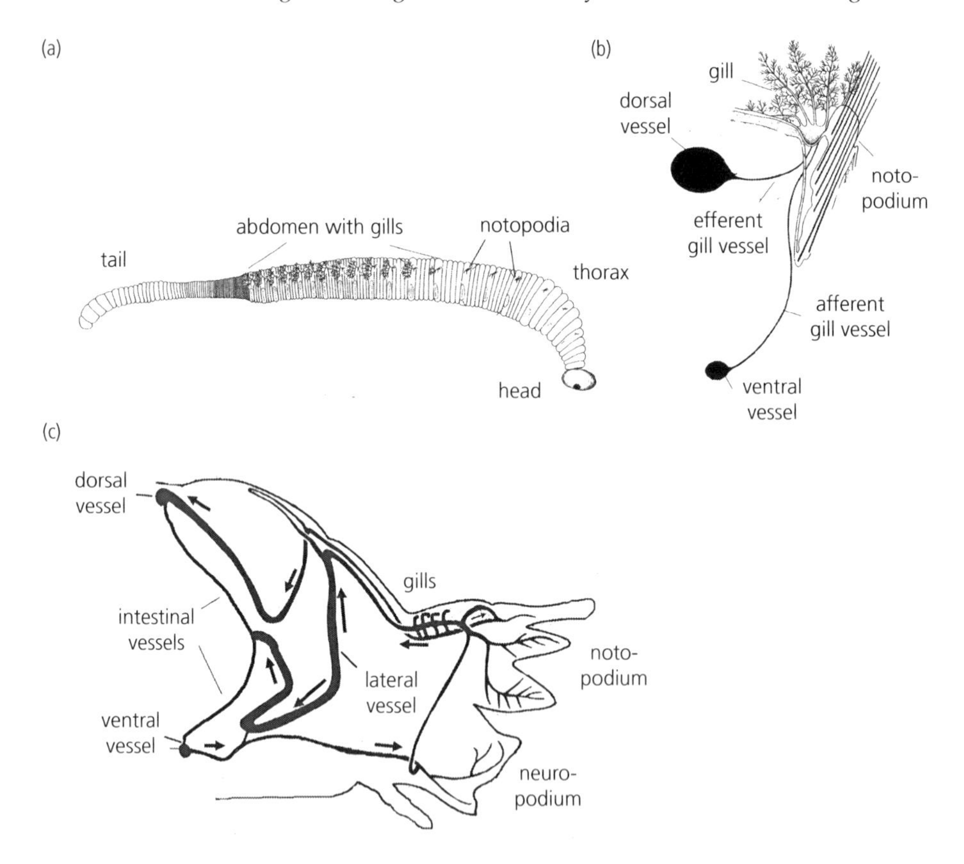

Figure 6.1 Typical lateral respiratory organs of polychaete annelids. Part (a) shows the location of the lateral gills in a lugworm *Arenicola*, shown in more detail in part (b). The circulatory system consists of single (in some species paired) often contractile dorsal vessels, which move the haemolymph from posterior to anterior and a ventral vessel with the opposite flow direction. These are connected by two sets of vessels: the often contractile lateral vessels supply the two-part parapodia, the ventral neuropodium and dorsal notopodium, the latter also giving rise to the gills, if present, and an interior network of vessels that supply the internal organs. Part (c) illustrates the haemolymph vessel system using the example of another polychaete, *Platynereis*. Modified after Westheide and Rieger, 2007.

can be ventilated either by the action of cilia or by mass body movement in vagile species.

Many species, for example, the Sabellidae (feather duster worms, or fan worms) and the closely related calcareous tube worms, Serpulidae, have, surrounding the mouthparts, a showy basket-like or spiral lophophore-like wreath that doubles in gas exchange and prey capture. Another example are the spaghetti-like gills of the burrowing *Amphitrite*. These dorsal protuberances of the anterior segments can be projected out of the burrow and provide a large surface area for gas exchange. This is also the case in the Siboglinida (beard worms, formerly called Pogonophora). In the familiar lugworm *Arenicola* of tidal flats, small gill bundles are concentrated on the dorsal part of the midbody. This species lives in tubes with two openings, which are used, respectively, for moving water in and out of the tube.

Looking now at the circulatory side of the respiratory faculty, the circulatory system is functionally closed in polychaetes and is very similar to the system in clitellates, such as the common earthworm that you may have dissected in your introductory biology class. And especially the afferent dorsal vessel and often also the pharyngeal loops are contractile, functionally serving as a heart.

And remember the red 'blood' in the earthworm? Three of the four known groups of respiratory proteins are found in annelids: only haemocyanin is lacking (see Chapter 4). Haemoglobin is most common, and a special green type called chlorocruorin is characteristic of the serpulid and sabellid worms. In *Amphitrite*, the coelomic haemoglobin has a higher affinity for oxygen than the blood haemoglobin, thus facilitating the passage of oxygen into the coelomic system that supplies the internal organs. Worms such as the lugworm *Arenicola* use their high-affinity respiratory proteins for oxygen storage in hypoxic periods during low tides, and some species such as *Euzonus mucronatus* can survive up to 20 days of anoxia.

The calcareous, tube-dwelling fanworm *Serpula* has both haemoglobin and chlorocruorin. In some annelids, the rare respiratory protein haemerythrin also occurs. The polychaete *Magelona papillicornis*, for example, has heterogeneous haemerythrin, both in non-nucleated cellular fragments, 'haemerythrocytes', which circulate in the vascular system and in the muscles (myohaemerythrins).

6.5.2 Clitellata

6.5.2.1 Oligochaeta

So slowly the pieces are coming together, but unlike polychaetes, the other major subdivision of annelids, the Clitellata, never evolved specialized respiratory organs. Instead, they use the well-vascularized skin. Having said that, there are a few exceptions, such as the genera *Dero* and *Aulophorus*, that have a few finger-like gills around the anus and the genus *Branchiodrilus*, in which long filaments (setae) extend from the lateral surfaces of each segment. Some other species might have dorsal or ventral filiform projections that they use as gills.

The oligochaete sludge worm *Tubifex* is an interesting case in point, illustrating how one can muddle through without a proper respiratory system. *Tubifex* inhabits highly eutrophied and oxygen-poor water. Convergent with larval dragonflies, it uses parts of its hindgut for oxygen uptake but it doesn't have a tracheal system to transport the oxygen. The animal burrows into mud or sludge, extending its posterior end into the water. Gut capillaries take up the oxygen from the water that is drawn into the hindgut via the anus by negative pressure. The less oxygen available, the further the animals reach into the water, with waving movements increasing convection. In this way *Tubifex* can survive at levels as low as about 1.3 kPa oxygen for long periods. Like other clitellates—both terrestrial and aquatic—*Tubifex* possesses haemoglobin that has an especially high oxygen affinity (see Chapter 4), giving the animals a red colour.

6.5.2.2 Hirudinea

Within the Clitellata, the leeches are terrestrial-to-aquatic, but only the aquatic Piscicolidae have developed gills as lateral leaf-like or branching outgrowths of the body wall. These gills are filled with coelomic fluid, which in leeches assumes the function of the circulatory system in gas transport, and a true circulatory system is lacking in most leech groups. Leeches without gills ventilate by undulating the body with only the posterior sucker attached: just the opposite to *Tubifex* (see section 6.5.2.1), which waves its posterior end in the water. Like *Tubifex*, the dorsal and the ventral vessels in the leech are connected by lateral vessels in each segment, but these are coelom vessels and not part of a separate circulatory system. Valves prevent backflow of coelomic fluid.

6.6 Molluscs: a universe parallel to arthropods and chordates

Of the approximately 100,000 extant molluscan species, most live in aquatic environments, and even terrestrial snails require a substantial amount of water to lay down the slime film they glide on. A mollusc in principle consists of four body parts: head, foot, visceral sac, and mantle. The mantle can produce the shell, which is characteristic but not necessarily present in all molluscs. But the important thing for respiration is that this body region usually lies protected, between the shell and foot, and can be quite thin-walled and well suited for gas exchange. And as we shall see, it can serve as the basic substrate for the formation of highly specialized respiratory organs.

Most aquatic molluscs—except for sea slugs and some intertidal and pulmonate snails—have gills. In basal radiations they are called ctenidia. The plesiomorphic condition is assumed to be one pair of ctenidia in the mantle cavity, but the number can vary considerably: Polyplacophora (chitons), for example, can have about 100 while within the Conchifera, monoplacophoran molluscs have only three to six pairs.

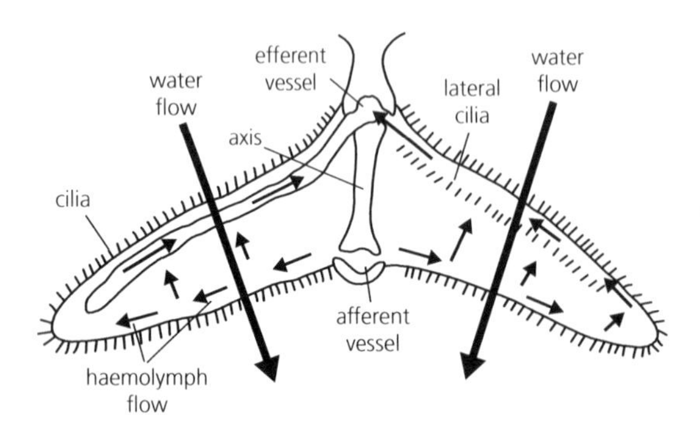

Figure 6.2 Schematic diagram of a typical bipectinate ctenidium. Note counter-current model. Original A.S.

Ctenidia (Figure 6.2) consist of a long, flattened axis which contains afferent and efferent blood vessels, muscles, and nerves. To each side of the axis, flattened, wedge-shaped filaments are attached which alternate in position with the filaments of the opposite side. The gills are held in position by ventral and dorsal membranes and are stabilized at the margin by a chitinous rod. The location of the gills divides the mantle cavity into an upper and a lower chamber. Water enters the lower chamber posteriorly, passes over the gills, enters the upper chamber, and exits the cavity. In this ancestral condition, the respiratory water stream is created by cilia located on the gills themselves. The afferent ctenidial vessels transport haemolymph into the gill and branch off into vessels perfusing the filaments before flowing back in an efferent vessel: a counter-current gas exchange model is approximated.

6.6.1 Gastropoda

Snails also show a fascinating progression in the respiratory faculty (Figure 6.3). The prosobranch condition with the gills anterior and water circulation pattern involving cleft or perforated shells (Archaeogastropda) is generally considered to be plesiomorphic. A single pair of gills in abalones (e.g. *Haliotis*) and keyhole limpets (e.g. *Diodora*), is referred to as the 'bipectiante state' and is presumed to approximate the ancestral condition. In *Haliotis iris*, the right gill alone is sufficient to meet the oxygen requirements at rest, but during activity the left gill, which receives little perfusion during rest, is also employed (Ragg and Taylor, 2006a, 2006b).

Now looking in more detail: within the Patellacea (limpets and turban snails), one or both of the original gills are completely reduced and secondary gills along the mantle groove, along the body or the mantle surface itself take over the function of gas exchange. Since these archaeogastropods tend to occupy intertidal zones where oxygen supply is usually not a problem, the mantle cavity alone can provide sufficient oxygen diffusing capacity.

Other marine snails are not restricted to rocky surfaces and well-oxygenated areas. They exploit diverse habitats and the gill structure reflects this flexibility. The gill axis is attached to the mantle wall: one side of the gill is completely reduced while the

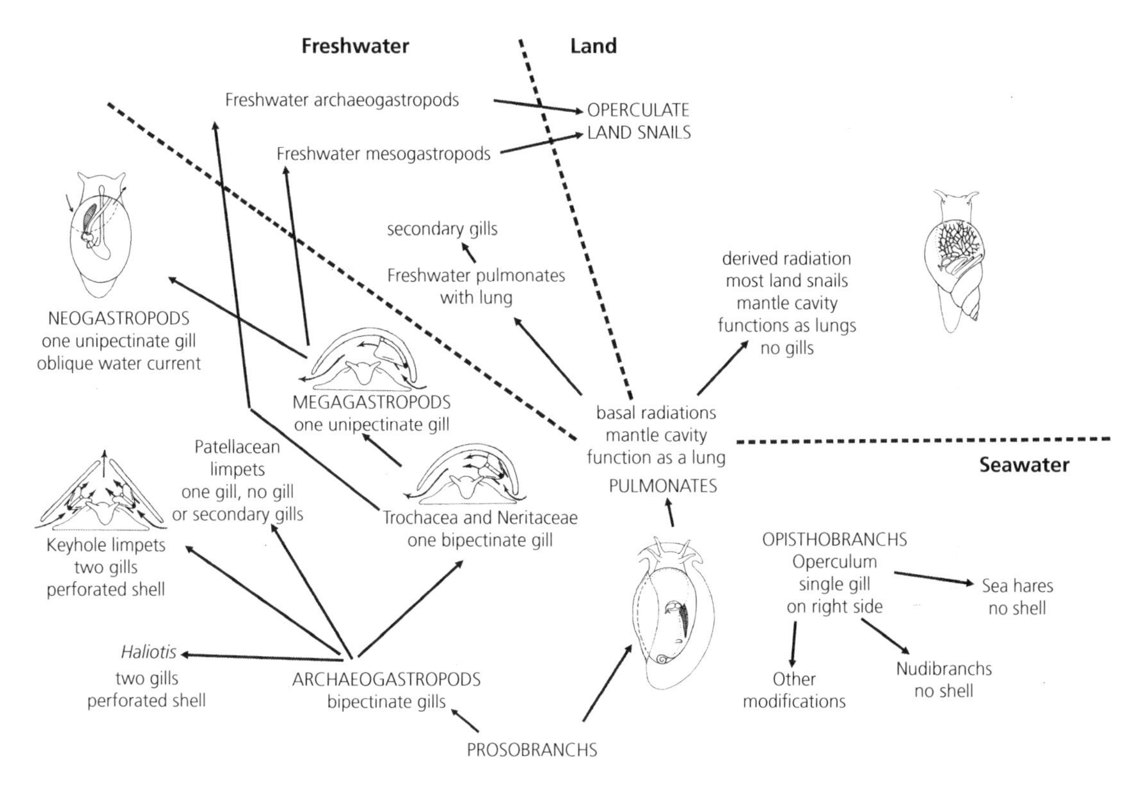

Figure 6.3 Respiratory organs of snails. The tendencies toward asymmetry and simplification of gill structure (e.g., bilateral, bipectinate → unilateral, bipectinate → unilateral, unipectinate) are clear but recent revisions in higher-level systematics make it less clear if these changes occurred once or several times. Note the multiple invasions of terrestrial habitats by non-operculate and operculate groups. Modified after Barnes, 1980.

other side of this 'unipectinate' gill projects into the mantle cavity. Together with this development, a siphon for the respiratory water current evolved. It is formed by an infolding of the mantle ridge, and enables the animals to have a burrowing or cryptic lifestyle since they are no longer dependent on external water currents for ventilation.

In keeping with the habitats and life histories, we see some snails (e.g. *Viviparus*) losing even these unipectinate gills while others form secondarily bipectinate gills (e.g. *Valvata*). In some caenogastropods, a siphon guarantees ventilation even while the snail is withdrawn in its shell with the operculum closed. The operculate Cyclophoridae and Pomatiasidae actually enter terrestrial habitats with the consequent loss of gills and the development of an air-breathing, vascularized mantle epithelium. And this occurred completely independently of the much more highly derived Panpulmonata (pulmonate snails), which evolved within the Heterobranchia but retained the right-side gill opening.

The opisthobranch condition, in which the gills lie in the posterior part of the body, developed when the entire mantle cavity and associated organs migrated around the right side. This may have involved complete loss of the gills and their later 'reinvention'. The gills of opisthobranchs are folded or 'plicate', rather than filamentous: a structure that may well challenge homology to the prosobranch condition.

Within opisthobranchs, there is a tendency to reduce the shell together with the loss of the original gills. Pteropods, anaspidians, and nudibranchs are good examples, the nudibranch sea slugs being the most spectacular. They have completely reduced the shell, mantle cavity, and gills, and developed a nearly perfect superficial secondary bilateral symmetry. The surface area of the dorsal body wall is greatly increased due to the formation of so-called cerata. The surface of these often showy, branched, club-shaped or finger-like structures functions as a secondary gas exchange area. Although some snails can have haemoglobin, particularly in the foot or in the musculature of the rasping tongue (radula), haemocyanin is the primary oxygen-transporting protein (see Chapter 4).

6.6.2 Bivalvia

In bivalves, which include the familiar clams, oysters, and so on, the gills are large, typically having the dual function of providing a water current that serves both respiration and food collection in filter feeding. The lining of the mantle cavity is important for gas exchange in most species and, as in snails, gives rise to the gills. Gills are on each side of the central foot and fill in the mantle cavity between foot and the two halves of the shell. In all bivalves there are more or less well-developed inflow and outflow pathways of water flow through the gills perfused with haemolymph, which is haemocyanin-free in non-protobanch bivalves (see later in this section) (Bergmann et al., 2007; Markl, 2013).

Different types of gills occur and are named according to their general structure (Figure 6.4). Protobranch gills are the simplest ones. They have many similarities with ctenidia, and probably have a major function in gas exchange. Gills are interconnected via ciliary tufts between the adjacent filaments. The modification for filter feeding is in general a lengthening and folding of the gills, producing filiform or lamellated structures. The number of filaments or lamellae tends to increase, leading to a very large surface area. Filiform gills are repeatedly folded in the mantle cavity with tissue connections between the folded filament halves, between adjacent filaments, and gills are connected to the mantle and foot for stabilization. Lamellibranch gills are ladder-like structures in which the connections between the filament halves are broadened, transforming the gills into sheet-like structures with a large surface area. The most highly specialized are the eulammelibranch gills in which the interconnections increase such that the lamellae consist of solid sheets of tissue. Further increases of the surface area result in plicate gills with an undulate surface. In lamellibranchs, the primary function of the gills has shifted from gas exchange to filter feeding.

Septibranchiate bivalves are deep-sea forms that have converted the gills into flat muscular septa, which horizontally separate a suprabranchial chamber and an inspiratory water chamber. Gas exchange takes place over the surface of the mantle. The chambers are connected by small slits through which the water is propelled. These species can even be carnivorous, the force of the muscular pumping septa being sufficient to carry small animals into the mantle cavity where they are seized and transported to the mouth, and the mantle has completely taken over the gas exchange function.

Since air-breathing bivalves are more of a curiosity than a major trend, they will be briefly discussed here rather than under air-breathing invertebrates. Mussels are the best studied group here. Like all animals inhabiting intertidal zones, mussels are subjected to enormous fluctuations in PO_2 and temperature. In most species the shells are kept closed during exposure at low tide and gills and mantle cavity are kept moist by the water retained in the shell. In *Mytilus galloprovincialis* from the Mediterranean Sea, haemoglobin and muscle myoglobin store oxygen and keep the metabolic rate constant in spite of the above-mentioned fluctuations (Barbariol and Razouls, 2000). And like other bivalves, they can also tolerate extended periods of hypoxia or anoxia anaerobically. But with all that oxygen in the air, why not use it? The ribbed mussel *Geukensia demissa*, which inhabits salt marshes on the Atlantic coast of North America, does just that. It is exposed to air for up to 70 per cent of the tidal cycle, and opening the shell is actively regulated in this species, allowing aerial respiration, which may even be obligatory (Huang and Newell, 2002).

6.6.3 Scaphopoda

The scaphopods, or 'tusk-shells', have a one-piece shell and a mantle which is open at both ends. In a way, the scaphopods represent a sort of a 'flow-through snail', if you will. But gills are lacking: instead cilia propel water through the tubular shell and over

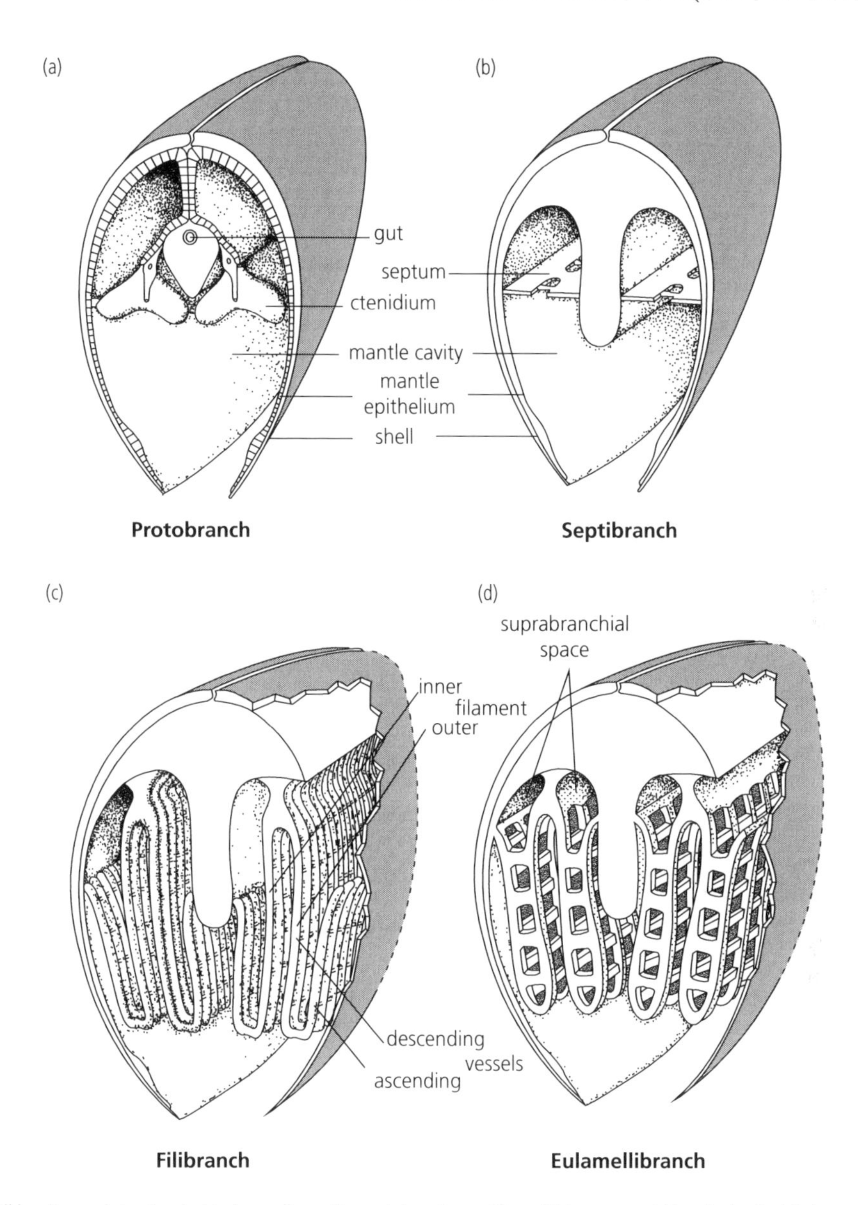

Figure 6.4 Gill location and structure in bivalve molluscs. The protobranch condition with two rows of bipectinate ctenidia becomes reduced in the septibranch condition or highly modified in the filibranch and eulamellibranch conditions. The gill structure tends to correlate with phylogenetic groups in bivalves. After Westheide and Rieger, 2007

the smooth surface of the mantle, where gas exchange takes place. Scaphopods lack the sequences that code for haemocyanin and it is not known if haemocyanin was always lacking in this group or if it was lost.

6.6.4 Cephalopoda

In cephalopods, the gills project into the mantle cavity, which forms a funnel (Figure 6.5). Ventilation is accomplished by muscular contraction of the funnel or undulation of the mantle edge. So here we see a siphon (ventilatory pump) turning up independently in a third molluscan group. Squids use the same water stream for ventilating the gills and for locomotion. Accordingly, they 'waste' a lot of potentially ventilatory water, and extract only 5–10 per cent of the dissolved oxygen. Cuttlefish, on the other hand, have successfully uncoupled respiration

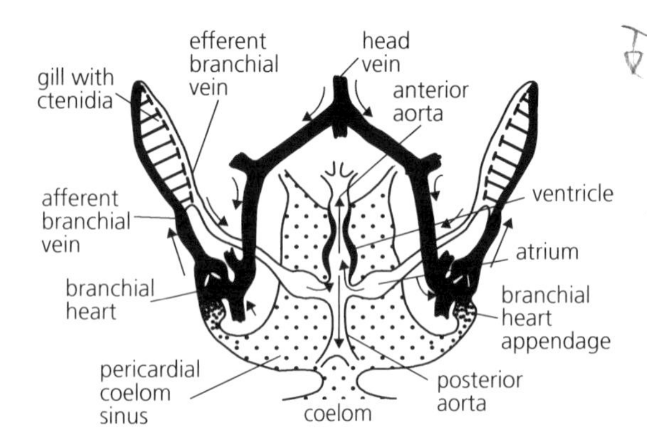

Figure 6.5 Schematic diagram showing the relative location of the gills, heart, and excretory organs in a cephalopod mollusc (*Sepia*) in dorsal view, anterior above. Haemolymph flows from head and body and is pumped by gill (branchial) hearts into the afferent branchial veins to the ctenidia of the gills, whence it enters via efferent branchial veins, and—now oxygen-enriched—the paired atria, and the ventricle of the heart for distribution to the body. The branchial hearts also provide haemolymph to a renal apparatus (branchial heart appendage) for ultrafiltration dialysis against coelomic fluid from the pericardial coelom sinus. After Westheide and Rieger, 2007.

and locomotion, using undulation of the lateral fin folds for swimming. They can extract up to 80 per cent of the dissolved oxygen from the ventilatory water, thus approaching the efficiency of teleost fish.

Regarding the structure of the respiratory system, nautiloid species (e.g. *Nautilus*) have two pairs of gills (hence the name Tetrabranchiata, which is sometimes applied to the group) while the remaining cephalopods, have only one pair. Cephalopod gills differ from those of other molluscs in that the large surface area is caused by infoldings and cilia are lacking. Within the gill filaments, haemolymph runs in closed capillaries, which contributes to the efficiency of the gills. Some species even possess special gill hearts.

Cutaneous gas exchange is also important in cephalopods. In *Octopus*, for example, between 3 and 41 per cent of the total oxygen uptake is directly through the skin (Madan and Wells, 1996). This may be considered an exaptation allowing these critters to escape from their aquaria in public zoos, as occasionally reported in the media. The metabolic cost of walking in *Octopus* is about three times the cost of swimming for a fish of similar size but only one-third the cost of tetrapod walking on land, since the animal does not have to support much of its weight. Maximum oxygen uptake during walking is 2.4 times resting values (Wells et al., 1983). Under hypoxic conditions, *Octopus* and *Nautilus* downregulate oxygen consumption to a PO_2 of 7–10 kPa while at the same time increasing ventilatory stroke volume (Wells and Wells, 1985, 1995).

6.7 Crustacea

Most aquatic crustaceans breathe with gills, which are part of the forked legs or of the adjacent body wall. Small crustaceans use the skin as the primary gas exchange organ and amphibious species as well as terrestrial crabs show a species-specific development of air-breathing structures. Most often the exites (lateral branch) of the legs form thin-skinned structures called epipodites. However, other parts of the legs or the thin-walled pleura may also serve as gills. So a huge variety of gill forms have arisen in keeping with systematic position, the structure previously present, and the metabolic needs in the crustacean group. Histologically, crustacean gills are made up of a haemocoel space covered by a thin inner epidermal and even thinner outer chitinous cuticular layer.

6.7.1 Decapoda

In decapod crustaceans (crabs, lobsters, etc.), the carapace not only protects the gills, it also aids in directing respiratory water current. In this group, water is moved through the gills by the beating movement of modified mouth parts called scaphognathites (Figure 6.6). In species with lamellar gills, such as crabs and lobsters, ventilatory water tends to flow opposite to the flow of the haemolymph, resulting in a counter-current model. Others, such as crayfish, have filamentous gills and the cross-current model would be a better approximation, but either gas exchange model is highly efficient. In the mangrove crab *Ucides cordatus*, for example, the cuticle is only 1.5–3 μm thick. The epidermal layer is 5–10 times thicker, but due to the vastly different Krogh diffusion constants of chitin and cellular tissue, the oxygen flux in these two layers is about the same.

6.7.2 Maxillopoda

The use of modified legs to create a ventilatory water current is common in crustaceans. Sessile species

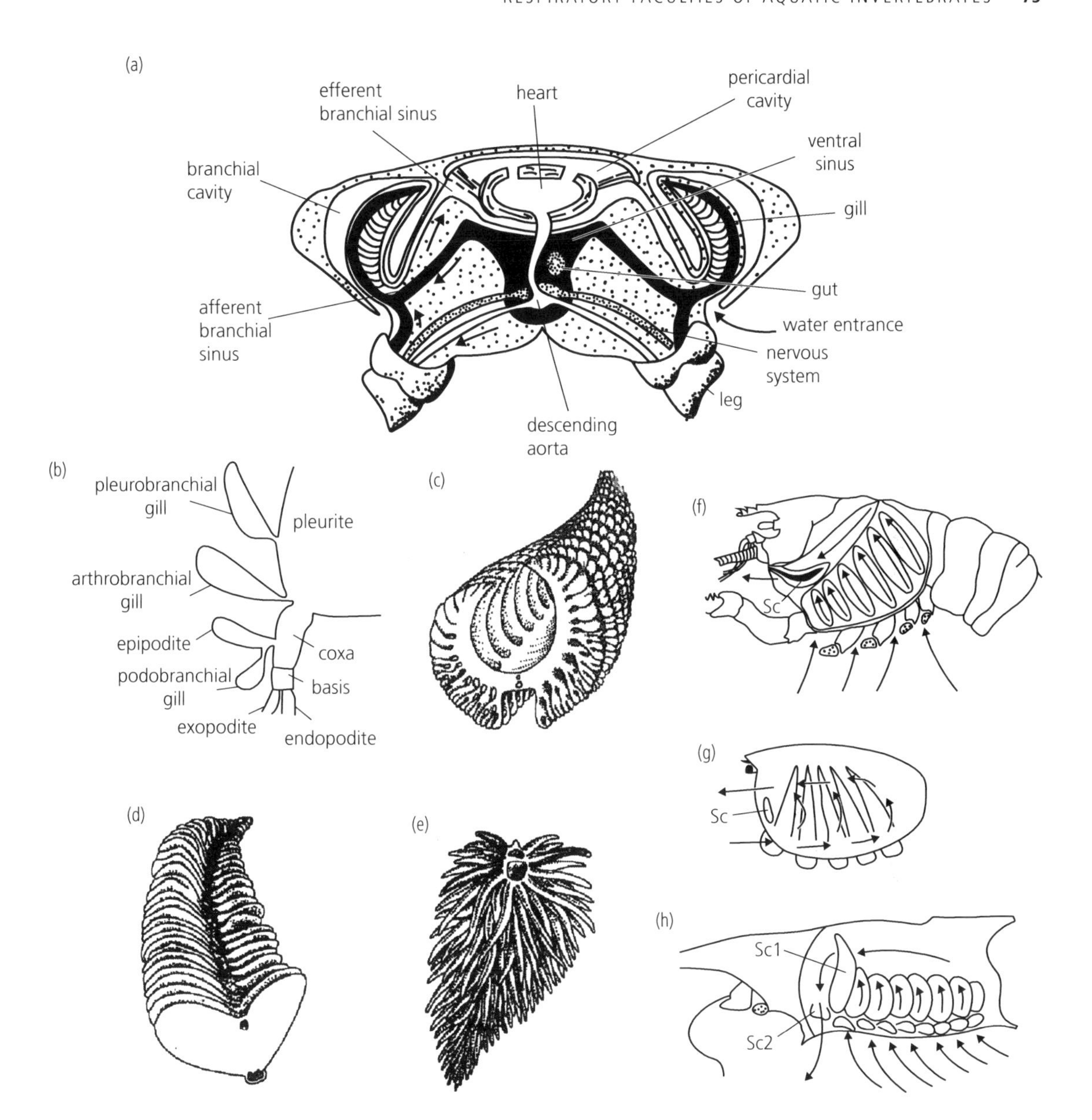

Figure 6.6 Gills of some large malacostracan crustaceans. Part (a), is a schematic cross section through a brachyuran crab showing the relative location of gills and heart, which is typical of many arthropods. Deoxygenated haemolymph flows into the gills via the afferent branchial sinus whence it is drawn via the efferent branchial sinus from the gills into the pericardial sinus, which surrounds the contractile heart. Unless otherwise designated, arrows in this illustration indicate the direction of haemolyph flow. Ventilating water is propelled by scaphognathites (see part (g)). Part (b) shows posssible location of gills in crustaceans and designations of these locations. Parts (c)-(e) show the gill structure in decapod crustaceans: (c), dendrobranch shrimps; (d), ctenidium-like phyllobranch structure of crabs and lobsters (compare part (a)); (e), filament-like structure of crayfish gill. Parts (f)-(h) show ventilatory water current caused by scaphognathites (Sc) in (f), lobsters (water entry at all leg attachment points); (g), brachyuran crabs (directed water with entry only anterior, see also part (a)); red mysid shrimp (water movement as in lobsters but propelled sequentially by two pairs of scaphognthites: Sc1 a modified exopodite of the 2nd maxilla as in decapods and Sc2, derived from the first thoracic leg). Arrows indicate the direction of ventilatory water movement. (a)-(d) and (f). After Westheide and Rieger, 2007, (g) and (h) after Paul et al., 1976.

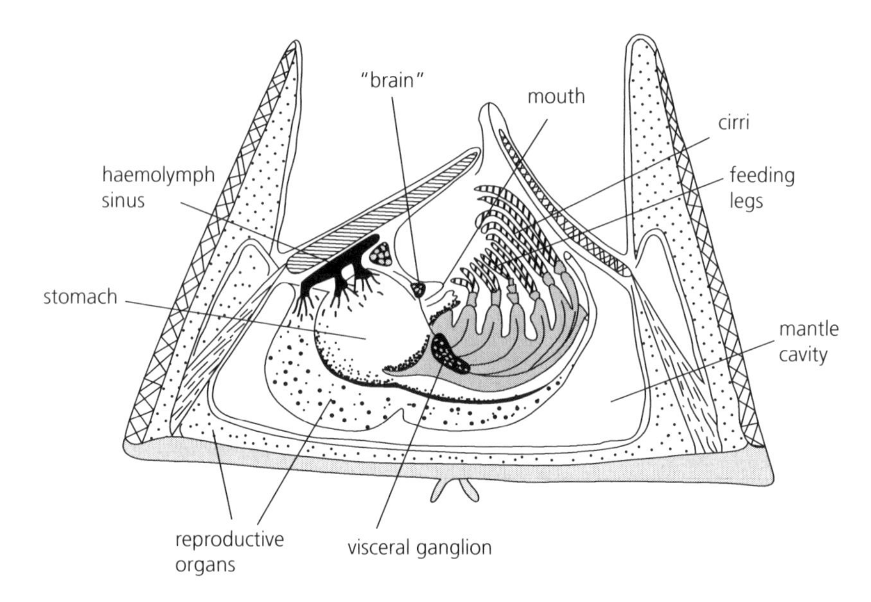

Figure 6.7 Semi-schematic vertical section through a generalized barnacle. The anterior three pairs of forked feeding legs are modified thoracic legs, which comb suspended particles from the cirri (see below) and transport them to the mouth. In most barnacle groups true maxillipeds (mouthparts) never develop. The more posterior 3 leg pairs, cirri, are also forked but are long and whip-like. They trap or propel suspended particles and ventilate the mantle cavity. After Westheide and Rieger, 2007.

such as barnacles (Maxillopoda: Cirripedia) are well known for this, and three of the six pairs of modified thoracic legs create a water current that in these suspension feeders provides both food and ventilation of the body surface (Figure 6.7). Gills and haemocyanin are lacking and the heart is reduced to a sinus: haemolymph is moved indirectly by beating of the legs. Gas exchange—in convergence with bivalve molluscs—is by diffusion across the lining (just to confuse matters, also called 'mantle') of the space where the body is suspended within the calcareous shell. Surprisingly, barnacles have a virtually closed circulatory system (haemocoel), which functions—in convergence with echinoderms and even spiders—as a hydroskeleton to extend the legs and, in gooseneck barnacles, the stalk. In that group, it is not unlikely that the oxygen stored in the haemoglobin of the stalk may be made available to the rest of the animal when it is exposed at low tide with the shell closed.

6.7.3 Branchiopoda

The water flea *Daphnia* (Branchiopoda) also breathes through the skin: but which skin? Perhaps not surprisingly given the course of this discussion, beating of the legs creates a ventilatory stream between the inner surface of the carapace and the lateral body wall, which serves as a cutaneous gas exchanger (Figure 6.8). But there is still more to be learned from the water flea. The laterally flattened rostral region has direct diffusive access to the first—most oxygen-rich—ventilated water. This can be advantageous during severe hypoxia, when the convective transport via the open circulatory system fails to supply enough oxygen to the sensory organs and nervous control centres. During periods of hypoxia, haemoglobin in the haemolymph aids gas exchange and transport. For more on this, see Chapters 4 and 8.

6.8 Chelicerata

6.8.1 Pycnogonida

The Pantopoda (sea spiders) are the only extant group within the Pycnogonida. In spite of the fact that there are more than 1300 extant species, little is known about them. These small, spider-like animals, rarely exceeding 1 cm in body length, inhabit coastlines or the deep sea, where they crawl slowly on substrate. Gas exchange is exclusively through the

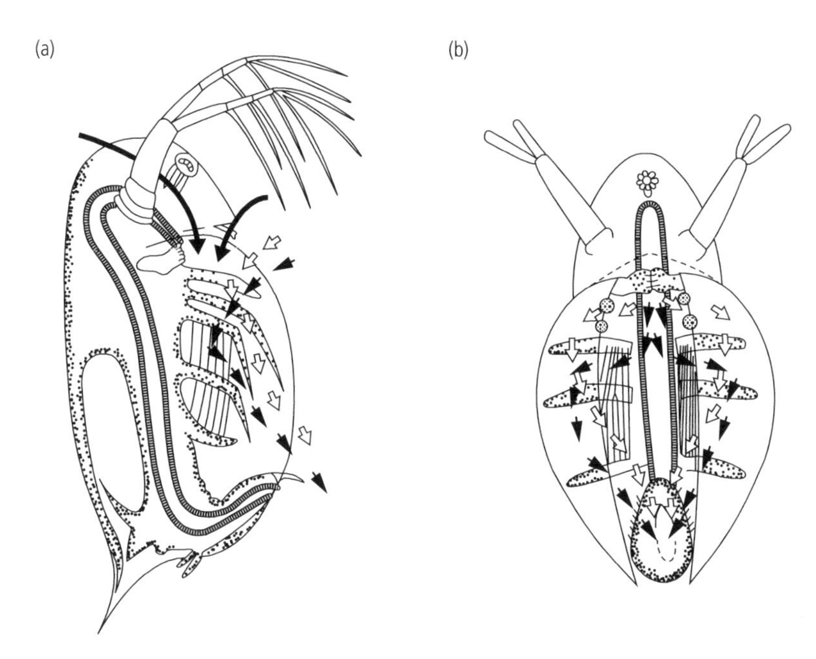

Figure 6.8 Sketch of a the water flea, *Daphnia*, a branchiopod crustacean. Beating of the 2nd antennae (1st antennae not shown) support the animal in the water column and also move water across the head (long arrows), whence it is further moved under the carapace by the broadened thoracic legs, ventilating the body wall (dark arrows) and the sub-carapacial space (light arrows). After Westheide and Rieger, 2007.

integument. There are no respiratory specializations in combination with the circulatory system except that the haemolymph does contain a simple, hexameric haemocyanin (Rehm et al., 2012).

6.8.2 Xiphosura

Today, only four species of horseshoe crab exist, the best known being the North American *Limulus polyphemus*. The other extant species are native to the Asian coast. Not only because of its position on a basally branching lineage among arthropods and chelicerates, but also because of its large size, often exceeding 50 cm in body length, *L. polyphemus* is one of the best investigated arthropod species, particularly regarding haemocyanin (see Chapter 4).

The respiratory organs are quite unlike those of crustaceans, and consist of five pairs of appendages from segments three to seven of the posterior body region (opisthosoma), each appendage forming a cuticular plate that covers leaf-like, lamellated gills (Figure 6.9). These are called book gills, because the lamellae lie one atop the other like the pages of a book. They are passively ventilated by movement of these appendages as well as actively by movement of a so-called flabellum, which is part of the fifth leg pair in the anterior body region (prosoma). The gills are covered with a thin (2 μm!) cuticle, and the haemolymph space is lined with an epidermal cell layer that is only about 0.15 μm thick. Pillar cells stabilize the haemolymph space (Reisinger et al., 1991).

6.9 Functional morphology of aquatic spiders and insects

Due to the high degree of convergence between arachnids and insects, we shall diverge from the systematic discussion at this point and talk about functional morphology in these groups. The important thing for our discussion of water-breathing invertebrates is that arachnids—from their very origin on—were and still are terrestrial. The same applies to insects.

6.9.1 Spiders and insects with an open, gas-filled respiratory system: breathing air under water

As stated previously, arachnids and insects—unlike annelids, crustaceans, molluscs, and chordates—are

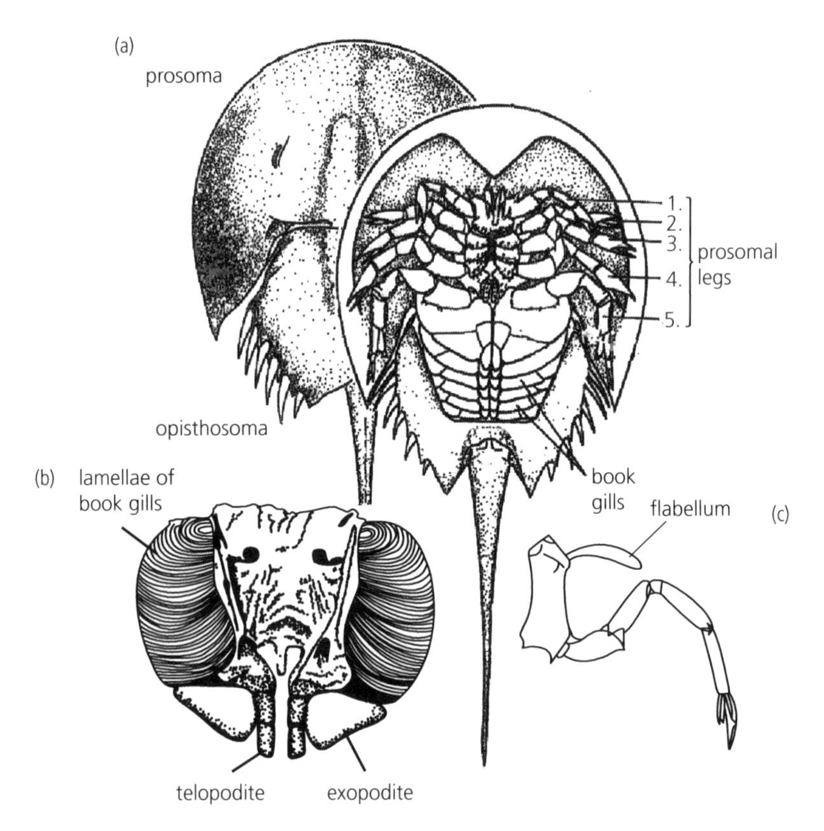

Figure 6.9 A horseshoe crab *Limulus* in dorsal (left) and ventral (right) view (a). In (b) the opisthosoma has been removed and is viewed dorsally, showing some of the book gill lamellae. Part (c) shows the 5th thoracic leg, removed. The flabellum aids in moving the gill plates, ventilating the book gills which they cover. After Westheide and Rieger, 2007.

plesiomorphically air-breathing groups. And as strange as it may seem, many spiders and insects that live under water continue to use the gas phase for breathing. They have secondarily adapted to under-water living in three different ways: (1) by trapping air in a submerged dwelling (diving bell); (2) by carrying air, to which the open stigmata of the lungs or tracheae have access; and (3) among insects, gills that contain closed tracheae.

6.9.2 The diving bell

The water spider *Argyroneta aquatica* is the only extant spider that really lives under water. Well, sort of. Actually it makes a kind of diving bell with its web and fills it with air. In spite of the fact that the oxygen content per litre of the water outside the web is much lower than in a litre of air at the surface of the water, the partial pressures are the same. And since gas can diffuse through the web, the air trapped inside the web has basically atmospheric levels. Such a good idea as a diving bell of course can also be used for emergency survival. Dysderidae (woodlouse spiders) and Clubionidae (sack spiders) can survive flooding for about 10 days using their dense webs as silken sacs to form a physical gill (Rovner, 1987; Messner and Adis, 1995).

6.9.3 The snorkel and siphon

For an insect living under water, the simplest way to get oxygen into an open tracheal system is to access the surface or to bore a siphon into the gas-transporting vessels (aerenchyma) of water plants (Figure 6.10). The latter can be really advantageous because the aerenchyma can connect to the oxygen-producing leaves, resulting in hyperoxic air. But umbilical-like connections also have disadvantages

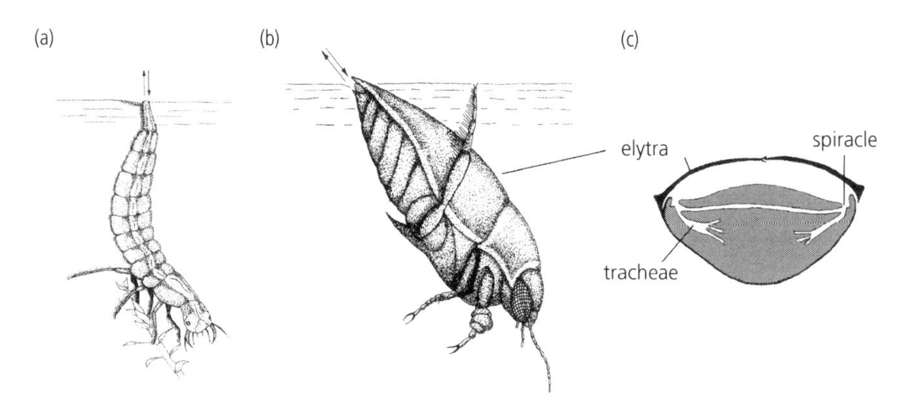

Figure 6.10 A representative aquatic insect, with open tracheal systems: *Dysticus*. the great diving beetle. Larva (a) and adult (b) breathing at the water surface. Part (c) is a cross section through the adult, showing the location of the captured air bubble and the tracheal spiracles. From Penzlin, 2005.

as they severely limit mobility underwater. A self-contained underwater breathing apparatus (SCUBA) is very much preferable as any diver will attest.

Many spiders and insects remain air-breathing even when foraging outside diving bells, or they never make a diving bell in the first place. So-called gas gills developed convergently in spiders and in numerous insect groups. For more on the evolution of these structures, see Chapter 9. There are two types of gas gills: compressible (gas trapping) and incompressible (plastron) ones.

6.9.4 The compressible gas gill

In the compressible gas gill, a gas bubble is bound to the surface of the animal and covers the open spiracles. The spiracles themselves are surrounded by hydropellic oil or hairs, but near the spiracles these structures may also be reduced.

When the water spider, for example, goes outside to hunt, it traps air in body hairs, and with this reservoir it can stay up to 3 days under water even at summer temperatures (Braun, 1931; Winkler, 1955). About two-thirds of the opisthosoma hairs are very short (5 μm) and serve in retaining the gas bubble while the remaining hairs are about twice as long and stabilize the bubble to form a thin film. The surface of the trapped bubble where it contacts the water is called a physical gill (Figure 6.11a). Of course this physical gill is not without its downside. It causes significant buoyancy problems and also the silvery sheen makes the spider an easy mark for predation by fish.

Argyroneta is also not the only spider to use body hairs for short-term survival in the water. Semiaquatic, hydrophilous spiders, including the genera *Dolomedes*, *Pirata*, and *Pisaura*, have surface structures (hairs or tubercles) especially on the ventral surface of the opisthosoma, where the spiracles of lungs and tracheae lie. In insects, compressible, temporary gas gills are bubbles captured under the elytra (wing covers) in bugs, or bound in the hairs on the ventral part of thorax and abdomen, particularly in beetles.

As you may know, compressible gas gills are stable only for a short time: near the water surface longer than at depth. But why is this so? Oxygen is consumed from the gas gill and roughly the same volume of carbon dioxide is released to the bubble via the spiracles, so the volume of the bubble should stay more or less constant, right? Wrong. Due to its high solubility and also its capability to chemically react with water, carbon dioxide gas leaves the bubble and remains only at the nearly negligible ambient level of 0.04 kPa. So the volume of the consumed oxygen is now missing from the bubble, meaning that the PN_2 increases proportionally, according to Dalton's law of partial pressures (see Chapter 3). Since the PN_2 is now higher in the bubble than outside in the water, nitrogen will diffuse out and oxygen will enter the bubble from the water because the PO_2 is higher outside than inside. Particularly, at the water surface ΔPO_2 is larger than ΔPN_2, so the loss of nitrogen is slow and a relatively large amount of oxygen will enter. The bubble eventually disappears, but slowly. At depth, the pressure in the gill increases by about 10 per cent per metre of depth. As the total pressure increases, the proportion of partial pressures stays constant but the absolute partial pressures of oxygen and nitrogen increase due to compression, meaning that the flow of oxygen into the bubble will slow down but the loss of nitrogen will increase. Oxygen continues to be taken up from the water but because of the large ΔPN_2 and a concomitant rapid nitrogen loss, the lifetime of the bubble decreases proportionally to the depth of the dive.

All of this means that spiders and insects with a compressible gas gill must periodically return to the air to avoid drowning, and the deeper they dive, the more often they must surface (Rahn and Paganelli, 1968; Mill, 1974, 1985). But given the tricks available (see Chapter 5), it would be surprising if aquatic insects did not take advantage of some of them. In back swimmer bugs of the genus *Anisops*, haemoglobin is present in some abdominal cells that are connected to the tracheal system. At the beginning of the dive, the insects are positively buoyant. Progressive oxygen consumption makes them neutrally and finally negatively buoyant. During the neutral buoyancy phase, the oxygen stored is released by the haemoglobin and stabilizes the air store (Matthews and Seymour, 2008).

6.9.5 Incompressible gas gills: plastron

Incompressible gas gills (plastron) are also called permanent gas gills, since their lifetime may extend to several days or even weeks. Some insects with a plastron even live permanently submerged. Hydrophobic, densely packed (10^4–10^6 per mm^2), curved hairs surround the spiracle and can also cover much of the body. Because of their shape and arrangement, the hairs cannot collapse, and the air volume in the gill remains constant at any water depth (Figure 6.11b).

Plastron gills are theoretically stable up to a pressure of about 500 kPa, which corresponds to a water depth of about 50 m. The pressure being the same in the bubble as the surrounding medium, nitrogen loss is minimal and the bubble is long-lived. Since oxygen is consumed by the animal, the PO_2 in the bubble is always slightly lower than in the water, meaning that oxygen constantly diffuses in from the water. In insects with normal and constant metabolic demand, plastron respiration is only possible in normoxic or hyperoxic waters. Under hypoxic conditions, the animal would consume oxygen more quickly than it can be replaced, and the bubble would eventually disappear (Thorpe and Crisp, 1947).

The best-known examples for plastron respiration are the benthic water bugs (Hemiptera: Aphelochiridae) of the genus *Aphelocheirus* (hair density $4.2 \cdot 10^6$ per mm^2) but others are also quite proficient. The European pygmy backswimmer bug *Plea minutissima* (Hemiptera: Pleidae), for example, switches from a compressible gas gill to plastron respiration in the winter, when it sinks to the bottom of the pond and stays there for months (Kovac, 1982). Independently, beetles have discovered the plastron, and the aquatic riffle beetles (Coleoptera: Elmidae) of the genus *Elmis*, with up to $1.6 \cdot 10^5$ hairs per mm^2 have a plastron that rivals that of the benthic water bugs. And at least three other major beetle taxa (Chrysomelidae, Hydrophilidae, and Curculionidae) also have a plastron.

And the list goes on: the pupae of black flies (Diptera: Simuliidae) and aquatic caterpillars of snout moths (Lepidoptera: Pyralidae) use a plastron late in development, but in tiger moths (Lepidoptera: Noctuidae) all caterpillar stages have plastron areas on the abdomen. And arching off at the very base of the Hexapoda, the intertidal springtail *Anurida maritima* survives high tide thanks to a plastron made up of hydrophobic, microscopic ornamentation of the epicuticle. Independently, this structure has also evolved among spiders, for example, in the semi-aquatic, hydrophilous species mentioned earlier and in *Phrynus marginemaculatus* (Amblipygi) where the plastron surrounds the book lung spiracles and enables the animals to stay under water for about a day (Hebets and Chapman, 2000).

6.9.6 Insect gills: insects with closed tracheal systems

Air trapping or binding previously described is necessary if the openings to the lungs or tracheae

(a) **Compressible gas gill**

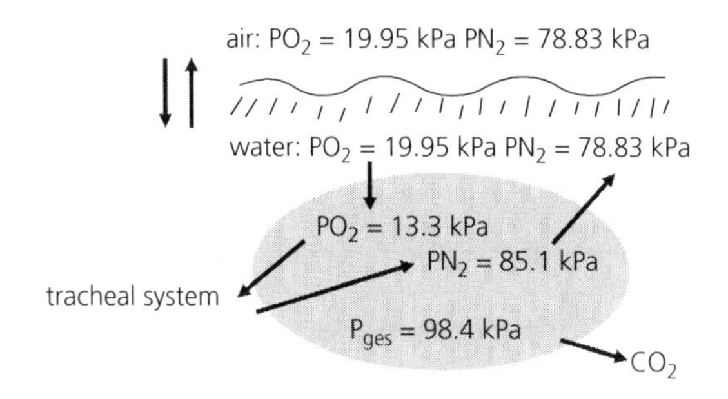

At water surface

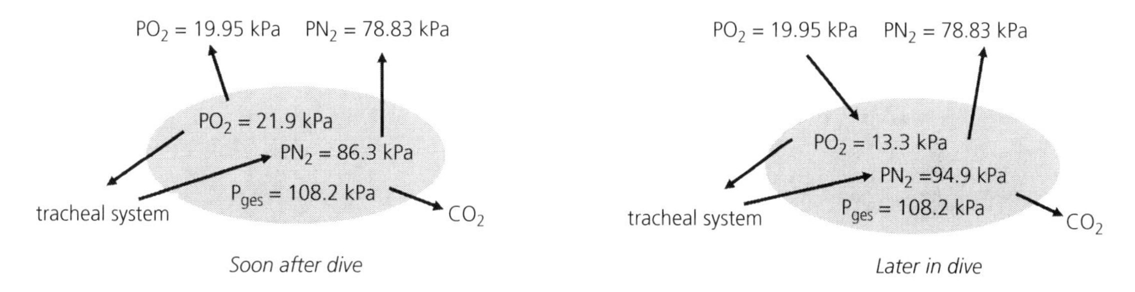

At 1 m water depth

(b) **Incompressible gas gill**

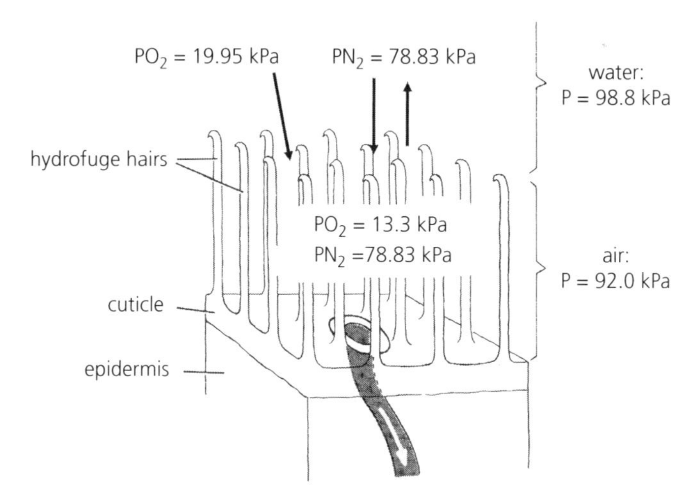

Figure 6.11 Compressible (a) and incompressible (b) gas gills. Part (a) shows partial pressures of gases initially at the water surface. Shaded area indicates the conditions inside the gas gill. Below, the partial pressures of oxygen and nitrogen inside and outside the gas gill at 1 m depth soon after a dive (left) and later in the dive (right). Note the increase in PN_2 and decrease in PO_2, leading to loss of nitrogen to the water as oxygen is consumed and to eventual collapse of the gill. Part (b) shows an incompressible gas gill with hydrufugal (non-wettable) hairs. Trachea shaded. Because the gill cannot collapse as oxygen is consumed the nitrogen partial pressure remains unchanged and oxygen diffuses in from the water. Similar conditions exist in a plastron (see text). (a), original AS; (b) after Eckert, 1998.

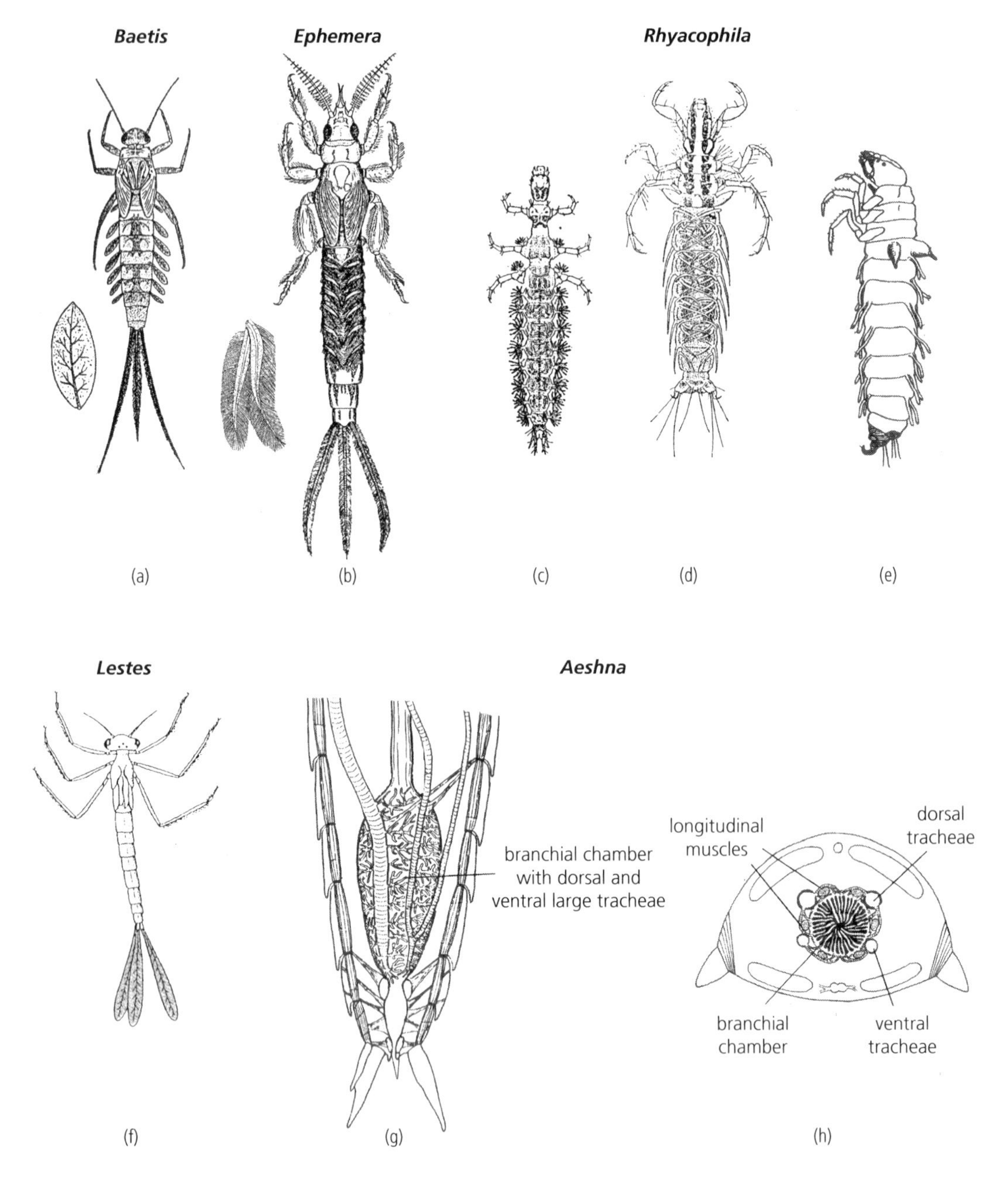

Figure 6.12 Aquatic insects with closed tracheal systems and gill-like structures: Ephemeroptera (a) and (b); Trichoptera (c)–(e); Zygoptera (f); water lung of a dragon fly in dorsal view (g) and cross section (h) Greven and Rudolph, 1973; Engelhardt 1972.

remain open and flooding with water must be prevented. But there is another alternative in tracheate arthropods: closed tracheae. Most species of wetland insects—in particular their larvae—have a closed system and very often, specialized body regions called tracheal gills are the site of gas exchange (Figure 6.12). Tracheal gills are leaf-like or filamentous evaginations of the body surface in which tracheoles are located directly under the external cuticle and usually arranged parallel to one another. Some examples of insects with closed tracheal systems are larvae of Ephemeroptera (mayflies) that have leaf-like gills along the abdomen and Trichoptera (caddis flies) with filiform gill appendages. Here, the large number of abdominal tracheal gills is an adaptation to the life in water with low oxygen content. The tracheoles are arranged just beneath the cuticle parallel to the longitudinal axis of the thread-like tracheal gills and lie enclosed in deep extracellular slit-like invaginations of the basal plasma membrane (Wichard and Komnick, 1971). In some species of Plecoptera (stoneflies), the tracheoles of the respiratory epithelium are extremely densely packed and the spaces between the tracheoles have a smaller diameter than the tracheoles themselves (0.2–1 μm). Larval Zygoptera (damselflies) have three long, leaf-like gills at the end of the abdomen. They are important under hypoxia: in normoxia, cutaneous gas exchange is sufficient (Wichard, 1979).

In such tracheal gills the boundary layer between the cuticle and the surrounding water is about 300 μm in standing water and down to 10 μm in moving water or during oscillating body movements. But ventilation is in general important for aquatic insects since they have all disadvantages of breathing in water compared to their terrestrial relatives or life stages.

Larval Anisoptera (dragonflies) are large, active predators that live completely submerged in water that is not always rich in oxygen. Their remarkable branchial chamber—probably better described as a water lung—in the hindgut, is undoubtedly the most highly specialized mechanism of aquatic respiration in insects (Figure 6.12h). In *Aeshna cyanea* (blue hawker dragonfly), the hindgut of the larva is widened and the anterior portion is modified into a branchial chamber. Six primary gill folds have transverse secondary folds on each side. Each secondary fold consists of two basal pads, each of which forms respiratory lamellae in its apical part. In the epidermal layer of the lamellae, just beneath the cuticle, a dense meshwork of tracheoles is found (Greven and Rudolph, 1973; Wichard and Komnick, 1974; Saini, 1977). Oxygen that enters the tracheolar space has to traverse the cuticular and the epidermal layers of the hindgut epithelium and, in addition, the epidermal and cuticular layer of the tracheole. The surface area for gas exchange is 44 $cm^2\ g^{-1}$. This compares with 28 $cm^2\ g^{-1}$ for a rat lung (Burri and Weibel, 1971). The thickness of the entire water–tracheole diffusion barrier, 0.27 μm, is also similar to that of a mammalian lung (Kohnert et al., 2004). But since lung tissue is about 20 times more permeable to oxygen than chitin, the rat lung would still have about 10 times as much diffusing capacity per unit body mass as the dragonfly larva, which is in keeping with the differences in metabolic rate. And once inside the tracheal system, the oxygen can diffuse at a speed comparable with that of blood movement in the rat but without need of an energy consuming circulatory system. Here the dragonfly may also have an added advantage, since the system is tidally ventilated by rhythmic filling and emptying the hindgut with water and this could cause some convection in the tracheae, too (Mill and Pickard, 1972; Pickard and Mill, 1974).

CHAPTER 7

Respiratory faculties of amphibious and terrestrial invertebrates

As pointed out in earlier chapters, the partial pressure of oxygen in water is the same as in the surrounding atmosphere but the oxygen content per unit volume is around 30 times less. Carbon dioxide, on the other hand, is freely soluble in water and also chemically reacts with it, not only in surrounding water but also in body tissues. Many of the problems of water-breathing can be better solved by leaving the water and breathing air on land.

In the present chapter, we shall talk about the respiratory faculties for air breathing that different animal groups have (see also Table 7.1), and in Chapter 10, take up the evolution of these faculties. To avoid confusion, we shall adhere to the order of discussion of different taxa that we used in Chapter 6 although this does not necessarily represent the order of appearance of air breathing in the different groups.

7.1 Annelida: Clitellata

In earthworms and leeches, gas exchange is through the skin, even in very large-bodied species. In each segment, major dorsal and ventral blood vessels are connected by lateral ones, which—at least in large animals—supply an extensive network of capillary-like spaces in the outer epidermal layer. Such a system is best developed in earthworms, for example, in *Lumbricus terrestris*, familiar to most anglers. Blood is moved from posterior to anterior in the dorsal vessel and in the opposite direction in the ventral vessel. In this group, extracellular haemoglobin plays an important role, and accounts for up to 50 per cent of the oxygen transported in the blood (Weber and Vinogradov, 2001).

Terrestrial leeches (Hirudinea: Haemadipsidae, e.g. *Haemadipsa* and some Pharyngobdellidae, e.g. *Cylicobdella*) do not keep the ancestral blood-vascular system just described. The septa separating adjacent segments have been reduced and the coelomic cavity is used as a hydraulic skeleton to support undulating movement. Branches of this hydrocoel approach the skin, which serves as the gas exchanger. Extracellular haemoglobin is found in only a few species.

7.2 Mollusca

Of the various molluscan groups only the snails and a few bivalves have air-breathing representatives. Air-breathing mussels have already been discussed as a special case of water-breathing bivalves in Chapter 6, and will not be taken up again here. Instead, we move directly on to snails.

Air breathing arose independently several times among snails: twice even among intertidal limpets and periwinkles (Henry et al., 1993). These snails use the mantle for gas exchange during land excursions. But the real experts are the pulmonate snails.

In the pulmonate snails (Figure 7.1), an orifice on the right side opens to a sac, which varies in

Respiratory Biology of Animals: Evolutionary and Functional Morphology. Steven F. Perry, Markus Lambertz, Anke Schmitz, Oxford University Press (2019). © Steven F. Perry, Markus Lambertz, & Anke Schmitz.
DOI: 10.1093/oso/9780199238460.001.0001

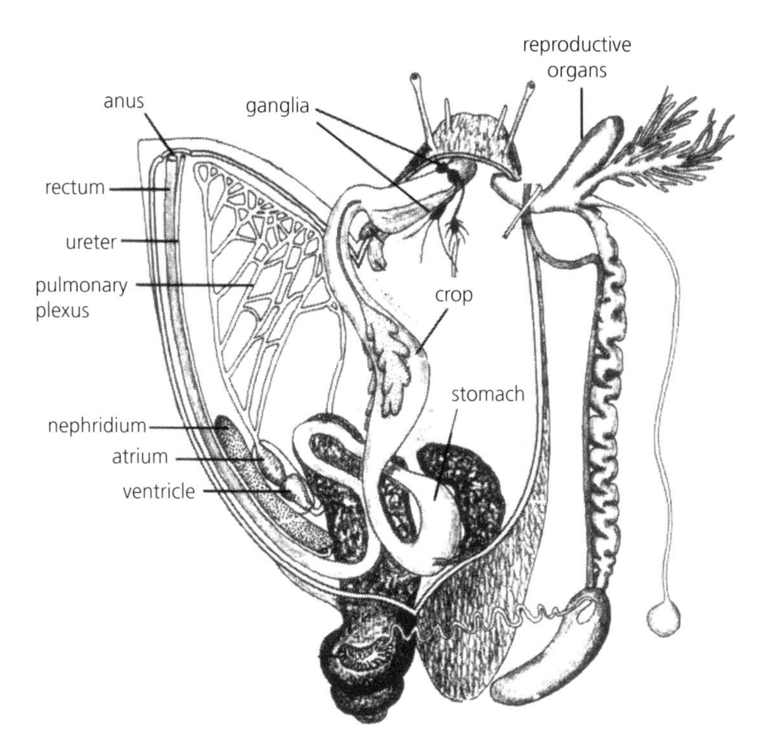

Figure 7.1 Dorsal dissection of the Roman snail, *Helix*, a eupulmonate, showing the lung with its network of closed vessels (pulmonary plexus), which supply oxygen-rich haemolymph to the atrium of the two-chambered heart for distribution to the body by the ventricle. Barnes, 1980.

structure from a smooth-walled bag to a lung with a rugose, highly vascularized surface. Many, such as the Roman snail *Helix pomatia* are purely terrestrial while others (e.g. representatives of the genus *Lymnaea*) lead an amphibious life, using inhaled air to help them survive in oxygen-poor water. The latter has been studied in detail and is even used as a model for better understanding the control of breathing in mammals, including humans (see Chapter 8). The respiratory faculty of air-breathing molluscs includes molluscan haemocyanin (see Chapter 4) and a two-chambered heart, which receives oxygenated haemolymph from the lung and pumps it to the rest of the body.

7.3 Onychophora

Velvet worms breathe exclusively with tracheae, which function as tracheal lungs. The animals need high humidity in the environment: otherwise water loss through the numerous, non-occludable spiracles of the tracheal system is life-threatening. The short atrium, lying behind the spiracle, branches into a tracheal tuft with the tracheae ending in the haemolymph. The tracheal volume of 7.3 $\mu l\ g^{-1}$ in *Peripatus acacioi* (Bicudo and Campiglia, 1985) is about ten times the value for the tracheal system of a jumping spider and in the same range as the tracheal system in the first instar of the stick insect *Carausius morosus* (Schmitz and Perry, 1999). Haemocyanin is present but its exact role in gas transport is poorly understood. Although tracheal lungs are present, the skin also serves in gas exchange. Breathing is continuous and decreases under hypoxia of a PO_2 of 10 kPa or lower. Thus animals are well adapted to their habitat under leaf litter or in rotting logs (Mendes and Sawaya, 1958; Woodman et al., 2007a, 2007b).

7.4 Myriapoda

Myriapods consist of Chilopoda (centipedes, including the groups Scutigeromorpha and Pleurostigmophora) and the Progoneata (millipedes, including

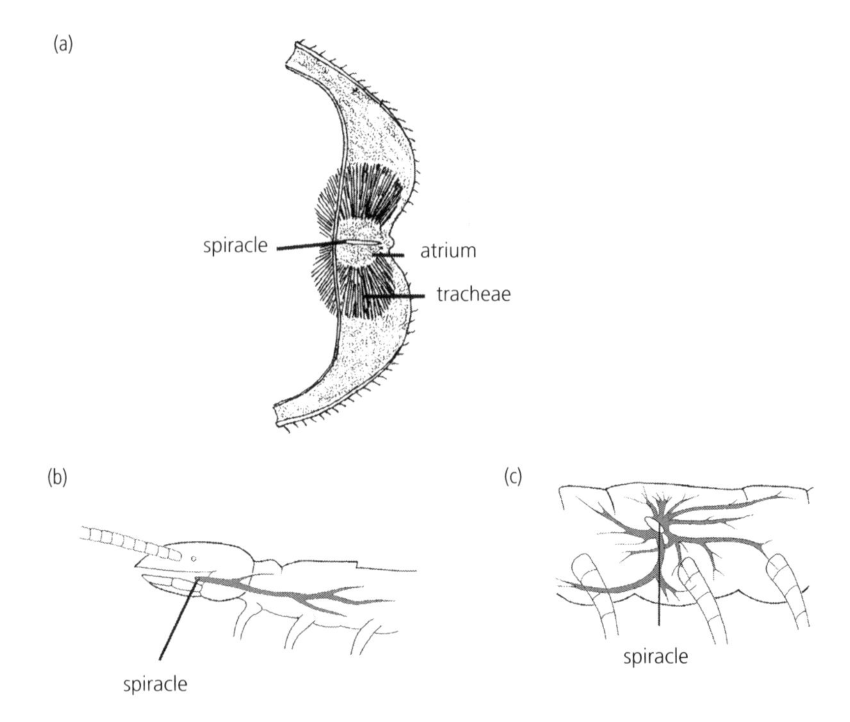

Figure 7.2 Tracheal lungs in myriapods. (a), Chilopoda, *Scutigera*; (b), Diplopoda, *Scutigerella*; (c), Chilopoda, *Lithobius*. Note the unusual position of a spiracle above the mandible in the pseudocentipede, *Scutigerella*. In (a) the back part of a tergite (dorsal plate) is shown from dorsal with spiracle and tracheal tuft. Anterior is to the left. After Westheide and Rieger, 2007 and Hilken, 1998.

the groups Symphyla, Diplopoda, and Pauropoda). Both groups, like velvet worms, have tracheae, but in some species their spiracles are occludable. This is a prerequisite for discontinuous breathing, which has been observed in some species (see also Chapter 8). The tracheae look like those of spiders and insects. The latter similarity was the nominal character for combining insects with myriapods to form the Tracheata, which is usually no longer recognized as a taxonomic entity. Haemocyanin has been shown to have a respiratory function in at least some species, but its exact role is still to be resolved: some millipedes have an insect-like tracheal system that supplies cells directly, while others have tracheal lungs (Figure 7.2).

Among the centipedes, the Scutigeromorpha (e.g. *Scutigera*) have unpaired, mid-dorsal spiracles near the posterior margin of the leg-bearing segments one to seven. The other segments lack spiracles. *Scutigera* has tracheal lungs, each spiracle opening into an atrium that gives rise to two large tracheal tufts with short tracheae, bathed in the haemolymph (Hilken, 1997). *Scutigera* is a fast-running animal with high oxygen demands, and its haemocyanin, which is a low-affinity protein with high cooperativity, is well suited for oxygen transport and release in the muscles (Mangum, 1986; Jaenicke et al., 1999). In the Pleurostigmophora, spiracles are situated on the sides of the animal (pleura) and the number of spiracles varies with the systematic group.

The 'true' millipedes, Diplopoda, have double segments, which arise by the fusion of two originally separate anlagen. Accordingly, these also have two pairs of legs and also two pairs of spiracles per segment. The spiracles open into atria from which numerous tracheae arise. These tracheae are similar to those of insects, either penetrating the organs or ending very nearby in the haemolymph. But, since haemocyanin is present (Kusche and Burmester, 2001; Kusche et al., 2003), some tracheal lung function certainly exists. The haemocyanin of *Spriostreptus*, in contrast to that of *Scutigera*, is a high-affinity protein with low cooperativity and probably serves primarily in oxygen storage. The animals live underground in a hypoxic environment and only come to the surface at night (Stewart and Woodring, 1973). A second millipede group, the Symphyla (e.g. *Scutigerella*), are very small but active animals. They have only one pair of spiracles in the head, supplying tracheae that serve the front end of the body. A third millipede group, the Pauropoda, are also very small. Tracheae are either lacking completely or locally present in the head and near to the base of the first leg pair (Hilken, 1998).

7.5 Chelicerata: Arachnida

The vast majority of arachnids are purely terrestrial. The plesiomorphic respiratory organs are book lungs, which are clearly derived from book gills (see Chapter 6). Among arachnids (Table 7.1), small-bodied species use the entire skin for gas exchange, whereas larger animals use lungs and/or tracheae. The so-called book lungs are diffusion lungs. Ventilatory mechanisms are lacking, and the spiracles function only as diffusion regulators (Paul et al., 1987). Lung lamellae are stacked like the pages of a book, in which the haemolymph-filled pages alternate with air spaces. Epidermis lines the inner surface facing the haemolymph, while the air space is covered with cuticle. Pillar cells reminiscent of those in fish gills keep the haemolymph spaces from overinflating, a function that is supported in the air spaces by cuticular struts and spines. The column-like struts bind the two sides of the air spaces with one another over about one-third of the lung area. Spines, which are present on the other two-thirds, extend from only one side and thus prevent collapse of the air spaces but also allow expansion.

Just to give an idea of the dimensions of such a book lung, here are a couple of examples. First the bird spider 'tarantula' *Aphonopelma hentzi*, which most frequently was referred to as *Eurypelma californicum* (see Nentwig, 2012). Here the haemolymph space is 60–70 μm across, while interlamellar air space is an order of magnitude thinner (Moore, 1976; Reisinger et al., 1990; Reisinger et al., 1991) and the diffusion barrier of this constant-volume lung is about the same thickness as in a bird lung: the epidermal layer is about 0.02–0.1 μm thick, while the cuticular layer is only slightly thicker: 0.03 μm. These dimensions are much less in smaller spiders such as the wolf spider *Pardosa lugubris*, and the jumping spider *Salticus scenicus*, both weighing about 20 mg. Here the haemolymph space measures 10–20 μm—comparable with the lamellae of a fish gill—but the air space is only 1–2 μm wide (Schmitz and Perry, 2000, 2002a). The air spaces can be thinner than the interlamellar spaces of fish gills because of the much lower density and viscosity of air. They are even thinner than the diameter of the smallest air capillaries in the bird lung. As explained previously, the interlamellar air spaces are supported by spines and struts and do not tend to collapse because the cuticle that covers them is dry and not subject to surface tension constraints.

7.5.1 Arachnids other than scorpions and spiders

A look at these chelicerate groups (Table 7.1) shows a mixed bag, ranging from some with relatively reduced lungs to some, such as the sun spiders (Solifugae, see later in this section) that have a tracheal system comparable to that of insects (Figure 7.3).

The microwhip scorpions (Palpigradi) are tiny (<3 mm long) cousins of whip scorpions (Uropygi, see below) that live in tropical sands and soils or in caves. Most of these tiny animals breathe through the skin but some species have three pairs of book lungs that look more like sacs than books. Nothing is known about the respiratory physiology of this group.

Two other arachnid groups, the Uropygi and the Amblipygi (whip spiders), breathe exclusively with book lungs and have haemocyanin in their haemolymph. Whip scorpions such as *Mastigoproctus giganteus* reach body lengths of up to 6.5 cm. Like mesothaelid spiders, they have two pairs of book lungs on the ventral side of the second and third opisthosomal segments. Whip spiders, with body sizes from 4 mm to 4.5 cm, are recognized by their huge and raptorial pedipalps and whip-like second legs. The book lungs are as in whip scorpions. In *Damon annulatipes* breathing is continuous and the metabolic rate is low, as in scorpions (Terblanche et al., 2004).

Both the Ricinulei (hooded tickspiders) and Pseudoscorpiones (pseudoscorpions) lack lungs and have a tracheal system, the ultrastructure of which has not yet been studied. Ricinulei reach

Table 7.1 Respiratory organs and circulatory system of air-breathing invertebrates

Taxon	Air-breathing organs (ABOs)	Circulatory system/respiratory protein
Mollusca		
Gastropoda Diverse taxa Panpulmonata (pulmonate snails)	Mantle cavity Lung	Open circulatory system Haemocyanin, haemoglobin
Annelida		
Clitellata (earthworms and leeches)	Skin	Open or functionally closed system Extracellular haemoglobin
Onychophora (velvet worms)	Tracheal lungs, skin	Open circulatory system Haemocyanin
Myriapoda		
Chilopoda (centipedes)	Tracheal lungs or tracheae, occludable spiracles	Open circulatory system Haemocyanin
Progoneata (millipedes)	Tracheae, occludable spiracles	Open circulatory system Haemocyanin
Arachnida		
Palpigradi (micro whipscorpions)	Skin or three pairs of book lungs	Not known
Uropygi (whipscorpions)	Two pairs of lungs	Open circulatory system Haemocyanin
Amblipygi (whip spiders)	Two pairs of lungs	Open circulatory system
Ricunulei (hooded tick spiders)	Tracheae	Partly reduced
Pseudoscorpiones (pseudoscorpions)	Tracheae	Open circulatory system with little development
Acari (ticks and mites)	Tracheae in large species, skin in small species	Open circulatory system
Opiliones (harvestmen)	Tracheae	Open system, little development Presumably no haemocyanin
Solifugae (sun spiders)	Tracheae, occludable spiracles	Open circulatory system No respiratory proteins
Scorpiones (scorpions)	Four pairs of lungs	Open circulatory system Haemocyanin
Araneae (spiders)	Two pairs of lungs in the opisthosoma or one pair of lungs and tracheae	Open circulatory system Haemocyanin
Crustacea		
Malacostraca		
Amphipoda (sand fleas etc.)	Gills	Open circulatory system. Haemocyanin
Isopoda: Oniscoidea (woodlice etc.)	Gills, pleopod lungs, and tracheae	Open circulatory system Haemocyanin

Table 7.1 (Continued)

Taxon	Air-breathing organs (ABOs)	Circulatory system/respiratory protein
Decapoda Anomura Coenobitidae (land hermit crabs)	Branchiostegal lungs, dorsal surface of abdomen	Open circulatory system Haemocyanin
Decapoda Brachyura Grapsidae and Gecarcinidae (land crabs)	Gills stabilized for air breathing, branchiostegal lungs (convergent with Anomura)	Open circulatory system Haemocyanin
Hexapoda		
'Entognatha' (springtails etc.)	Tracheae or skin	Open circulatory system, little development
Insecta (insects)	Tracheae, occludable spiracles Ventilatory air sacs	Open circulatory system Haemocyanin in some species Extracellular haemoglobin

body lengths of only 5–10 mm and have an exceptionally thick and stable cuticle, which excludes cutaneous gas exchange. The circulatory system is reduced and nothing is known about their respiratory physiology. Pseudoscorpions are in the same size range as tickspiders and live under bark, stones or leaf litter, or as commensalists in the nests of mammals. *Chelifer cancroides* can also be found in houses. Two pairs of spiracles are on the ventral side of the third and fourth opisthosomal segments, and breathing is discontinuous.

Within the Acari (ticks and mites), both cutaneous and tracheal respiration are seen but book lungs are lacking. Most mites breathe with tracheae, the number of spiracles varying in number from one to four pairs, but all are in the anterior part of the body. In some species the respiratory organs are completely lacking. Also, the circulatory system has undergone reduction. Many mites are only a couple of hundred microns long, so studying their respiratory physiology is tough, but the centimetre-long giant red velvet mite, *Dinothrombium magnificum*, has been looked at (Lighton et al., 1993; Fielden et al., 1994). When not burrowing, it shows continuous respiration. Its standard metabolic rate per unit body mass is in accordance with expected values, but when the animals are burrowing, it is low, suggesting discontinuous breathing.

Opiliones (harvestmen such as 'daddy longlegs') have tracheae and no vestiges of book lungs. On the ventral side of the prosoma behind the fourth leg pair, the tracheal system has two spiracles that are opened and closed indirectly by muscles attaching at the tracheal wall and the atrium behind the spiracular opening. Also, movement of the coxa—the most proximal leg segment—influences the spiracular opening, since it forms one wall of the atrium. The spiracles are deeply countersunk and have densely packed bristles (trichomes), both of these characters representing adaptations that reduce water loss. The two stem tracheae of the ventral spiracles branch into smaller tracheae and tracheoles, but most tracheae are limited to the prosoma. Air sacs are lacking. The thickness of the tracheal walls ranges from 1–2 μm in the largest tracheae to 0.2 μm in the tracheoles. Most organs are not penetrated by tracheae but the nervous system and—in some species—also the muscles profit from a direct supply. Long-legged harvestmen possess additional spiracles and tracheae in the tibia of the legs.

As omnivores, feeding on living and dead animals and also on plant material, harvestmen need a continuous food supply. Their metabolic rates are greater than in spiders but less than in insects. The short-legged, forest-dwelling harvestman *Nemastoma lugubre*, with a body mass of 1.5–4.0 mg, has a

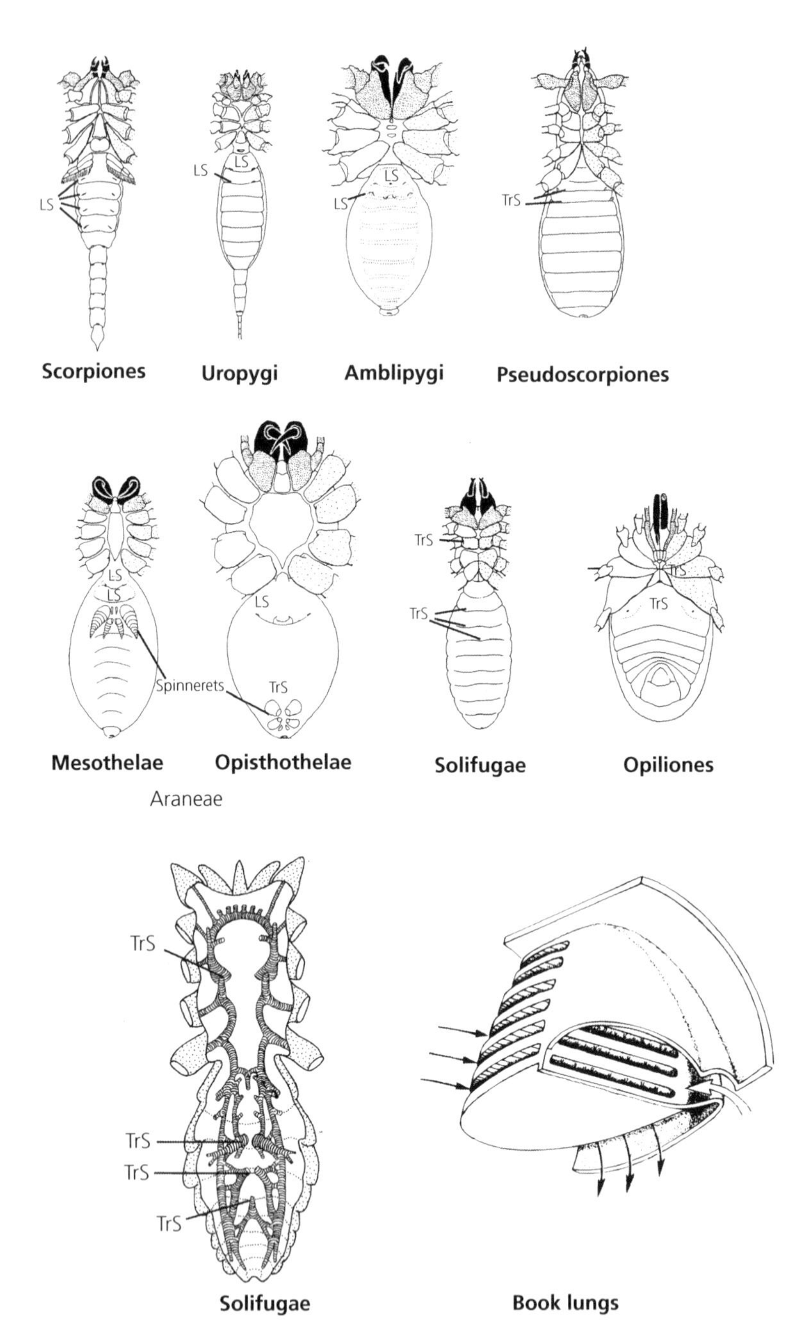

Figure 7.3 Ventral view of diverse terrestrial chelicerates showing lung stigmata (LS) and tracheal stigmata (TrS). Note the posterior relocation of stigmata and spinnerets in Opisthothelae compared with Mesothelae. The posterior stigmata open to tracheae, lacking in the Mesothelae, which possess two pairs of book lungs. The highly developed tracheal system (shaded) of Solifugae, with 3 pairs of stigmata and 1 unpaired one, is reminiscent of insects. All arachnid book lungs (final image) have a similar structure, clearly derived from book gills. Air (white arrow) enters an atrium, whence it diffuses in between vascular lamellae, through which haemolymph flows (black arrows). The oxygenated haemolymph then flows, usually through closed vessels, to a heart similar to that of other arthropods for distribution to the body. After Westheide and Rieger, 2007.

diffusing capacity (DO_2) of 10–25 μl min^{-1} g^{-1} kPa^{-1} (Hoefer et al., 2000; Schmitz and Perry, 2002b). This compares favourably with 12–16 μl min^{-1} g^{-1} kPa^{-1} in active hunting spiders such as wolf spiders and jumping spiders. Breathing is continuous, both at rest and during spontaneous and forced treadmill walking. Spontaneous walking raises the resting $\dot{V}O_2$ up to threefold while constant running on a treadmill raises the value about fivefold. Their thick-walled tracheae serve mainly in gas transport while the tracheoles carry out terminal diffusion as in insects. Haemocyanin has been reported, but in lower concentration than in spiders. This would lead to the hypothesis that gas exchange takes place as a combination of lateral and terminal diffusion. However, there is some doubt about the presence of haemocyanin, so it appears that the last word has not been spoken.

But the champions of tracheal respiration among the arachnids are the Solifugae (sun spiders). Here is an arachnid that 'thinks it is an insect'. These tropical and subtropical relatives of pseudoscorpions can reach up to 10 cm body length and are fast runners. The large size and oversized chelicerae together with their quickness makes them formidable predators and they can even inflict painful wounds in humans. Their completely tracheal respiratory system has seven occludable spiracles: one pair in the prosoma, two in the opisthosoma, and an unpaired one in the opisthosoma behind the paired ones. As in insects, air sacs aid in ventilation, the tracheae are branched, and tracheoles penetrate the epithelia of the organs (Gruner et al., 1993). Solifugans also are similar to insects regarding their breathing patterns. They show discontinuous gas exchange during dry periods and during starvation or rest (Lighton, 1996; Lighton and Fielden, 1996), but breathing is continuous during activity, and active ventilation by contractions of the prosoma, supports respiration (Nogge, 1976; Slama, 1995). Haemocyanin has not yet been found (Markl, 1986) and tracheal gas exchange seems to be the same as in insects, with an emphasis on terminal diffusion.

7.5.2 Scorpiones

Scorpions have in the opisthosoma four pairs of book lungs and small, slit-like spiracles. The heart lies in the mesosoma (equivalent to the anterior part of the opisthosoma in spiders) and discrete arteries supply internal organs, appendages, and the nervous system (Farley, 1990; Kamenz et al., 2005). Although newly hatched scorpions initially use cutaneous gas exchange, after the first moult, the book lungs grow along with the growth of the instars and take over gas exchange completely (Farley, 2008). That scorpions have a larger respiratory surface area than similar-sized spiders is not surprising since they have twice as many lungs.

Like spiders, scorpions have a low metabolic rate (Paul and Fincke, 1989), and when forced to exercise, the anaerobiosis leads to accumulation of D-lactate in the leg muscles (Paul and Fincke, 1989; Prestwich, 2006). But the larger respiratory surface area also leads to much greater carbon dioxide release than in spiders during activity and recovery (Paul and Fincke, 1989). Carbon dioxide plays a particularly important role in the acid–base control of scorpions. In the emperor scorpion, *Pandinus imperator*, after running, 62 per cent of the additionally released carbon dioxide is used for buffering, compared with 34 per cent in the North American bird spider tarantula, *Aphonopelma hentzi* (Paul and Fincke, 1989; Paul et al., 1989; Paul, 1991, 1992).

7.5.3 Araneae

In spiders, two pairs of book lungs in the second and third opisthosomal segments are ancestral and lungs remain important respiratory organs in most species. In many species, however, tracheae are developed and book lungs and tracheae complement one another (Figure 7.4). In jumping spiders, which are stalking predators, for example, 25–30 per cent of the entire diffusing capacity lies in the tracheae. Another example are the Uloboridae (hackled orb-weavers), which preen their webs. In this taxon, the extensive tracheal system extends into the legs and is best developed in those species that constantly use their legs for web maintenance. Generally speaking, when tracheae are well developed, lungs are less developed and vice versa. Most frequently, the second pair of book lungs becomes reduced, but occasionally also the anterior one or even both. However, some spiders rely on book lungs only (e.g. *Pholcus*—in fact only maintaining a

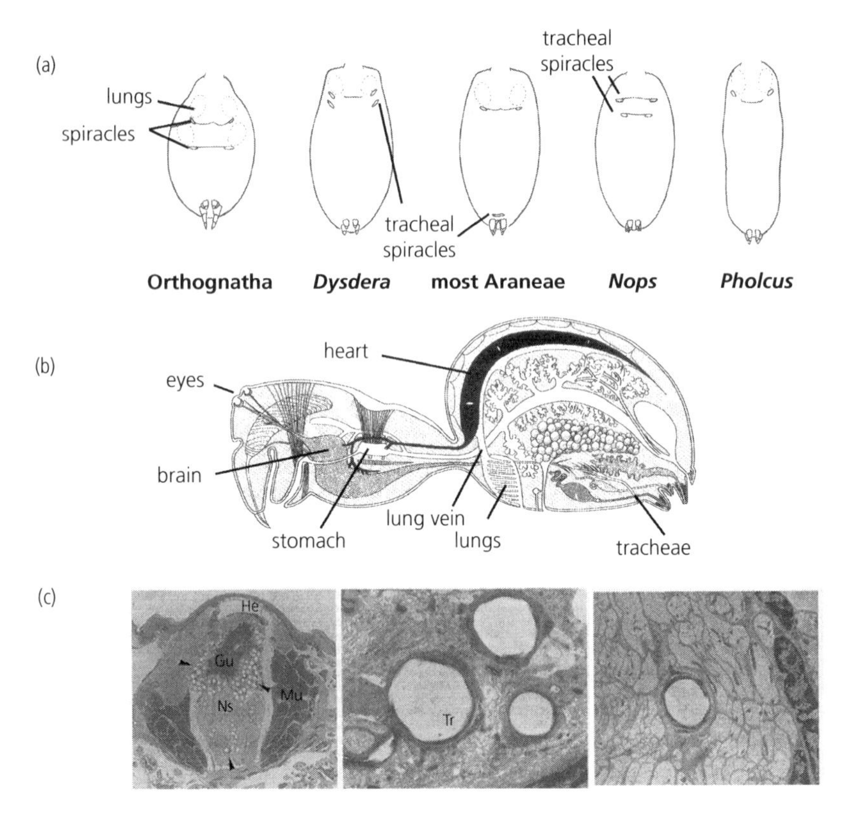

Figure 7.4 The respiratory system of spiders. Part (a) shows ventral views of the opisthosoma and the location of the lungs (outlined with dotted lines), spiracles for lungs and tracheae. Unless otherwise labeled, the anterior (upper in the images) spiracles are for the lungs and the posterior ones, for the tracheae. Part (b) is a schematic longitudinal section through a typical spider (see also "most Araneae" in (a), above) showing the location of lungs, tracheal stigmata and heart in the opisthosoma, while the metabolically demanding main parts of the central nervous system, the stomach and legs (not shown) are in the prosoma, reached through the narrow pedicel. In the left-hand electron micrograph note tracheae and tracheal bundles indicated by arrow heads and tracheal penetration of the nervous system. The following two, higher power micrographs micrographs show large and smaller tracheae in cross section in the gut, and in the nervous system, respectively. Gu, gut; He, haemolymph vessel; Mu, musculature; Ns, nervous system; Tr, trachea. After Westheide and Rieger, 2007; Original electron micrographs A.S.

single pair), whereas others breath exclusively via a tracheal system (e.g. *Nops*) (Braun, 1931; Millidge, 1986; Bromhall, 1987).

Tracheae in spiders are invaginations of the body wall, supplied by a spiracular opening, and up to four tube-like tracheae can be present. The lateral two (so-called primary) tracheae are modified book lungs, whereas the medial two (so-called secondary) originate from other internal structures such as cuticular tendon insertions (apodemes) (Purcell, 1895; Lamy, 1902; Purcell, 1909, 1910; Levi, 1967; Forster, 1980; Ramirez, 2000). Most arachnid tracheae look very much like those of insects, being tubular structures, or for that matter like miniature vacuum cleaner hoses. The tracheal spiracles of arachnids can be closed but the occlusion mechanisms never reach the sophisticated level seen in insects. Primary and secondary tube tracheae in the entelegyne spiders often remain simple tubes, restricted to the opisthosoma, as in the common house spider, *Tegenaria domestica*. But in some groups the tracheae enter the prosoma via the pedicel that connects the opisthosoma with the prosoma, and even penetrate muscles, the nervous system, or the epithelia of other organs (Kaestner, 1929; Schmitz and Perry, 2000, 2001, 2002a; Schmitz, 2004, 2013). Haplogyne spiders, on the other hand, develop a special kind of tracheae called sieve tracheae, which are bundles of tube-like projections considered to be derived from rounded book lung lamellae.

The thickness of tracheal walls is between 0.2 and 0.6 μm and is about equally divided between cuticle and epidermis. The smallest tracheae have about the same barrier thickness as the book lungs (Schmitz and Perry, 2001, 2002a), which is comparable with that of a mammalian lung. But having said that, the mammal would still have a greater diffusing capacity per unit surface area because of the extremely low KO_2 of chitin. The tracheae usually form tracheal lungs, ending in the haemolymph, and use the entire surface for gas exchange. Tracheal lungs increase the respiratory surface area and the effectiveness of respiration considerably because gas exchange is no longer restricted to the second and third opisthosomal segment where the lungs are located, and can extend through the pedicel into the prosoma.

The diversity in lung morphometrics of spiders becomes clear when comparing the agelenid house spider *Tegenaria domestica* with the wolf spiders and jumping spiders. The morphological diffusing capacity in *Tegenaria* (4–9 μl min^{-1} g^{-1} kPa^{-1}) is clearly less than that in the other two groups (12–16 μl min^{-1} g^{-1} kPa^{-1}), mainly due to the diffusion barrier, which is nearly twice as thick in *Tegenaria* (Strazny and Perry, 1984; Schmitz and Perry, 2001, 2002a).

In some groups the tracheae combine lateral and terminal diffusion models: the tracheal walls may be thin enough to allow gas exchange with the haemolymph but the tracheae also penetrate some organs. Lateral diffusion may be discouraged either by thick diffusion barriers, as in jumping spiders, or by grouping the tracheal branches into bundles that run together through body regions in which lateral diffusion is not necessary. In other words, oxygen loss over the length of the tracheae is restricted but not prevented.

7.6 Crustacea

Among the crustaceans, semi-terrestrial or completely terrestrial forms are found only within the Malacostraca: in Isopoda, Decapoda, and Amphipoda. Life on land in crustaceans required a transitional phase in which animals possessed varying competences for water and air breathing. This can be still seen in amphibious crabs that have species-specific morphological and physiological adaptations to their lifestyle, with a transition of using the gills as air-breathing organs up to complete adaptation to air breathing with respiratory organs that are similar to lungs. The semi-terrestrial amphipod *Talitrus saltator*, for example, uses its gills for air breathing. It lives on the beach of European coasts and takes up oxygen via its gills on the pereiopods.

7.6.1 Isopoda: Oniscidea

Oniscidea is the only isopod group that is completely terrestrial; about 4000 species exist. They live in nearly all terrestrial habitats, the most extreme being deserts. As an adaptation, isopods developed lung-like or even trachea-like structures, but many species also retained their gills (Hoese, 1982; Schmidt and Wägele, 2001). The legs of the posterior region of the animal, the pleon, can be modified for gas exchange in air. Both lungs and tracheae are part of the exopodites of the pleopods (Figure 7.5), whereas gills are derived from the endopodites.

In ventral uncovered pleopod lungs, gas exchange takes place at the thin-walled ventral surface of the exopodites. This lung type can be found in certain basally branching representatives of the Oniscidea, as well as within some Crinocheta, which is a subordinated taxon of the Oniscidea. Within the Crinocheta, however, also dorsal uncovered pleopod lungs with only small surfaces exist, whereas other species display covered lungs with increased surfaces made by tubules and wrinkles. The most complex respiratory structures are completely internalized pleopod lungs with spiracles and large surface areas. Both the covered lungs, especially the internalized ones, are especially suitable for arid environments (Paoli et al., 2002). One extreme example of such a specialization is the xeric genus *Periscyphis* (Oniscidea, Eubelidae). The lungs start at occludable spiracles on the posterior margin of the pleopodal exopodite with a short tubular entrance, extend into the sacciform atrium and end in an enormous number of tubular, tracheal-like structures, distributed to the pleon, and in some species filling a large volume of this body part (Ferrera et al.

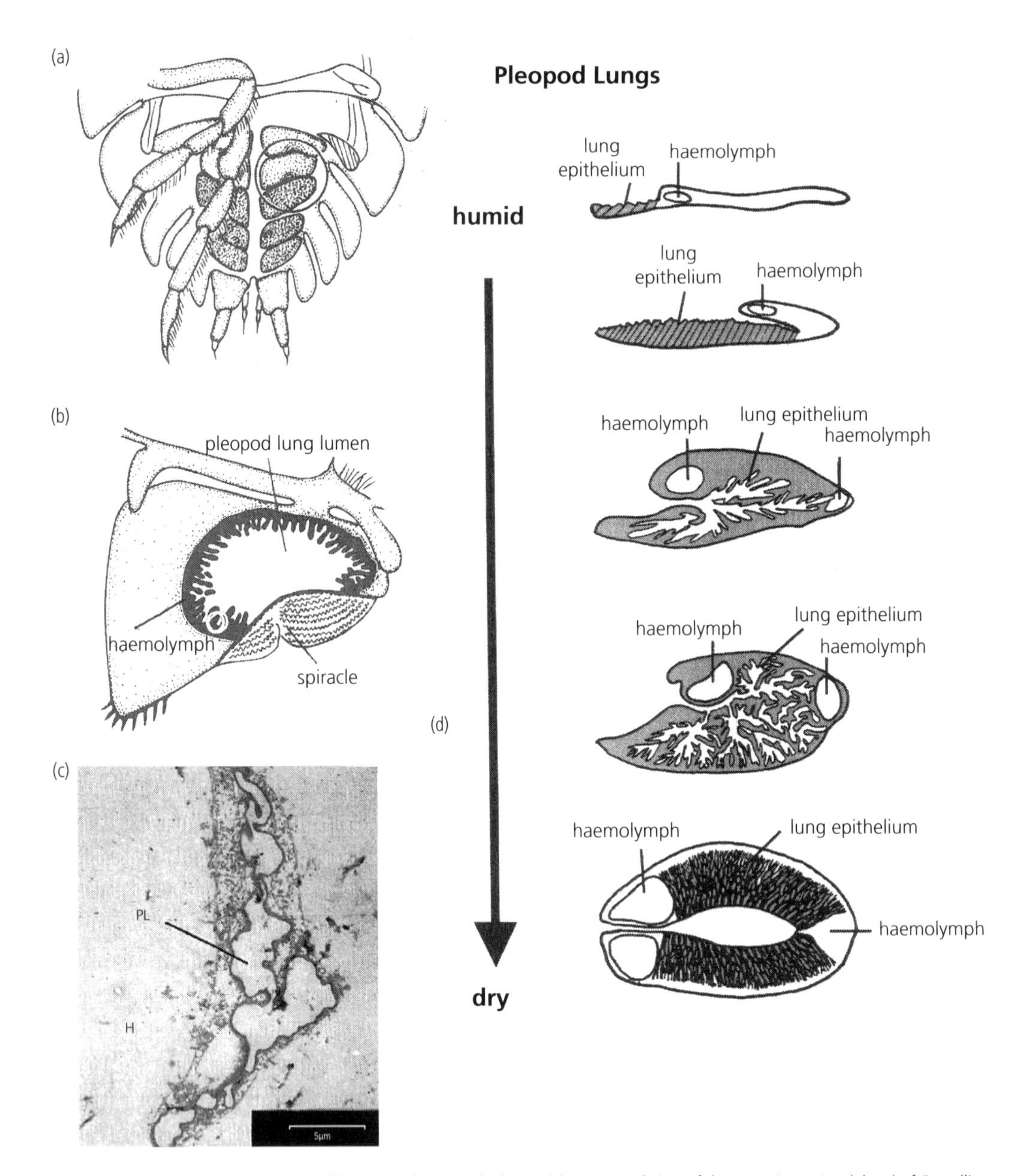

Figure 7.5 Pleopod lungs in terrestrial wood lice (Isopoda: Oniscidea). Part (a) is a ventral view of the posterior region (pleon) of *Porcellio* showing the pleopods (shaded). Encircled area is shown in detail in (b). Part (c) shows an electron micrograph of a portion of the air-filled pleopod lung, PL, surrounded by haemolmyph, H, similar to encircled region in part (b). Scale bar is 5μm. Part (d) shows a progression of pleopod lung types, from flat, plate-like structures to closed lungs with an occludable spiracle, seen as one progresses from species that inhabit humid habitats to desert species: upper to lower sketch *Oniscus, Trachelipus, Porcellio, Perscyphis, Hemilepistus*. The gas-exchange structures in the lungs receive haemolymph from the open circulatory system and release it to efferent vessels (haemolymph). In highly derived desert species, air-filled tracheal lungs (not shown) extend out of the pleopod lungs into the posterior region (pleon) of the animal. Parts (a) and (b) After Westheide and Rieger, 2007; part (c), courtesy of Suaad Bennour, part (d), Hoese, 1982.

1997). Another desert genus, *Hemilepistus*, can survive thanks not only to its complex pleopod lungs, but also to a sophisticated social behaviour and inhabiting burrows. In terrestrial isopods, respiratory water loss is compensated for by water vapour absorption at the ventral pleon (Wright and Machin, 1990; Wright and Machin, 1993).

7.6.2 Decapoda

Two separate lines of decapods have successful terrestrial forms: the Anomura (hermit crabs), with the almost completely terrestrial coconut crab *Birgus latro*, and the Brachyura (short-tailed crabs), which include the familiar fiddler and ghost crabs, Ocypodidae. At least five other major lineages of brachyurans can also survive prolonged exposure to air, but the most highly terrestrial belong to the Grapsidae, where the halloween crab *Gecarcinus ruricola* is completely terrestrial (Bliss, 1968).

One problem with gill breathers is that the gills collapse due to surface tension problems when air enters them. One way of circumventing the collapsing gill problem is to use the gill chamber rather than the gill itself as the gas exchange organ (Figure 7.6). We have seen something analogous to this in molluscs that reduce the gills and use the mantle epithelium as a lung. Since the metabolic rate remains unchanged whether the animals are in water or on land, and air has 30 times more oxygen per unit volume than water, theoretically they can get along with the modest surface area that the gill chamber offers when breathing air. If the habitual metabolic rate increases, the gill chamber can be modified with ridges or depressions (Diaz and Rodriguez, 1977; Greenaway et al., 1996). This way the crabs can have the best of both worlds: they have an air-breathing organ but keep their gills. Since the gills are often reduced, terrestrial crabs are poorly equipped for water breathing (Innes and Taylor, 1986) but at least they won't drown. At least most of them won't—*Birgus latro* will.

Another alternative is to modify the gills themselves for air breathing. The gill lamellae become short and stiff, which reduces respiratory water loss over the now smaller surface area and also makes the lamellae stiff, preventing collapse. Collapse is also often prevented by spike-like projections, analogous to those in the book lungs of arachnids. Comparing now: aquatic crabs have a gill surface area between 400 and 1400 $mm^2\ g^{-1}$; intertidal species, 500–900 $mm^2\ g^{-1}$; bimodal breathers, 35–325 $mm^2\ g^{-1}$; and terrestrial species, only 12–500 $mm^2\ g^{-1}$. The diffusion barrier thickness is similar in aquatic (1–10 μm) and bimodal breathers (1–15 μm), but the so-called branchiostegal lungs in terrestrial species have diffusion distances of only 0.2–1 μm (Henry, 1994). In this way a terrestrial crab can end up with the same diffusing capacity as an aquatic one with only one-tenth of the surface area. A third way to

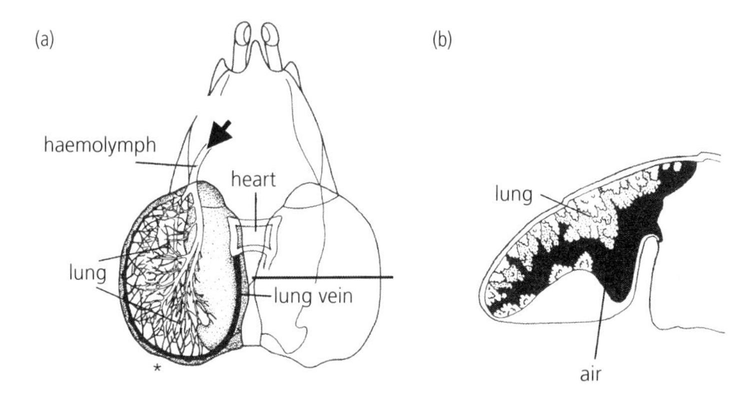

Figure 7.6 Terrestrial crabs. The left lung of the coconut crab, *Birgus latro* in dorsal view (a) and in cross section (b). The level of the cross section is indicated by a black line. Air in the lung cavity is shown in black in part (b), and the approximate place of entry is indicated in part (a) by the asterisk, opening not visible in the illustration. The arrow in part (a) shows the direction of haemolymph flow into the lung. Deoxygenated haemolymph is shown here in black; oxygenated, white. Penzlin, 2005.

circumvent the collapsing gill problem is to fill the gill chamber with water and carry the water along on land.

7.7 Hexapoda

Tracheae of hexapods are tube-like invaginations of the integument. Large main tracheae arise at the spiracles and divide into dorsal, visceral, and ventral branches. These branches subdivide further with decreasing diameter and end in the finest branchings, the tracheoles, which terminate in or near the epithelia of the organs. In most insects except the basally branching and primarily flightless groups, tracheae between the segments are interconnected and there are also anastomoses between the left and the right sides of the body. In addition, air sacs of varying sizes and locations are found in the tracheal system of numerous groups (Figure 7.7). They are best developed in Diptera (flies) and Hymenoptera (bees and close relatives), but are lacking completely in other groups, such as

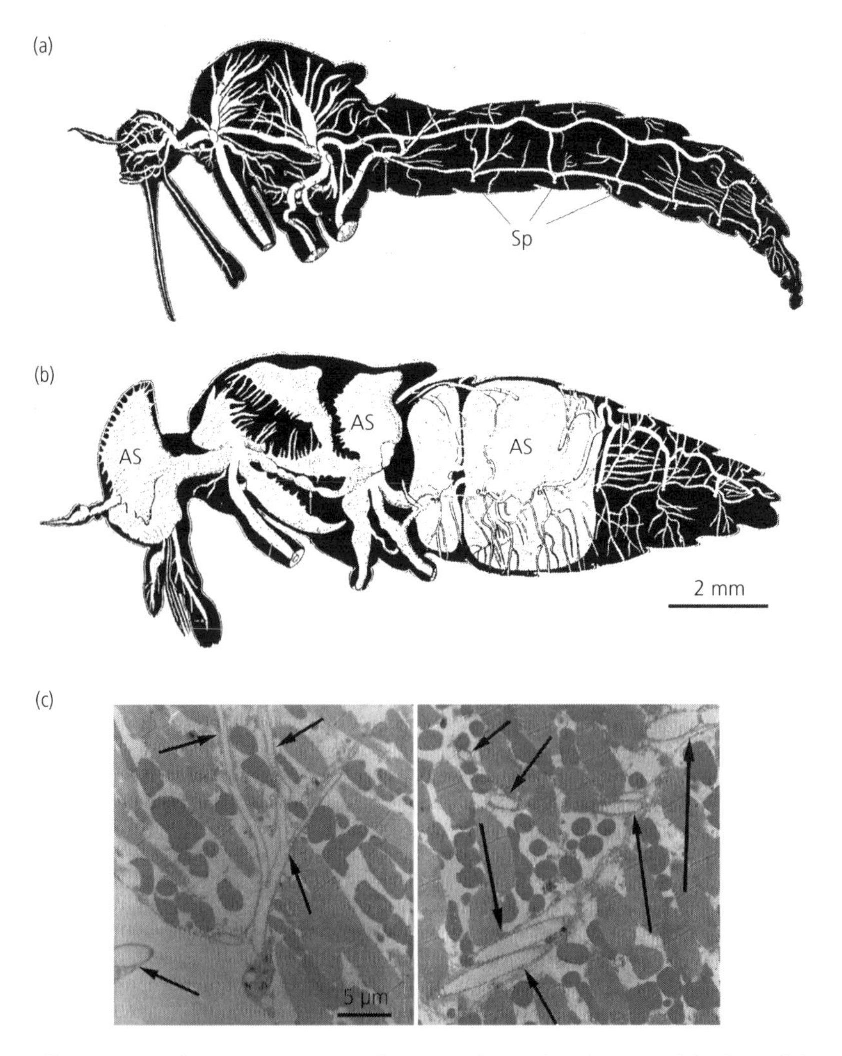

Figure 7.7 Drastic difference in air-sac formation in two species of Diptera in schematic lateral projection: (a), a dagger fly (Empididae); (b), a horse fly (Tabanidae). Tracheae in white. Part (c) shows fine branches of tracheae (arrows) in the musculature of *Drosophila*. AS, Air sac; Sp, Spiracle. (a),(b) From Faucheux, 1974; (c), Original A.S.

Blattodea (roaches). Air sacs serve in ventilation and increase the volume of the tracheal system considerably, thereby decreasing the overall density of the animal.

Histologically, tracheae are composed of special 'tracheal cells'. The terminal part of a given trachea—the tracheole—is composed of a single terminal tracheal end cell that is extremely attenuated, such that the false impression may be created that the air spaces invading the tissue are intracellular. The intima of the tracheae and also of air sacs consists of cuticle which, in turn is made up of a soft endocuticle and a more solid, epicuticle. Exocuticle is found only in the taenidia: hardened flexible threads disposed in a coil that stabilizes the trachea (Wigglesworth, 1983a).

Two examples show how well suited for high performance tracheal morphology is. In the dragonfly genus *Aeshna*, for example, tracheoles do not penetrate the muscle fibres, but the diameter of the muscle fibres is only about 20 μm, which means that the maximum trachea-to-mitochondrion diffusion distance is 10 μm. The theoretical physiological maximum for the diffusion distance during flight with an oxygen consumption of 2 ml g^{-1} min^{-1} is 12 μm: a perfect match of anatomy with demand, including a good safety margin. In wasps and bees, on the other hand, tracheoles actually penetrate the muscle fibres and lie so close to the mitochondria that a trachea–mitochondrion diffusion distance is not even measurable (Weis-Fogh, 1964a, 1964b).

7.7.1 Terminal versus lateral gas exchange models

Looking at the tracheal system as a whole, one gets the impression that they are very functionally built (Weis-Fogh, 1964a). Unlike lungs, tracheae transport the oxygen directly to the organs without involving blood or haemolymph, and can deliver oxygen very rapidly because of the high diffusion rate of oxygen in air compared with fluids (see Chapter 3). In 1920, August Krogh published the first of his first famous papers on insect respiration (Krogh, 1920a, 1920b, 1941) whereby the importance of gas phase diffusion in insect gas exchange was documented (Weis-Fogh, 1964b; Pickard and Mill, 1974; Scheid et al., 1982; Wasserthal, 1996). In the meantime, however, we know that convective gas movement is also important, particularly in large and/or active insects (Kestler, 1985; Slama, 1988; Wasserthal, 1996; Westneat et al., 2003). More about this later.

In his experiments, Krogh (1920a, 1920b) assumed that the area available for diffusion was the combined cross-sectional area of the spiracle openings and the diffusion distance was the length of the trachea from the spiracles to the metabolizing tissue (see Eq. 3.18). In *Rhodnius prolixus*, the bug that transmits Chagas disease, it turns out that cross-sectional area of branching tracheal system remains constant from spiracle to organs. So the total cross-sectional area of the spiracles is actually a pretty good estimate of the respiratory surface area for diffusion.

In general, insect tracheal systems carry out terminal gas exchange, which takes place almost exclusively in the tracheoles (see Chapter 3) and the tracheae simply contain and conduct gas. Even if tracheae are located where gas exchange could occur, most of the oxygen will be released in the tracheoles. So, knowing the maximal cross-sectional area of all tracheae taken together and the distance to the flight muscles, it was possible for Krogh to predict the maximal oxygen consumption of an active insect. This would not be the case if oxygen were to be lost on the way to the tracheoles.

Using stereological morphometric methods, Schmitz and Perry (1999) developed another model that allowed estimation of the diffusing capacity over the entire tracheal system: the so-called lateral gas exchange model (see Chapter 3). It turned out that even in a flightless, 'lethargic' stick insect, *Carausius morosus*, more than 70 per cent of the entire diffusing capacity lies in the tracheoles, and it can be expected that it would be much more in a fly or a bee. However, lateral diffusion does play a considerable role, especially in many non-hexapod tracheated arthropods.

7.7.2 Spiracle closing mechanisms

One key to the extremely high respiratory performance of insects is the highly developed tracheal system, but a second key development is the spiracular closing mechanisms. The plesiomorphic condition is ten pairs of spiracles: two pairs located in the lateral wall of the meso- and metathorax and the remaining ones in the first eight abdominal segments. But many insect groups reduce the number of spiracles (Nikam and Khole, 1989).

With the exception of xeric species, respiratory water loss is only a small proportion of total water loss (Chown, 2002), and many species even inhabit hot deserts and high mountains. But a closing apparatus is also a precondition for discontinuous respiration and the basis for unidirectional air flow, which is realized in large flying insects such as grasshoppers.

As further discussed in Chapter 8, spiracle closing is always active while opening can be active or, more often, passive. Both the closer and opener muscles attach at the cuticular spiracular valve, but passive opening is accomplished by elastic fibres, which also attach at the valves. Alternatively, the elasticity of the valve itself closes the spiracle when the closing muscles relax.

There are basically two types of closing systems: the so-called internal and the external closing mechanisms (Mill, 1985). In the external closing apparatus, which is characteristic of, for instance, cockroaches or grasshoppers, but also occurs in other groups such as in the metathorax of butterflies, the spiracular valves are part of the body wall. The valves are usually thickened cuticular lips where the muscles attach. The trachea either begins directly at the body wall or opens into an atrium, which is often covered with a thick cuticle that has hairs or bristles.

In the internal closing apparatus, the cuticular valve system is located behind the atrium, at the beginning of the main trachea. The spiracle is usually surrounded by a hair-bearing area: the so-called peritreme. In many insects, such as bees and butterflies, there is one stationary valve and a second moveable valve where the muscles and ligaments attach (Beckel and Schneiderman, 1956, 1957; Nikam and Khole, 1989; Schmitz and Wasserthal, 1999). Many spiracles have hairs/bristles or sieve plates. In addition, deep countersinking of the closing apparatus, cuticular elytra or wings can protect the spiracles. In aquatic insects with open tracheal systems, glands secrete fine surpellic film that keeps water from entering the spiracles (see Chapter 6).

7.7.3 Physiology and breathing mechanisms

Over the past century, virtually every aspect of insect metabolism, including respiratory physiology, has been intensively studied, with a particularly important flurry of classical works by such investigators as John B. Buck (1912–2005), Howard A. Schneiderman (1927–1990), Vincent B. Wigglesworth (1899–1994), and others in the mid 1900s. Due to the extremely high oxygen consumption, the physiology of flying insects has attracted much attention. More about this later.

Gas exchange at rest can be continuous, cyclic, or discontinuous, and the cost of transport during terrestrial locomotion (walking/running) depends on number and length of the legs and the load carried. Terrestrially locomoting insects respond quickly to exercise, rapidly arriving at dynamic steady-state oxygen consumption rates. Usually this condition is maintained aerobically, and the animals are called 'marathoners'. Compared with large mammals, which modify their gait to achieve staged steady states as velocity increases, in insects, oxygen consumption for terrestrial locomotion increases linearly with speed, and the minimum cost of transport is relatively high compared with craniotes.

In contrast, other terrestrial arthropods such as spiders and crabs are categorized as 'sprinters', running short distances at high speeds. Under these conditions anaerobic metabolism is important and an aerobic steady state is achieved only during slow locomotion. In the long term, however, all 'pedestrian' animals except spiders, where the long-term oxygen consumption rate is lower than predicted, have similar mass-specific costs of transport, mainly depending on the number and length of legs (Harrison et al., 1991; Krolikowski and Harrison, 1996).

Per unit distance covered, flight is much more economical than walking or running. But per unit time elapsed, insect flight is the most energy-demanding exercise known, with aerobic scopes around 100 or even higher in bees, flies and moths. A hovering orchid bee (Euglossini), for example, has a $\dot{V}O_2$ of up to 154 ml g^{-1} h^{-1} (Withers, 1981; Casey et al., 1985; Harrison, 1986), or approximately twice that of a humming bird (68–85 ml g^{-1} h^{-1}) (Hargrove, 2005). Flight muscles are fully aerobic

(Weis-Fogh, 1964a) and during flight they account for more than 90 per cent of the total $\dot{V}O_2$.

Insect flight muscles have an extraordinary mitochondrial volume density of up to 43 per cent of muscle fibre volume (Casey, 1992). High flux rates are achieved because of high respiratory enzyme content per unit muscle mass and because these enzymes operate at high fractional velocities. The respiratory enzymes occupy the mitochondrial cristae, which are present in a very high surface density, such that enzyme densities per unit cristae surface area are similar to those found in mammals. But during flight in insects, the oxygen consumption rates per unit cristae surface area are much higher than in mammals, because of higher electron transfer rates per enzyme molecule. In this system, the cytochrome C oxidase operates close to its maximum catalytic capacity. Insects use mainly carbohydrates as fuel for short flight. In locust grasshoppers, the initial stage of flight is served by carbohydrates, but fat is used during migration (Suarez et al., 1996; Wegener, 1996; Suarez, 2000).

Flying insects have mass-specific oxygen consumption rates between 3- and 30-fold greater than in the same insects while running. This relation is much higher in insects than in mammals and birds in which the ratio is maximally twofold. These values should not be confused with aerobic scope: the ratio of maximum-to-resting oxygen consumption rates. The mass-specific $\dot{V}O_2$ during flight decreases with body size, but mass-specific power output is constant or increases, implying an increase in efficiency with size.

In moths, scaling of $\dot{V}O_2$ during flight is almost identical to that of bats and birds, and their aerobic scope reaches 300. Flapping flight in moths is the most energetically demanding mode of locomotion per unit time. Since the mechanical efficiency of the flight muscle is only about 20 per cent and energy can neither be created nor destroyed, the remaining four-fifths of the energy expenditure appears as heat. Some of this heat is retained in the thorax and most moths do have a fur-like integument, so body temperatures can be as high or higher than those of birds and mammals (Bartholomew and Casey, 1978). This also means that moths can be active at lower temperatures than many other insects.

Aerodynamic theories predict that the mechanical power required for flapping animal flight varies with speed according to a 'U-shaped' curve: at intermediate speed there is a pronounced power minimum, typically half the power required for hovering flight. These theories apply to bats and birds. But for bumblebees in free, controlled, forward flight over a speed range of 0 to 4 m s^{-1}, the metabolic rate does not vary significantly with flight speed, suggesting that this model may not apply to the bumblebee, and maybe not to insects at all (Ellington et al., 1990). In order to minimize the energy use in flight, *Drosophila* uses an elastic flight motor, with elastic storage of energy in the flight apparatus. Flight muscles show only 10 per cent efficiency in these animals (Dickinson and Lighton, 1995), the rest being stored in elastic loading or lost in heat production.

Even a representative of a relatively basal branch of winged insects, the dragonfly *Erythemis simplicicollis*, is completely insensitive to ambient oxygen levels at rest, indicating that there must be a large safety margin inherent to its tracheal system. This would be expected in a more highly specialized flier, such as the honeybee, *Apis mellifera*. But then during flight in the bee, the tracheal system is highly exploited, the safety margin is small, and decreasing ambient PO_2 becomes rate limiting (Harrison and Lighton, 1998). At a PO_2 below 10 kPa (approximately 4000 m elevation), the wingstroke frequency in the honeybee decreases, at 8 kPa (approximately 7000 m elevation) it can no longer support a metabolic rate consistent with flight and at 5 kPa (equivalent to 2000 m above the top of Mount Everest) bees cannot fly at all (Joos et al., 1997).

So in insects we have the perfect flying machine in terms of matching of structure with function in high performance combined with adaptability and an ability to function well also at lower levels of activity. A striking characteristic of this system is that it is to a large degree self-limiting, the muscular regulation of spiracles, for example, having a regulating or fine-tuning function rather than an existential one (see also Chapter 8).

CHAPTER 8

Control of breathing in invertebrates

As pointed out in Chapter 3, if actively metabolizing tissue is more than a few millimetres from its external oxygen source, an aerobic animal cannot survive by direct diffusion alone and will need some way of distributing respiratory gases that enter and leave through contact surfaces with the surroundings. So, knowing the physical and chemical properties of oxygen and carbon dioxide, we shall now look at which variables can be manipulated physiologically, then see what mechanisms actually exist for doing this in various animal groups, and finally take a look at how these mechanisms are regulated by the nervous system.

To begin with, we must realize that in the Bilateria, body cells actually are seldom in direct contact with the external medium. In the exchanger (skin, gill, or lung), oxygen is transferred from the external environment to an internal transport medium and carbon dioxide is liberated from the internal transport medium to the external environment. The opposite takes place in the tissue. In insects and a few other arthropods that have tracheae that enter the tissue, the transport medium is the internalized air itself.

Looking at the equation for morphological diffusing capacity in Table 8.1, we find in the exchanger, tissue constants (expressed as KO_2,CO_2), the physical properties of which cannot be changed, at least not in the short term. But surface area (S) and harmonic mean barrier thickness (τ) can be influenced by stretching, collapsing, or by only partially perfusing the respiratory organs. The latter would, for example, very quickly reduce the effective S and/or increase τ dramatically and within seconds. Now looking at the equation for physiological diffusing capacity, we find ΔPO_2, CO_2. This can be quickly changed by factors that are under hormonal, central, or autonomic nervous control: namely ventilation and perfusion of respiratory organs and selective perfusion of internal organs.

Finally, as homeotherms we sometimes forget that the regulation of respiration can also be a passive phenomenon. $\dot{M}O_2,CO_2$, KO_2,CO_2, and $\Delta PO_2,CO_2$ all increase, for example, as the temperature increases. Looking from this side of the mirror, as it were, we can also see that to begin with, respiration is at least in part a self-regulating process. Assuming an infinite pool of oxygen, if $\dot{M}O_2,CO_2$ decreases, $\Delta PO_2,CO_2$ will automatically go down, simply because less oxygen is being removed from the 'animal' side of the gas exchange surfaces.

The complexity of the control of breathing can be appreciated when we realize that all of the variables listed in Table 8.1 can influence all of the others. Given the result of this interaction, the organism adjusts those parameters under its control, usually without conscious intervention.

In non-tracheate animals—that is, in ones in which a liquid internal transport medium is used—the most important immediate physiological goal is to achieve an optimal ventilation/perfusion relationship for providing the tissues with adequate oxygen and removing the carbon dioxide.

Anatomical properties that would be just sufficient for oxygen diffusion would probably release

Respiratory Biology of Animals: Evolutionary and Functional Morphology. Steven F. Perry, Markus Lambertz, Anke Schmitz, Oxford University Press (2019).
DOI: 10.1093/oso/9780199238460.001.0001

Table 8.1 Major variables involved in gas exchange and their location

	Exchanger	Internal transport medium	Tissue
Anatomy (structure) **Morphological diffusing capacity** $DO_2,CO_2 = KO_2,CO_2 \cdot S/\tau$	Surface area Barrier thickness	Pump (heart), vessels Separation of oxygen-rich and oxygen-poor blood Content of Hb/Hc	Diffusion distance Mitochondria
Physiology (function) **Physiological diffusing capacity** $DO_2,CO_2 = \dot{M}O_2,CO_2 / \Delta PO_2,CO_2$	Driving pressure Ventilation Temperature Barometric pressure	Properties of Hb/Hc Perfusion rate Temperature Barometric pressure	Driving pressure pH Temperature Barometric pressure

DO_2,CO_2, diffusing capacity for oxygen or carbon dioxide; KO_2,CO_2, Krogh's diffusion constant for oxygen or carbon dioxide in diffusion barrier; S, surface area of exchanger; τ, harmonic mean thickness of diffusion barrier; $\dot{M}O_2,CO_2$, rate of oxygen consumption or carbon dioxide release; $\Delta PO_2,CO_2$, driving pressure.

too much carbon dioxide and result in a condition that can only work properly in the presence of respiratory proteins such as haemoglobin or haemocyanin. In addition, since animals are not always operating at their metabolic maximum, the respiratory system does not need to be going full tilt all the time. Rather, it needs to be regulated according to the demand. It is this regulation that we shall be discussing in this chapter.

8.1 Sensing the surroundings and the behavioural regulation of respiration

Probably the easiest way to regulate metabolism and respiration at the same time is to move to somewhere where the external conditions are different. In most cases, this means finding conditions where the temperature, moisture, and oxygen content are closer to optimal for a particular animal at any given time, season, and life phase. This means that these conditions need to be sensed by the organism, and if the sensors are in a single location, the signal must be stored in order to compare it to the situation after moving. Alternatively, the sensors may be located at different locations around the body and the signals directly compared. In any case, sensing organs or cells are necessary. Mechanisms for temperature sensing and in particular moisture estimation are all sciences in and of themselves and the reader is advised to consult appropriate sources. Carbon dioxide/pH are also common internal indicators of metabolic performance and pH-sensitive regions in the brain have been precisely located (Nattie, 1999), but the structure of these sensors has eluded us to date. We shall come back to this. Oxygen sensing is better known.

All animals—indeed, perhaps all living organisms—from single-celled critters to the most complex insects, birds, and mammals can detect levels of ambient oxygen and react appropriately. Entire organisms react in complex behavioural ways as illustrated by the following examples. Aerobic protists migrate from oxygen-poor sediment into the water column of lakes (Finlay and Esteban, 2009). Parasitic nematodes sense the fall in oxygen in netted herring and bore through the intestinal wall into the muscle, much to the economic detriment of the fishing industry unless the fish are gutted on the spot. But response to ambient oxygen tension also applies to tissues and cells. Mammalian endothelial cells, for example, move into areas of low oxygen tension during wound healing. Although the exact way in which cells detect hypoxia is still an area of active research, once a changed oxygen level has been detected, the pattern of response from the molecular to the behavioural level has been demonstrated in many organisms. For more see HIF (hypoxia-inducible factors) and HSP (heat shock proteins) in Chapter 5.

A simple but instructive model organism is the free-living nematode *Caenorhabditis elegans*, which shows a strong preference for ambient oxygen levels between 5 and 12 per cent. Above this range, specialized oxygen-sensing cells in the nervous system bind molecular oxygen to the specific soluble guanylate cyclase homologue *gcy-35* haem domain in the DNA and cause the release of 3′,5′-cyclic guanosine monophosphate (cGMP), a common second messenger in sensory transduction. This gene is also present in *Drosophila* (Gray et al., 2004), where it could be considered an exaptation, since flies are capable of internal respiratory control. But in *C. elegans*, expression of *gcy-35* ultimately results in fast behavioural responses. Among other things, when oxygen exceeds the preferred levels, the GCY-35 protein that the *gcy-35* gene codes for mediates oxygen sensing in four neurons that regulate social feeding behaviour and causes the worms to remain in unexposed locations (Gray et al., 2004).

Respiratory systems are completely lacking in lineages more basally branching than Bilateria. But even within basally branching bilaterian groups, respiratory systems are recognizable only in combination with other faculties such as locomotion or feeding, since effective circulatory systems are lacking. In more complexly organized animal life forms, highly aerobic animals have evolved, maintained and refined sensing capabilities for oxygen, a substance that could represent a limiting factor in metabolism in an aquatic environment. Let us now move on to the animals that create conditions compatible with life within their own bodies by regulating their respiratory faculties.

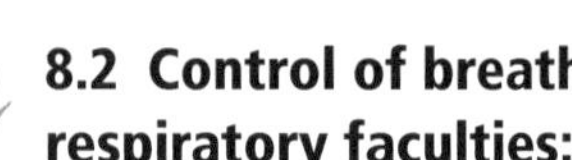

8.2 Control of breathing in animals with respiratory faculties: general concepts

In animals that have a respiratory faculty, gas exchange can occur either as a continuous or a discontinuous process. In the first case, the response in terms of behavioural change (e.g. relocation of the organism to a more favourable environment) or of changes in the ventilation and perfusion of the respiratory organs need only be regulated with respect to metabolic demand and ambient availability (usually PO_2), whereas in the latter case the respiratory system also needs to be turned on and off. The 'hardware' necessary in either case is similar. The organism needs oxygen receptors in a peripheral location to detect ambient availability and in a central location to estimate the need for more or less oxygen in the animal. These receptors were probably already present long before the respiratory control mechanisms evolved, and were an important prerequisite that allowed this evolution to occur in the first place.

Once the need has been ascertained, the respiratory system—usually in concert with the circulatory system—responds in two ways: (1) by increasing or decreasing the ventilation of the respiratory surfaces and (2) by increasing or decreasing the circulatory perfusion of the respiratory organs, and at least some portions of the body. When the ventilatory apparatus has been activated (if it was turned off), most animals with respiratory faculties have some sort of cyclic pumping mechanism for moving the respiratory medium across the surfaces of the gas exchanger. The volume of medium moved per unit time can be regulated by changing the ventilatory frequency, the stroke volume (i.e. the amount of medium moved per ventilatory cycle) or both. Similarly, the concomitant circulatory heartbeat frequency or stroke volume is also usually centrally regulated.

8.3 Mollusca: Gastropoda

Snails originated in the water where the majority still are found (Bouchet et al., 2005). One of the most exceptional groups is the Panpulmonata, the air-breathing snails. This group is the most intensely studied with respect to the control of breathing.

The great pond snail *Lymnaea stagnalis* lives in oxygen-poor, fresh-water habitats. Although its specialized respiratory proteins (see Chapter 4) allow it to survive submerged in hypoxic conditions using cutaneous respiration, when at the surface it breathes air using a region of the mantle cavity that is modified to serve as a lung. Air breathing in snails is a complex process consisting of (1) detection of internal hypoxic conditions, (2) a behavioural response causing it to surface, and (3) breathing. The breathing process, in turn, involves opening of the pneumostome (the opening to the lung cavity), expulsion of air, cyclic ventilation of the lung, and finally closure of the pneumostome.

The hypoxia-induced drive originates in oxygen receptors at the periphery and is conveyed to the central pattern generator (CPG) neurons via the right pedal dorsal neuron number 1. The CPG, which coordinates air breathing, consists of three mutually inhibitory interneurons, which cause the pneumostome to open and close and to initiate the respiratory rhythm (Taylor and Lukowiak, 2000). The similarity of this simple neuronal network to the CPG in tetrapods is remarkable, and recent advances in cultivating the CPG of *Lymnaea* on a silicon chip and measuring its activity directly has opened the door to the promising possibilities of doing the same in mammalian—indeed even human—cell cultures (see Py et al., 2010).

Respiration in *Lymnaea* is subservient to the whole-body withdrawal response and is altered by adaptive responses to hypoxia during water breathing (e.g. depression of metabolism) (Taylor and Lukowiak, 2000). Although nervous input from PCO_2-sensitive neurons in the osphradium (a sensory organ connected to one of the visceral ganglia in snails) to the CPG exists (Syed et al., 1990; Syed et al., 1991; Wedemeyer and Schild, 1995), a direct synaptic connection between oxygen-sensing peripheral chemoreceptor cells and a CPG neuron may be more important (Bell et al., 2008), mediating an increase in ventilatory frequency when the environment becomes hypoxic (Inoue et al., 2001). Below temperatures of 12.5°C, the respiratory CPG becomes quiescent, regardless of the ambient PO_2 (Sidorov, 2005). So *Lymnaea* appears to use both oxygen and carbon dioxide for driving its complicated breathing response.

Juvenile *Lymnaea* rely more on cutaneous, aquatic respiration than do adult snails, which employ more pulmonary, aerial gas exchange. There are also significant age-related differences in the synaptic connectivity within the CPG and in peripheral inputs to the CPG, which favour a more rhythmic activity in the adult CPG. Life-phase differences in respiratory biology are not unusual and we shall encounter them also in other taxa. In *Lymnaea* it correlates well with the decreased dependence on cutaneous respiration in the adults (McComb et al., 2003).

In more terrestrial pulmonate snails such as *Helix aspersa* and *H. pomatia*, the pneumostome does respond to mild hypoxia as in pond snails, but ventilatory regulation is mainly carbon dioxide driven: again, in remarkable convergence with terrestrial craniotes, as we shall see in Chapter 13. Intact suboesophageal ganglia and an intact anal nerve are necessary to mediate the carbon dioxide response. Carbon dioxide chemoreception is restricted to a discrete region of the medial margin of the visceral ganglion and changing the pH of the haemolymph alone is enough to elicit a response from these 'central' chemoreceptors (Erlichman and Leiter, 1993, 1997).

8.4 Crustacea

8.4.1 Malacostraca: Decapoda

Probably based on their economic importance, availability, and reasonable body size, the best studied crustaceans are Homarida, Astacida (lobsters and crayfish), Anomoura (hermit crabs), and Brachyura (crabs). The four groups differ mostly in the structure of the 'tail', being long and useful in swimming in lobsters and crayfish, modified for fitting into captured shells in hermit crabs, and reduced and permanently tucked under the cephalothorax (the combined head and thorax body regions) in most other crabs. The ventilatory pump in all adult decapod crustaceans is the scaphognathite. This is a highly modified, paddle-like appendage, which was originally part of the feeding apparatus. By sweeping back and forth it draws water into the gill chamber and forces it across the respiratory surfaces.

Water-breathing crustaceans possess a mainly oxygen-driven ventilatory control with carbon dioxide playing only a minor role. In the crayfish *Astacus fluviatilis*, a non-spiking oscillatory CPG cell, which appears to react to ambient PO_2, is located in the suboesophageal ganglion and drives the muscles of the scaphognathite. Additional oxygen sensors in the scaphognathites themselves and/or in the gills provide feedback to the CNS. So some regulation of ventilation may occur at the level of the CPG, but central and peripheral oxygen sensors are necessary to mediate the ventilatory response to hypoxia.

Bradycardia is a frequent response to hypoxia, but appears to be the direct result of a lack of oxygen at the level of the cardiac ganglion. This local metabolic process is reversed by restoring normoxia to the region. But the stroke volume of the heart is often increased during hypoxia and therefore is assumed to be independently regulated. Neurohormones appear to influence this process by a direct effect on the heart or indirectly via the CNS, but more research is necessary to clarify this interesting phenomenon.

In amphibious crabs, both oxygen and carbon dioxide are possible respiratory modulators, but which of the two predominates depends on whether water breathing or air breathing is the main focus of the species. In other words, the highly terrestrial Caribbean hermit crab *Coenobita clypeatus* remains far more sensitive to oxygen than to carbon dioxide even when on land, and it breathes air with its stiff gills, using water trapped in the branchial chamber as an interface (McMahon and Burggren, 1988). This is also a pretty 'clever' way of passively reducing the diffusing capacity in keeping with the higher level of oxygen on land. Other species, such as the Atlantic blue land crab *Cardisoma guanhumi*, are able to adjust scaphognathite beat frequency using both gases, the relative importance switching when the respiratory medium is changed: greater sensitivity to ambient carbon dioxide than to oxygen when breathing air, but to oxygen when breathing water.

At least five only distantly related brachyuran species and two anomourans have evolved complete terrestriality. Some even have evolved lungs in the branchial space or have lung-like modifications of the carapace. One of the best studied hermit crabs, the coconut crab *Birgus latro*, shows an enhanced carbon dioxide sensitivity, convergent with terrestrial snails and amniotes (McMahon and Burggren, 1988).

8.4.2 Malacostraca: Cladocera

Another well-studied group of crustaceans are the water fleas (Malacostraca: Cladocera), primarily the genus *Daphnia*. These tiny limnic animals are an important food source for fish and due to their ability to make haemoglobin they are able to survive in large numbers even during warm periods when the waters are hypoxic.

Ventilation in *Daphnia* is an oxygen-controlled process that regulates the beat frequency of paddle-like peraiopods, which have the dual function of food acquisition and ventilation. Leg beat frequency is increased at high temperature in a food-rich environment even when the oxygen level in the water is very low. In addition, hypoxia-induced increase of heart rate and perfusion rate during short-term hypoxia occur. The amount of haemoglobin is increased after a few days under hypoxia. In haemoglobin-poor animals, circulatory convection covers less than half of the oxygen demand, but the contribution is much greater in haemoglobin-rich animals (Bäumer et al., 2002). Also the type of haemoglobin is flexible and genetic control elements and oxygen conditions near the haemoglobin synthesis sites determine which types of subunits are produced. Either more macromolecules of unchanged P_{50} or an increased proportion of macromolecules with greater oxygen affinity are synthesized (Paul et al., 1997; Pirow et al., 1999; Paul et al., 2004; Pirow and Buchen, 2004). Although not regulated by breathing control mechanisms, direct supply of organs with the highest metabolic rate—namely, the sensory organs and the nervous system—is achieved in these minute animals by direct diffusion through the cuticle.

8.5 Hexapoda

The control of breathing in insects involves central pacemakers and/or central and peripheral chemoreceptors, which sense carbon dioxide, pH, and oxygen. What is regulated is the opening and closing frequency of the spiracles as well as pumping of the air sacs, primarily by movement of the abdomen. To learn more about this, read on. But first, a little joke.

Imagine two men on a dark night, one kneeling under a street light and the other standing next to him. The standing man is saying, 'If you lost your keys on the other side of the street, why are you looking here?' The kneeling man answers, 'Because the light is better here'. Of course the man who lost his keys will not be able to find them under the street light, but he may become more familiar with the terrain and may be able to recognize his keys faster when he looks where he thinks he lost them. By the same token, we shall proceed to describe what is known about the control of breathing in selected groups that have been best studied and then try to piece together a reasonable evolutionary scenario for insects in general.

8.5.1 Opening and closing the stigmata

The stigma closure muscles receive excitatory and inhibitory impulses from branches of the median nerve, which contains axons of motoneurons (Miller, 1969). These axons branch to innervate muscles on the right and on the left side of the segment and interneurons connect the motoneurons of the stigmata with the CNS.

Opening and closing the stigmata represents a compromise involving two conflicting physiological processes. Keeping the stigmata open would of course benefit rapid gas exchange. But most insects are small and if ambient humidity is below the saturation point for water vapour, it would diffuse through the open stigmata, desiccating the animal within seconds.

To release carbon dioxide, an insect may (1) open the stigmata, (2) increase ventilation, (3) reduce the fluid filling in the terminal branches (tracheoles) of the tracheae—which increases the surface area for carbon dioxide to diffuse into the tracheal system—or any combination of these three mechanisms. The relative importance of these various mechanisms differs among species.

The muscles of the stigmata are under CNS control, responding to the internal pH or PCO_2. In addition, muscles that are responsible for abdominal pumping movement or other ventilatory muscle activity respond to central PCO_2 receptor stimulation and are innervated by the segmental lateral nerves and by neurons, the axons of which are located in the median nerve.

Historically, the first work on the influence of carbon dioxide on the stigma muscles in lepidopterans (butterflies/moths), grasshoppers, dragonflies, and dipteran flies dates back to the late 1950s (Case, 1956a, 1956b), but is based on the earlier classical studies of the Dutch comparative physiologist Engel Hendrik Hazelhoff (1900–1945) in the 1920s (Hazelhoff, 1927). In lepidopteran pupae, oxygen affects the CNS, resulting in inhibitory and excitatory neurogenic regulation of stigma muscles, which are also under myogenic control (Case, 1957a, 1957b; Hoyle, 1959; Burkett and Schneiderman, 1974). High carbon dioxide levels applied to the tracheal system cause the stigmata to open (Burkett and Schneiderman, 1974), and since there is no evidence for a peripheral carbon dioxide chemoreceptor, it appears that the stigma muscle itself responds to changing gas levels. But on the other hand, lowering the ambient carbon dioxide alone does not cause the stigmata to close (Beckel and Schneiderman, 1957). The effector might be the inhibition of neuromuscular synapses or changing membrane properties. Multipolar neurons innervate spiracular muscles and the cuticular closing apparatus of the stigmata in Lepidoptera (Schmitz and Wasserthal, 1999). In the tsetse fly *Glossina morsitans*, a multipolar neuron lies in the vicinity of the stigmata with one process extending to the stigma muscle (Finlayson, 1966). Such mechanoreceptive neurons might control the opening status of the stigmata in addition to the chemoreceptive control and work as a feedback system.

8.5.2 Air flow and breathing movements: discontinuous breathing

Many tracheate arthropods breathe discontinuously. An important prerequisite for discontinuous respiration is closable spiracles. As pointed out in Chapter 7, this is the case in some arachnids and centipedes as well as in insects. Most studies have been undertaken in insects, and six hypotheses (see Chapter 10) for the origin of discontinuous gas exchange cycles (DGC) have been in discussion since this phenomenon was first discovered in lepidopteran pupae (Heller, 1930). The first systematic investigations were carried out a bit later (Punt, 1944, 1950; Buck et al., 1953; Buck and Keister, 1955; Punt, 1956a, 1956b; Punt et al., 1957; Buck, 1958; Buck and Friedman, 1958; Buck and Keister, 1958; Buck, 1962), and this continues to be an area of active research.

A typical DGC (Figure 8.1) can be divided into three parts: in phase 1, the closed-spiracle 'C' phase, in which external gas exchange is negligible. Because of oxygen consumption in the tissue, the internal oxygen level decreases continuously while approximately the same amount of carbon dioxide is released. But since the carbon dioxide is buffered in the haemolymph and tissues and does not necessarily stay as a gas, the pressure in the tracheal system goes below atmospheric (Bridges and Scheid, 1982). A critical PO_2 (usually 4.5 kPa) triggers phase 2, the flutter, or 'F' phase, in which spiracles open and close in rapid succession. Because of the negative pressure, air enters and the oxygen it contains is consumed as fast as it comes in (Chown et al., 2006). With every opening of the spiracles, the pressure difference between tracheal system and environment decreases. Also small quantities of carbon dioxide

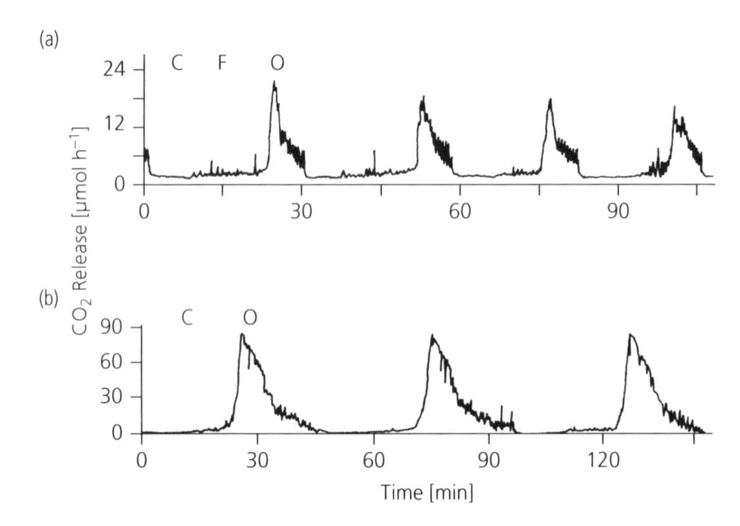

Figure 8.1 Discontinuous breathing pattern in an insect, measured on hand of carbon dioxide release. Part (a): C, spiracle closed, no carbon dioxide released. F, flutter phase of rapid opening and closing of the stigma; O, carbon dioxide rapidly released as stigma opens and then ceases as the stigma is stutteringly closed. In part (b) a discontinuous breathing pattern is depicted in which the F phase is missing. Kestler, 1985.

leave the tracheal system, but the driving pressure for carbon dioxide is much less than for oxygen (Hetz and Bradley, 2005), and since only about 25 per cent of the carbon dioxide produced is released during the flutter phase, it continues to accumulate. A critical carbon dioxide level triggers phase 3, the open, or 'O' phase, also referred to as the ventilation, or 'V' phase. Carbon dioxide is released and oxygen is taken up in a burst-like fashion. The carbon dioxide is retrieved from the buffer system and nearly the entire air content of the tracheae is exchanged, resulting in a rise in PO_2 to near the ambient value. Accordingly, the three phases of DGC have been referred to as CFO (closed–flutter–open) or CFV (closed–flutter–ventilation) respiration. The control of breathing in insects remains an area of active research and gas exchange patterns may be much more diverse than previously recognized (Quinlan and Gibbs, 2006).

8.5.3 Grasshoppers: a case in point

A great deal of research on respiratory control in insects was carried out on grasshoppers (Orthoptera) between the 1960s and 1990s, in particular by J.F. Harrison and colleagues P.L. Miller and M. Burrows (Harrison, 2001). So, in the spirit of the man looking for his keys, let us take a closer look at this group.

The control of breathing in grasshoppers is particularly interesting because several different states of metabolic demand exist: rest, quiescent but alert, and locomoting, whereby locomotion can be in the form of slow walking, hopping, or flying. The control of breathing under these conditions differs. When at rest, grasshoppers tend to respire discontinuously, the stigmata opening and closing in response to changes in haemolymph carbon dioxide, and trans-spiracular resistance limits gas exchange. In alert but quiescent grasshoppers, abdominal pumping also occurs and gas exchange is regulated by a combination of spiracular opening and abdominal pumping. Bulk air flow is predominantly unidirectional (front to back or back to front). But during hopping, bulk gas flow is caused by resulting cuticular deformation rather than by abdominal pumping, and airflow is mainly tidal. Internal oxygen stores and increased ΔPO_2 support the increased oxygen demand. After hopping, increased abdominal pumping restores tracheal gases, but this process appears to be a 'hard-wired' behavioural response, since it is affected neither by ambient PCO_2 and PO_2 nor by pH of the haemolymph. During flight, tidal thoracic auto-ventilation drives bulk gas flow to the flight muscles, while abdominal pumping supplies the remainder of the body. Tracheal PCO_2 affects spiracular opening directly and also indirectly via the CNS, while the tracheal PO_2 affects spiracular opening only by means of a central nervous mechanism. Tracheal PO_2 and PCO_2 in addition influence abdominal pumping via the CNS, while haemolymph pH has no effect on stigmata and ventilation (Harrison, 1997).

So grasshoppers appear to have a mosaic of mechanisms that control breathing, ranging from discontinuous breathing and direct response of the stigma muscles to carbon dioxide during rest to various combinations of direct and indirect control of ventilation and stigma opening during locomotion. This host of control mechanisms is not surprising taking into account the very different anatomical conditions and ventilatory mechanisms that correlate with rest and locomotion.

8.5.4 Grasshoppers and beyond

Let us now inspect the control of gas exchange in grasshoppers even more closely and compare it with what is seen in other insects. But please keep in mind that we are only using grasshoppers as a case in point and that this is only the very superficial sampling of the vast literature on this subject, selected to demonstrate the types of control of breathing that have evolved. In particular, great advances have been made studying roaches and adult and larval lepidopterans and dipterans, which are hardly mentioned here. For further study the reader is referred to recent reviews (Lighton, 1996; Chown et al., 2006; Lighton, 2007; White et al., 2007; Chown, 2011).

In grasshoppers during rest, tracheal PO_2 is maintained at about 18 kPa in the large longitudinal trunks, and the PCO_2 is 2 kPa. Teleologically speaking, these high oxygen levels 'make sense' in terms of repaying the oxygen debt after hopping, which can happen suddenly and without warning, but how are they maintained? Ventilation is controlled by ganglia in the metathorax and in the anterior part of the abdomen. These pacemaker ganglia initiate the rhythm that drives abdominal pumping in grasshoppers. Modification of the pacemaker rhythm is accomplished by stretch receptors in the abdomen, whereas opening and closing of the stigmata is controlled either by the CNS, by the corresponding ganglia, or locally. In metathoracic stigmata with single closer muscles, local carbon dioxide control is dominant, while central control modifies the response to local carbon dioxide. In the prothoracic and abdominal stigmata, which have both closer and opener muscles, central nervous control of the muscles dominates, and carbon dioxide acts on the CNS. Local ganglia and metathoracic ganglia are connected via interneurons, which coordinate opening and closing of the stigmata with abdominal pumping.

At high carbon dioxide levels, synchronization fails and all the stigmata open, at least theoretically maximizing cellular access to oxygen. But in some insects fluid level in the tracheoles also limits oxygenation of the tissue. This fluid level depends on the metabolic demands of a given tissue and might also be under the control of neurohormones (Wigglesworth, 1983a). In response to locomotion, for example, hormones such as octopamine can stimulate ventilation. In the locust grasshopper *Schistocerca gregaria*, the closer muscle in the mesothoracic stigma receives input from two excitatory motoneurons and processes of a peripherally located neurosecretory cell. These muscles possess specific octopaminergic receptors that increase the cyclic AMP level. This is consistent with the hypothesis of central or peripheral octopamine containing neurons that provide the neurosecretory inputs to the muscle (Swales et al., 1992).

Oxygen—or, better said, lack of oxygen—is in and of itself also a powerful central nervous stimulator. As mentioned before, in *Drosophila*, genes that code for soluble guanylyl cyclases, which catalyse the synthesis of low levels of cGMP under normal atmospheric oxygen levels, are strongly activated by anoxia. The increase is dose dependent for ambient oxygen concentrations of less than 21 per cent. Guanylyl cyclases are co-expressed in a subset of sensory neurons, where they are ideally situated to act as oxygen sensors (Morton, 2004).

The response of chemoreceptors in insects differs among species. For example, in cockroaches such as *Periplaneta*, receptors respond to a change in pH of the haemolymph, whereas in grasshoppers this could not be shown. Also in cockroaches, abdominal movements associated with rhythmic tracheal ventilation elicit sensory feedback. Impulses from abdominal mechanoreceptors modulate ongoing pacemaker rhythms during tracheal ventilation, coordinating rhythmic neural and mechanical activity (Farley and Case, 1968).

Grasshoppers appear to have a different mechanism for regulating ventilation. Neuronal or humoral responses associated with locomotion can stimulate ventilation by means of a feed-forward regulation that is assumed to increase receptor sensitivity, resulting in a more precise regulation by negative-feedback mechanisms already present. Teleologically speaking, this again 'makes sense' because local acid–base conditions in muscle may differ significantly from those in the CNS.

In addition to these peripherally modulated signals driving rhythmic activity, a CPG can be present in the metathoracic ganglion (Farley et al., 1967; Bustami and Hustert, 2000; Marder and Bucher, 2001; Woodman et al., 2008) or (in the Cuban burrowing cockroach *Byrsotria*) in the first abdominal ganglion. This ganglion also responds to increasing levels of carbon dioxide by increasing the frequency of neuronal bursting (Myers and Retzlaff, 1963). The rhythmic impulses are transferred to other segments, where ganglia distribute them to the muscles involved in abdominal pumping and spiracular control.

But how does this CPG respond to changes in oxygen levels? An oxygen receptor is necessary to allow an unequivocal response of the CNS to changes in PO_2. Such a receptor has been demonstrated in the horse lubber grasshopper *Taeniopoda eques*, in which isolated ganglia clearly respond to changes in PO_2 (Bustami et al., 2002).

In dragonflies, rhythmically firing ventilation centres exist in the abdomen, but the brain and thoracic ganglia also have a considerable influence over ventilation. Synchronized spiracular movements result in unidirectional air flow, which ventilates the flight muscles and then the abdomen (Miller, 1962). Factors that act on the spiracular closing muscle (dragonflies lack a spiracle opening muscle), are carbon dioxide, desiccation, hydration, and hypoxia. Complete inhibition of the closer muscles has also been demonstrated during flight or ventilation in some species (Miller, 1964).

To briefly bring together some of the more important elements of respiratory control in insects: a host of mechanisms including a central nervous pattern generator, modulator neurons in the CNS, and local responses of muscles to low oxygen and carbon dioxide levels as well as associated local receptors for oxygen and carbon dioxide have been demonstrated. The sophistication of the ventilatory apparatus in dragonflies as well as the 'ancient' origin of this group may suggest that such mechanisms were already in place early on in the evolution of winged insects. A clear evolutionary sequence, however, still eludes us.

Undoubtedly the most far-reaching insight to come from the study of the control of breathing in insects was made only recently (Hetz and Bradley, 2005). These studies on the mechanisms controlling DGC in moth pupae and in grasshoppers suggest that tracheal PO_2 is not influenced by ambient PO_2 in the flutter phase, but that the main function of the closed phase appears to be to reduce the PO_2 around the cells and protect them from oxygen toxicity. To place this in an evolutionary context: the most important aspect of the control of breathing could be described as maintaining the PO_2 in the tissue at levels approaching those that existed when aerobic life originated. The plethora of respiratory control mechanisms we have seen in this chapter illustrates a phenomenon that reoccurs in this book: a crucial physiological state is preserved or improved upon while the mechanisms that bring it about change dramatically.

CHAPTER 9

The evolution of water-breathing respiratory faculties in invertebrates

Since animal life originated in the water it should come as no surprise that the first mechanisms of gas exchange deal with the acquisition and elimination of dissolved oxygen and carbon dioxide. The mechanisms for gas exchange in sponges, cnidarians, comb jellies, flatworms, and roundworms have been discussed in Chapter 6. Some remarkable specializations surfaced. But since internal circulatory systems aside from a gastrovascular cavity or a fluid-filled body cavity are lacking and no dedicated gas exchange surface other than the outside of the animal exists, it is moot to speak of evolutionary directions of the respiratory faculty in these animals. The same applies to 'lophophorates' and echinoderms. We see specializations concomitant with the lifestyle these animals pursue, such as tentacle-like organs that can double as respiratory surfaces, papillar gills that supply local needs, and even some unique situations such as counter-current conditions in sand dollars or water lungs in sea cucumbers, but these are isolated occurrences and well-developed respiratory faculties are lacking. So again, we can hardly speak of recognizable evolutionary trends, let alone cascades.

9.1 Annelida

When we come to annelids and their allies, we see an amazing variety of respiratory proteins and gas exchange organs on various parts of the body, combined with an essentially closed circulatory system. But annelids seem to have spread out along an evolutionary plateau and 'appear to be quite content' with their existence as worms. Somehow they never achieved the big breakthrough with a central heart and externally ventilated gills. Or maybe their relatives did, and after that quantum jump we now call them molluscs, as the larval anatomy would suggest, or even arthropods.

9.2 Mollusca

With molluscs we appear to have a universe parallel to that of arthropods and craniotes, in which we see an amazing number of convergences but also some unique solutions. When speaking about the evolution of the respiratory faculty it only makes sense when we compare phylogenetically related groups, the closer the relationship the more meaningful the comparison. Among molluscs this group is Conchifera.

The conchiferans start out as bilaterally symmetrical animals with a protected mantle cavity containing paired gills. Basally branching conchiferan lineages can have up to ten pairs of gills. From this condition there is a tendency to release the bilateral symmetry, reduce the number of gills often on just one side of the animal, then on both sides, finally eliminating them entirely. We don't know exactly when this reduction occurred during the phylogenetic history of the clade

Respiratory Biology of Animals: Evolutionary and Functional Morphology. Steven F. Perry, Markus Lambertz, Anke Schmitz, Oxford University Press (2019). © Steven F. Perry, Markus Lambertz, & Anke Schmitz.
DOI: 10.1093/oso/9780199238460.001.0001

but it would be worth investigating the hypothesis that it coincided with geological periods of high atmospheric oxygen.

Within the Conchifera, gastropods repeated this sequence. Beginning with bilaterally symmetrical forms, basally branching lineages such as limpets eliminated the gills, and air-breathing species even evolved that use only the mantle lining as a gas exchange surface. Later, a more derived heterobanch lineage gave rise to highly terrestrial pulmonate snails, some of which used their lungs, their giant haemocyanin molecules, and refined, centrally controlled breathing mechanisms to allow them to invade oxygen-poor aquatic realms where they combine cutaneous gas exchange with the snail's version of breath-hold diving. Another lineage, the Opisthobranchia, following a complete torsion of the body cavity, appear to have 'reinvented' the gill, now at the posterior end of the animal. Finally, the opisthobranch subgroups Nudibranchia (sea slugs) and Sacoglossa (sap-sucking sea slugs) use surface elaborations of their digestive glands—rather than the mantle—as a gill and survive entirely by cutaneous gas exchange. The sacoglossid genera *Elysia* and *Plakobranchus* carried this evolutionary direction one step further: by kleptoplasty of chloroplasts from the algae they have eaten, and endosymbiosis of the chloroplasts in the skin, they have become truly photosynthetic bilaterian animals (Green et al., 2000; Wägele et al., 2011), capable of providing their own oxygen and carbohydrates.

Bivalves are really the professional filter feeders among animals. Accordingly, we see a stepwise evolution of the filter-feeding mechanism in keeping with different life histories. Most bivalves use cilia to move water over the gills, which serve the double function of filter feeding and gas exchange, the most highly derived form being the familiar lamellibranchiate one. But the dependence on tiny cilia for feeding and breathing represents a trade-off and also automatically limits the size of the animals. In another group of bivalves—the septibranchiate molluscs—the water flow is muscle powered, gas exchange returns to the mantle, and what once was a gas exchanger is again used in feeding—even prey capture.

Although the idea of a predatory clam may seem bizarre, the evolutionary interrelationship between filter feeding, respiration, and predation as seen among bivalve molluscs is something that will turn up again in chordates. There we shall see how the filter-feeding mechanism gave way to respiratory function of the gills, displacing the function of food acquisition to the mouth and later the jaws. Both the bivalve molluscs and also the hypothetical transitional basal craniotes—each in its own way—broke the trade-off gridlock and changed from cilia-powered to muscle-powered ventilation, allowing the evolution of predatory forms and the evolutionary cascades that can accompany this lifestyle.

Cephalopod molluscs, on the other hand, seem to have been predatory from the beginning. Like bivalves, cephalopods first appeared in the Cambrian and experienced two further periods of evolutionary radiation during the late Palaeozoic and the Mesozoic. The best known cephalopod groups are the nautiloids, squids/cuttlefish, and the octopus. Indeed, the largest known extant invertebrates are cephalopods: the giant squid *Architheutis* reaches a length (including the tentacles) of over 10 m. Most cephalopods are pelagic, but octopus are secondarily adapted to a more secretive and benthic lifestyle.

The respiratory system in cephalopods was not compromised with filter feeding as it was in bivalves and chordates, and (sometimes even giant) highly competitive, pelagic carnivores evolved. The functional morphological convergence with aquatic craniotes includes a heart (complete with atria and a ventricle), a highly effective oxygen-transport protein in the internal circulation, a closed circulatory system in the respiratory faculty, and a muscle-powered ventilatory apparatus. In the two extant lineages Nautiloida and Coleoida, we see a now familiar progression: an apparent reduction in the number of gills and also of cardiac atria from four pairs in *Nautilus* to two pairs, respectively, in the Coleoida. The gills are ventilated by contractions of a muscular mantle, which ends in a mobile siphon.

Regarding the siphon, we see quite different evolutionary directions. It is not only used for ventilation: it is also used for jet propulsion, which can be so strong as to even propel so-called flying squid (Ommastrephidae) out of the water to glide for tens of metres. The double purpose of the siphon obviously compromises the respiratory faculty, particularly in *Nautilus* and squid.

Cuttlefish sidestep this trade-off by using an undulating fin for browsing locomotion, but octopus found another solution. They either walk on their tentacles or swim by adducting all the tentacles at the same time, using a web at their base to force out trapped water. But there is still another alternative: both octopus and cuttlefish circumvent the trade-off by using cutaneous gas exchange, which in the octopus can account for more than 40 per cent of the total oxygen consumption in a resting, well-ventilated animal (Madan and Wells, 1996).

9.3 Crustacea

Most aquatic crustaceans breathe with gills, which develop as part of the forked legs or on the adjacent body wall. Very small crustaceans use the skin as the primary gas exchange organ, whereas amphibious species as well as terrestrial crabs show a species-specific development of air-breathing structures (see Chapter 6).

The best-studied crustaceans—the Malacostraca—include such diverse lineages as Decapoda (crabs and lobsters), Lophogastrida (lophogastrid shrimp), Amphipoda (amphipods), and Isopoda (isopods). Branchiopoda (including fairy shrimp and water fleas) represents another group whose respiratory faculty has been well studied. So these will be the groups we choose to trace the evolution of water-breathing crustaceans: not because other groups are not potentially revealing but simply because more information is available. The 'light' is better there.

Looking first at evolutionary lines within the decapods, in addition to the gills themselves, the most important prerequisite is the carapace, which not only protects the gills, it also allows a directed respiratory water current. As explained briefly in Chapter 6, in all decapods the beating movement of a modified second maxilliped (jaw leg) called a scaphognathite (shovel jaw) sucks ventilatory water into the carapace cavity from openings around or behind the legs, and below the gills and moves it in a ventrocaudal to dorsocranial direction. The biophysical results differ among various groups of decapods depending on carapace morphology and gill anatomy. In species with lamellar gills, such as crabs and lobsters, ventilatory water tends to flow counter-current to the flow of the haemolymph. Others such as crayfish, for example, have filamentous gills and the cross-current model is a better approximation. But either gas exchange model is highly efficient and this respiratory configuration is present even in the most basal branches of decapods.

Although some decapods can swim and many have a powerful, backwards fleeing behaviour, most are benthic walkers. The respiratory faculty described earlier, together with haemocyanin as a respiratory protein, is completely sufficient and no remarkable evolutionary directions are evident. The gill surface area of the active, purely aquatic blue crab *Callinectes sapidus*, with 13.7 $cm^2\ g^{-1}$, is similar to that of a pelagic fish and diffusion barrier thicknesses of 1.5–3 μm for the cuticle, and 10–15 μm for the epidermal layer, are typical (Martinez et al., 1999). Due to the vastly different Krogh diffusion constants of chitin and cellular tissue, these values result in an oxygen diffusing capacity approximately equal in each of these two layers, but nearly an order of magnitude below that of the fish.

As opposed to predominantly benthic crabs, lobsters, and crayfish, shrimp tend to be pelagic. Although most shrimp are smaller than crabs and lobsters, an exception shows what is possible. The pelagic lophogastrid red mysid shrimp *Gnathophausia ingens*, with up to 35 cm of body length, is arguably the biggest shrimp going. This species, which was already discussed in Chapter 5 because of its exceptional abilities as a metabolic regulator, inhabits poorly oxygenated waters, which of course is a chicken-and-egg question: can it live there because it already has superior respiratory ability or did this ability develop as an adaptation to the hypoxic environment? But once evolved, the animal never looks back. It just uses its aerobic abilities whenever and wherever it can.

Now, the giant shrimp just described is not a crab or a lobster, but it has evolved a very similar ventilatory mechanism. In addition to using the second maxillary leg as a scaphognathite, it uses a modified first thoracic leg to serve the same purpose. A minor function may become a major one. We call this exaptation. In the earlier example a leg pair that was once used in feeding in decapods or walking in shrimp becomes a ventilatory motor.

To sum up the evolutionary directions found in water-breathing crustaceans, larger and more active forms have gills, effective haemocyanin, and mechanisms for creating a ventilatory current, usually caused by leg movement. Within each of the numerous groups, adaptive radiation including the respiratory faculty occurs, but only rarely does this result in a 'high-performance' animal.

9.4 Chelicerata

As discussed in Chapter 6, there are two primarily aquatic—actually marine—extant chelicerate groups: Pycnogonida (sea spiders) and Xiphosura (horseshoe crabs). Both groups are misnomers since sea spiders are not spiders and horseshoe crabs are not crabs, but then, nobody is perfect.

Since the chelicerates represent an ancient group probably dating back to the Cambrian and their colonization of dry land occurred long before the first Jurassic record of sea spiders, it is not known if the latter became secondarily aquatic or if they always lived in the sea. And since they have no recognizable respiratory organs, the only thing that we can really say that is relevant to the evolution of the respiratory faculty in sea spiders is that they have a presumably plesiomorphic homo-hexamer haemocyanin.

Horseshoe crabs, on the other hand, have complex haemocyanin and book gills (see Chapters 4 and 6), which not only gave rise to the book lungs of terrestrial arachnids, but were also present in the now-extinct eurypterids (sea scorpions). Like their Palaeozoic cousins the horseshoe crabs, the 'sea scorpions' also have a misleading name, since they did not live in the sea, but were coastal marine and fresh water inhabitants and they were are not scorpions either. But they were formidable animals and were top predators in the coastal realm during the Palaeozoic. Like horseshoe crabs, sea scorpions possessed five pairs of book gills and they may have been capable of breathing both in water and in air. Some species even may have been amphibious, emerging onto land for at least part of their life cycle.

9.5 Convergence of Arachnida and Hexapoda

Arachnida (scorpions, spiders, mites and ticks, etc.) are unequivocally terrestrial, possibly evolving from something similar to Palaeozoic eurypterids that entered littoral and/or periodically drying freshwater habitats. Similarly, Hexapoda were from the very beginning terrestrial and evolved the tracheal system to the most effective of all respiratory faculties. But some species adapted secondarily to life under water and many can at least survive flooding and continue aerobic metabolism using air as an interface and transport medium. All of those species kept their air-breathing respiratory organs but developed additional structures that allow them to continue breathing while submerged.

The chitinous cuticle has the ability to form superficial structures that have a huge variety of functions from wings to adhesive feet, but particularly relevant for respiration are the tracheae, and elaborations that make the surface non-wettable and bristles that can trap air. Just as there is no evidence of the homology of tracheae in the diverse lineages that have them, we find no reason to assume that the ability to repel water and trap air cannot have originated repeatedly and independently.

So plastrons—like compressible gas gills—may have evolved many times among small, non-calcified arthropods, notably spiders and especially insects. Note also that they appear to be lacking in larger-bodied groups and also in crustaceans (with the exception of some isopods), although some are certainly small enough to profit from them. The explanation could lie in the fact that crustaceans are basically well-adapted, water-dwelling animals that have evolved some terrestrial forms, whereas spiders and insects were terrestrial from the outset and have evolved convergent mechanisms for surviving or even living in the water using their similar chitinous cuticle. Accordingly, we find unrelated spiders, bugs, and beetles that affix air layers to their bodies when under water and other insects that developed closed tracheal gills of widely deviating structure and located on different parts of the body.

Although we cannot draw any meaningful lines for the evolution of water breathing in arachnids and insects, we do find support for one statement that reoccurs in this book: once something has evolved and certain genetic programmes are present, the animal in question never looks back. This also applies to the use of chitin as a building block of a respiratory faculty.

CHAPTER 10

The evolution of air-breathing respiratory faculties in invertebrates

As pointed out in Chapter 3, for active, aerobic animals it 'makes sense' to breathe air, primarily because a litre of air contains at least 30 times as much oxygen as the same volume of water. In Chapter 7, Table 7.1 gives an overview of the types of aerial respiratory organs seen as well as the respiratory proteins present. Over a million species of terrestrial arthropods can't be wrong: tracheae must be fantastic respiratory organs. But book lungs are the ancestral organs of arachnids and together with the mantle cavity in air-breathing snails, they represent the second-most widely distributed form of aerial gas exchanger among invertebrates. The 'who has what' topic has already been dealt with in more detail in Chapter 7, so let's take a look at how it may have come about.

10.1 Annelida

The oldest fossil records of terrestrial earthworms (Clitellata: Oligochaeta) are from the Carboniferous (Selden, 1990). Gas exchange is through the skin, even if species are pretty big such as certain South African and Australian ones. Their adaptations to life on land, as seen in earthworms and terrestrial leeches, are modest to say the least and well suited only for amphibious or fossorial terrestrial life. So it is very unlikely that annelids did—or for that matter ever will—serve as a platform for the origin of an extensive line above ground.

10.2 Mollusca

Of the numerous molluscan groups, only the gastropods have ventured onto dry land, but they did a pretty good job of it. About 50,000 extant species of air-breathing pulmonate snails are known, and the Panpulmonata—to be sure the largest—is still only one of several groups of air-breathing snails.

With a lung, haemocyanin, and a two-chambered heart, the respiratory faculty does not seem to be a limiting factor for terrestriality in snails. This is more likely to be their reliance on adequate water in particular for gliding locomotion on a mucous layer. In addition to pulmonates (see later), unrelated, operculate air breathers such as the Helicinidae, Cyclophoridae, and Pomatiasidae independently became terrestrial. Most of them are tropical. They lack gills and lungs, and gas exchange takes place across the well-vascularized wall of the mantle cavity. In some species, a breathing tube or a notch in the shell aperture allows air breathing even when the operculum is closed. And then there are the limpets (Patellogastropoda: Patellidae). These representatives of a basally branching lineage are marine and often inhabit intertidal zones. They lack lungs and breathe air using the mantle, which is often modified to form modest gills.

Within the Panpulmonata, species occupy biotopes ranging from aquatic to completely terrestrial. The oldest fossil terrestrial snails are known from the

Respiratory Biology of Animals: Evolutionary and Functional Morphology. Steven F. Perry, Markus Lambertz, Anke Schmitz, Oxford University Press (2019).
DOI: 10.1093/oso/9780199238460.001.0001

late Carboniferous (Selden, 1990). They are assumed to derive from operculate (those that can retract completely into the shell and close it with an operculum) heterobranchs (see Chapter 6) with a single gill (Barnes, 1980). Possibly air breathing developed among ancestral snails that lived in tidal pools or other temporary bodies of water. The ontogenetic sequence reveals first a relatively conventional mantle cavity, and the lungs develop by remodelling the cavity and/or the roof lining with a highly vascularized epithelium. In pulmonate snails, the gills are primarily completely reduced and the edges of the mantle cavity become sealed, leaving only a small opening, the pneumostome, on the right-hand side. In some species. the surface area of the mantle cavity is further increased by ridges and blind-ending tubules. Ventilation is accomplished by flattening and arching of the floor of the lung cavity, the pneumostome either remaining open or opening and closing with ventilatory cycles (see Chapters 6 and 8).

One evolutionary direction that separate lines of pulmonate snails have independently taken is a secondary aquatic-to-amphibious life history, whereby aquatic forms such as the Siphonariidae (false limpets) have even developed secondary gills (Hodgson, 1999). The gills of these intertidal snails are either remnants of the ancestral structures or completely new, secondary gills.

The best-studied pulmonate snails—the relatively basally branching Basommatophora—also fit this description. *Lymnaea stagnalis* (the great pond snail), for example, is a bimodal breather, gas exchange being mainly (>60%) through the skin under normoxic conditions. But in hypoxia, the rudimentary lungs take over. The pneumostome opens and closes at the water surface, expiration being accomplished by contraction of mantle muscle while inspiration occurs due to the elasticity of the mantle. The lung is inflated passively and the pneumostome is closed by the pneumostome muscle. *Lymnaea* uses haemocyanin as respiratory protein while the great ramshorn snail, *Planorbis corneus*, has haemoglobin. Accordingly, *P. corneus* exploits its pulmonary oxygen store better than *L. stagnalis* and can stay under water longer. Some deep lake lymnaeids have actually secondarily completely abandoned air breathing and fill the mantle cavity with water, while other species evolved secondary external gills on various parts of the body, often near to the pneumostome, as in planorbids, mentioned earlier.

Regarding the evolutionary sequence in the origin of air breathing in gastropod molluscs, we propose the following: first to evolve was the behavioural response of surfacing in response to hypoxic water conditions. Due to its relatively thin epithelium, the mantle cavity probably served as a primary site for aerial gas exchange from the very beginning, and this is also where the gills are found in aquatic and marine snails. So the next phase was the stepwise modification of the mantle cavity to form a lung-like organ. Then, those individuals presenting a large surface area as well as the ability to open and close the mantle cavity in a coordinated manner, were better able to survive adverse (i.e. hypoxic) conditions, and air breathing evolved.

Many stylommatophorans, such as the common terrestrial slug genera *Arion* and *Limax*, are shell-less, while others such as the edible 'escargot' snail *Helix pomatia*, the grove snail *Cepea nemoralis* and the hairy snails *Helicella* do have shells. In another group of terrestrial slugs, the Systellommatophora, the leatherleaf slugs (Veronicellidae) have completely reduced the lung cavity and gas exchange is entirely cutaneous. But even within the Systellommatophora, some slugs have adopted an at least partially aquatic lifestyle. The intertidal Onchidiidae, for example, have developed a new posterior lung sac and pore, which remains closed during submersion, and gas exchange is cutaneous.

Typical for pulmonate snails is not the evolution of high-performance forms, which for a snail is really an oxymoron anyway, but rather to cover the entire breadth of aquatic-to-terrestrial life and to keep the options open. As any amateur gardener can attest, snails exhibit a strategy in which they survive and reproduce successfully on land. But as we have seen previously, the option of secondarily adopting aquatic life remains.

10.3 Initial terrestrialization(s) of arthropods

Marine onychophoran-like animals are reported from as far back as the early Cambrian. But since they consist almost completely of soft tissue, their fossil record is at best uncertain. The first certain crown-group fossils are from Cretaceous amber. Assuming that the older fossils are in fact related to extant velvet worms, it is presumed that the transition to terrestrial life and possibly the origin of onychophoran tracheal lungs took place during the Ordovician and Silurian: between 490 and 430 mya (Monge-Najera and Hou, 2002).

The fossil chelicerate group Eurypterida (so-called sea scorpions) were probably the first arthropods to occasionally call land their home. These formidable predators, such as the 1.8 m long *Pentecopterus decorahensis*, were among the largest arthropods that ever existed, and date back to the upper Ordovician, some 467 mya. One late Silurian specimen demonstrated both book-gills like their relatives the horseshoe crabs (e.g. *Limulus polyphemus*) and arachnid-like book lungs (Manning and Dunlop, 1995), suggesting that at least some eurypterids had an amphibious lifestyle. From the Silurian (444–419 mya), true scorpions—which also may have been semi-aquatic (Kjellesvig-Waering, 1986; Dunlop and Webster, 1999)—and also spider-like trigonotarbid arachnids as well as centipedes are known (Jeram et al., 1990; Shear et al., 1996). Since extant scorpions, spiders (Kamenz et al., 2008), and centipedes have either book lungs or tracheal lungs and are terrestrial, we can assume that by the end of the Silurian terrestrialization had to be taken seriously. But did terrestrialization of chelicerates take place once, with a common ancestor of scorpions and spiders coming on land and evolving book lungs, which they passed on to their heirs (Scholtz and Kamenz, 2006), or did these groups evolve their virtually identical book lungs separately by closing the plates covering the book gills in the same way, leaving spiracles for diffusion (Jeram, 1990)?

As they say in German: *Der Teufel steckt im Detail* (The devil hides in the details). As attractive as the idea of a single terrestrialization is, it has some problems: book lungs in scorpions, spiders, and other book-lung breathers are located in different opisthosomal segments. So scorpions and one or more other arachnid lineages are likely to have come on land independently and developed lungs convergently (Dunlop and Webster, 1999). Scorpions and other arachnids must then have diverged while still aquatic (Dunlop and Webster, 1999; Dunlop et al., 2007). Molecular data show that the same genes are expressed in gills of horseshoe crabs, in book lungs, lateral tracheae, and (in spiders) the spinnerets (Damen et al., 2002). But unfortunately this still does not answer the question of whether scorpions and spiders developed their lungs from gills separately or if they inherited them as lungs from a common terrestrial ancestor.

Terrestrial Chilopoda (centipedes) and Progoneata (millipedes) are known from the Devonian (Shear and Selden, 2001; Dunlop et al., 2004). Although the respiratory organs of the fossil forms have not yet been found, the extant centipedes have tracheal lungs and some millipedes even have true tracheae that penetrate tissue. In addition, the first wingless hexapods and, within the Arachnida, the Scorpiones, Opiliones, Amblypygi, free-living Acari, and Pseudoscorpionida, contributed to the Devonian terrestrial fauna. This colonization of dry land was presumably exacerbated by an increase of oxygen in the atmosphere when the content exceeded 20 per cent (Ward et al., 2006). This meant that animals with respiratory systems adapted to breathing water could survive longer on land and eventually evolved specialized air-breathing organs (ABO).

After a temporal gap—presumably caused by a fall in oxygen during the latter half of the Devonian (Berner et al., 2007)—a new evolutionary change occurred, and a surge of terrestrial arthropod groups appeared during the Carboniferous and early Permian, 345–250 mya (Ward et al., 2006). Atmospheric oxygen content in this period is thought to have reached 30–35 per cent. The mesothelic spiders appear to have been the first extant group on land during this surge, followed by the other arachnids, most insect groups and the remaining centipede and millipede groups. Even though elevated PO_2 would have increased the diffusive flux in the tracheal system by about 67 per cent (Dudley, 1998), diffusion alone would not have been sufficient to supply oxygen to the organs of the giant terrestrial arthropods that existed back then: some ventilatory mechanism must have been present.

Many arachnids possess tracheae that look very much like those of insects. But since the last common ancestor of insects and chelicerates may date back to the Ordovician, before any animals lived on land and certainly long before those arachnid groups that have tracheae originated, these two groups must have developed tracheae independently.

Hexapods were long thought to have evolved from a common ancestor with terrestrial myriapods, and therefore—as insects—would never have confronted terrestrialization (Selden, 1990). But more recent phylogenetic studies suggest a closer relationship between hexapods and crustaceans (Regier and Shultz, 1997; Glenner et al., 2006). In this newer scenario, hexapods are presumed to have originated during the drought-prone early Devonian, some 410 mya within the Malacostraca—perhaps the branchiopods—from a common ancestor with the group to which the water flea, *Daphnia*, and the fairy shrimps belong. Or perhaps from amphipods. None of these animals have anything resembling tracheae. So insects, centipedes, millipedes, and arachnids developed tracheae independently (Glenner et al., 2006). But how did animals from within groups that otherwise have no trace of tracheae suddenly develop them? Usually when new structures appear it is a case of exaptation, a functional anatomical paradigm change, in which a secondary function of a previously existing structure becomes primary, but very rarely does something originate completely *de novo*.

Genetic studies give us some insight in to this question. The placodes that give rise to tracheae and leg primordia in *Drosophila* arise from a common pool of cells. Whether the cells become tracheae or legs depends on the local activation state of a gene called *Wingless*. In crustaceans, we see a similar close relationship in signalling pathway and its role in the development of gills and appendages. Remember: crustacean gills often are modified exopodites of the legs. That these similarities are more than coincidental is supported by the fact that homologues of tracheal inducer genes are specifically expressed in the gills of crustaceans. In other words: crustacean gills and insect tracheae may have a common genetic origin (Franch-Marro et al., 2006). Clearly all arthropods have the raw material to construct trachea, but since they can occur in different parts of the body, the question of homology of tracheae in the classical sense is not really relevant, except in closely related species. The question is whether the same genetic programme or programmes are involved in trachea formation in different groups, and what turns the programmes on and off. Further genetic studies on non-model organisms such as centipedes, wood lice, and spiders will hopefully shed more light on this fascinating subject.

10.4 Origin of spider tracheae

The Araneae (spiders) are remarkable in that several groups have independently supplemented the book lungs with tracheae. Although the presence of more than one respiratory system is not unusual—think of snails, crabs, and lungfish—spiders are exceptional in the huge diversity of different combinations and configurations realized.

The first fossil records of true spiders, Mesothelae, are from the late Carboniferous to early Permian around 295 mya. However, the fossil record of other tracheate arachnids dates back much further, with, for instance, the harvestmen appearing in the early Devonian about 410 mya, and an independent origin of their tracheae appears likely. Also other arachnids—such as sun spiders—might have evolved tracheae independently. One feature supporting the hypothesis of a repeated and convergent evolution of tracheae within arachnids is that the number and position of tracheal spiracles are different in each major group.

But even within the Araneae alone, the situation is far from conclusive. Starting with two pairs of book lungs in the second and third opisthosomal segment, we see virtually every combination realized: reduction of one pair of lungs, replacement of the second pair of lungs by tubular tracheae, replacement of the first pair of lungs by tubular tracheae, modification of book lungs to sieve tracheae, and almost everything in between, including the complete absence of book lungs.

Particularly well-developed tracheal systems have evolved in the two most species-rich taxa: the jumping spiders (Salticidae) and the sheet-web or canopy weavers (Linyphiidae). In the six-eyed spiders (Segestriidae) and their close relatives the woodlouse hunters (Dysderidae) as well as in such phylogenetically diverse groups as crab spiders (Thomisidae), the cribellate orb-weavers (Uloboridae and Dictynidae), and the water spider *Argyroneta*, the lungs even are so reduced that the tracheae are the main respiratory organs (Braun, 1931; Millidge, 1986; Bromhall, 1987). Given the long evolutionary history of spiders and the extreme variation and combination in book lung and tracheal configuration, it is difficult to say at this juncture to what extent there is any overarching phylogenetic imprint and the problem appears to be best approached from a functional point of view.

10.5 Evolutionary respiratory physiology of spiders

Spiders are uncompromisingly terrestrial and have apparently been so from the very beginning. But, as just mentioned, one of the most puzzling chapters in the phylogenetic history of spiders is why they appear to have repeatedly replaced at least half of a perfectly good respiratory system—namely, the book lungs—with tracheae. And since, unlike insects, they do have a reasonable alternative to tracheae, spiders provide an optimal testing ground to study the advantages of direct, tracheal, gas exchange as opposed to indirect gas exchange involving the lungs, circulatory system, and respiratory proteins.

Simple school geometry tells us that a system of tubes provides much more surface area than parallel plates, so replacing a book lung by sieve tracheae results in an immediate advantage. Extending the tracheae into the haemolymph sinus adds even more surface area and further extension through the pedicel into the prosoma represents a quantum jump in gas exchange efficiency without requiring changes in the open circulatory system.

How does this translate into the physiology of everyday life of a spider? During low-activity phases, the partition between aerobic and anaerobic metabolism depends on the ATP requirement: as the ATP demand rises, spiders quickly deploy anaerobic metabolism, whereby D-lactate, the major by-product, accumulates in the legs and the prosoma (Prestwich, 1983a). Most spiders are completely exhausted if chased around for 1 or 2 minutes. Repaying the oxygen debt in small-bodied species takes about half an hour, while in larger spiders like tarantulas it can take hours, during which time the animals can hardly move (Anderson and Prestwich, 1985; Paul, 1986; Paul and Fincke, 1989; Paul et al., 1989; Paul, 1991, 1992). So spiders—even species that are free-ranging predators—tend to avoid strenuous exercise. Watch wolf spiders running in very short spurts or jumping spiders ambushing prey after sneaking up slowly and convince yourself.

Treadmill experiments give more insight into spiders' metabolic strategy. Running at high or exhaustive speeds leads to a considerable oxygen debt. Aerobic scope during running is between 3 and 10 times in most species, but can be up to 17.8 in the wolf spiders *Geolycosa* (McQueen, 1980; Culik and McQueen, 1985), which has only modestly developed tracheae. Now, this aerobic scope is comparable with that of a humming bird, but we need to remember that the wolf spider has a very low resting metabolic rate to begin with compared with that of a bird. For jumping spiders, in which tracheae are well developed and extend into the prosoma, the tracheae support aerobic metabolism at intense activity, supporting the hypothesis that they have evolved together with increased aerobic needs in some spider groups in which the circulatory component of the respiratory faculty cannot keep pace with the aerobic needs of the species (Prestwich 1988a; Prestwich, 1988b; Schmitz, 2005).

Let's now have a look at the circulatory component. Spiders have an open circulatory system. They do have major vessels connecting the book lungs with the heart and even major vessels leaving the heart, but capillaries are always lacking. In four-lunged spiders such as tarantulas, the anterior and posterior circulation are separated, so that haemolymph from the prosoma enters only the anterior lung pair while that from the opisthosoma enters the posterior lungs (Paul et al., 1989). But during fast running when metabolism in the leg muscles becomes anaerobic, prosomal perfusion is interrupted by a muscular valve at the anterior end of the pedicel (Paul and Bihlmayer, 1995). So haemolymph flow and pressure can be functionally adapted even without capillaries (Paul et al., 1991; Paul et al., 1994b; Paul et al., 1994a). In the same species, the arterial oxygen partial pressure (P_aO_2) of 3.7 kPa remains constant during rest and walking, but increases during the recovery phase, reaching a maximum of about 10 kPa at the end of recovery (Angersbach, 1978).

In tarantulas, which lack tracheae, a crucial factor in oxygen transport in the haemolymph is the arteriovenous PCO_2 difference ($\Delta P_{av}CO_2$). During rest, spiracles are nearly closed and the $\Delta P_{av}CO_2$ is small. During recovery from an exhaustive run, the $\Delta P_{av}CO_2$ increases and the spiracles open, causing an increase in P_aO_2. The $\Delta P_{av}CO_2$ falls and oxygen can be released to the tissue. Together with an increase in heart rate, this results in a more intense use of haemocyanin in respiration. Araneomorph spiders with two lungs and a prosomal tracheal system have significantly lower maximum heart rates than those with tracheae limited to the opisthosoma. Return to normal heart rates in recovery phases after running are also faster in species with prosomal tracheae (Bromhall, 1987a). In general, the heart rate correlates inversely with body size as does metabolic rate, but correlation with hunting/prey-capture style remains inconclusive (Carrel and Heathcote, 1976; Greenstone and Bennett, 1980; Carrel, 1987).

Against this background let us take another look at the selective advantages of tracheae over lungs. One factor is the better stigmatal anatomy and control of water loss in tracheae (Ellis, 1944; Levi, 1967; Anderson and Prestwich, 1975; Levi, 1976). Certainly small spiders are threatened by evaporation at low humidity, but this does not hold true for all spiders and conversely, those with a well-developed tracheal system and reduced lungs such as saxicolous and crepuscular six-eyed spider *Dysdera*, are not even diurnal let alone exposed to a desiccating environment.

Another reason, at least for replacing lungs, could be the curious hydraulic leg extension mechanism of spiders and separation of the prosoma from the opisthosoma. Breathing with book lungs alone will result in local hypoxia in the prosoma when spiders are walking, and tracheae reaching into the prosoma will solve that problem. But most spiders run in short spurts and haemolymph exchange between pro- and opisthosoma can happen between sprints. Also, not all spiders that walk a lot (e.g. wolf spiders) have tracheae in the prosoma and vice versa: not all spiders with prosomal tracheae are free-ranging hunters (e.g. crab spiders normally just sit and wait in flowers). In addition, this argument applies only to spiders, since other arachnids and tracheate arthropods do have leg extensor muscles. So let's have another look.

High local oxygen requirement is best served by tracheae, whereas a general oxygen demand—particularly at rest—is well served by book lungs. Jumping spiders can support relatively high aerobic metabolic rates better than comparable spiders with a less-developed tracheal system, such as wolf spiders (Prestwich, 1983a; Schmitz, 2004, 2005, 2016). In support of this hypothesis, tracheae indeed tend to supply exactly those organs that need a large amount of oxygen compared with the surrounding organs. The Uloboridae, for example, actively groom their web elements with the third leg pair during prey capturing, and also have a better tracheal supply to the muscles of these legs than to other legs (Opell, 1987, 1990; Opell and Konur, 1992). And in jumping spiders, which rely mainly on their huge eyes for prey capture and protection against predation, tracheae supply predominantly the nervous system and sense organs in the prosoma (Schmitz, 2004, 2005). Also, the haemocyanin shows high affinity but lower concentration in spiders with a well-developed tracheal system (Schmitz and Paul, 2003). This point needs further investigation but is certainly consistent with the hypothesis that haemocyanin serves primarily in oxygen storage in tracheated spiders, but is better suited for oxygen transport and release at the tissues in those dependent on book lungs.

10.6 Crustacea

Terrestrial and semi-terrestrial crustaceans are found only among the malacostracan groups Isopoda (Oniscoidea), Decapoda (Anomura and Brachyura), and Amphipoda. The first terrestrial isopods and decapods emerged independently from aquatic ancestors probably in the late Eocene (only about 40 mya) but amphipods never advanced beyond the semi-terrestrial stage. The semi-terrestrial amphipod 'sand flea' *Talitrus saltator*, for example, uses pereiopod (walking leg) gills for air breathing. But what is most exciting, in contrast to arachnids or insects, in which the initial terrestrialization took place millions of years ago, it is still going on among crustaceans. So they can serve as a present-day analogue of what happened long ago in other groups: a veritable time machine, as it were, for respiratory scientists.

10.6.1 Isopoda

Oniscoid isopods, with their reduction of a calcareous exoskeleton and highly developed pleopod lungs, have become fully terrestrial and thrive even in deserts, where they spend their days underground. In several lineages of wood lice, the pleopod lungs have even given rise to tracheal lungs. Here the convergence with spiders is remarkable. As we remember, spiders inherited internalized book lungs, which were presumably derived from the exopodites of extremities located in the posterior part of the body of their aquatic ancestors. Independently in some groups all or part of the book lungs were converted into tracheal lungs. In isopods, the endopodites of the pleopods form gills and, as in arachnids, the exopodites develop into lungs that are not unlike single leaves of arachnid book lungs. These become either dorsally or ventrally closed and give rise to tracheal lungs. There seems to have been a transition of species living in wet to dry environments and a parallel change from gills to pleopod lungs or even trachea-like structures. Examples for this evolutionary scenario are *Oniscus* that has a simple and uncovered lung, *Porcellio* that has lungs invaginated into the exopodite and *Hemilepistus* and *Periscyphis* with highly developed tubular lungs (Hoese, 1982; Schmidt and Wägele, 2001).

10.6.2 Decapoda

But crabs are another story. Actually, aside from having good walking legs, decapods got dealt a weak hand for becoming terrestrial: they have a heavy, often calcified exoskeleton, a relatively inefficient open circulatory system, a scaphognathite ventilatory system (which is okay for water but is poorly suited for moving air through the gills), and gills that can fill with water or collapse on leaving the water. Each group has achieved its terrestriality separately and nowhere do we see a series of synergistic changes that release an evolutionary cascade. Instead, decapod crustaceans appear in general to have spread out along a plateau in which the animals possess varying competences for breathing both water and air. This can be seen in species-specific morphological and physiological adaptations to a lifestyle extending from using gills as air-breathing organs up to complete adaptation to air breathing with respiratory organs that resemble lungs (see Chapters 7 and 8).

At least partial terrestrialization has evolved and is still happening independently in several groups (Burggren, 1992; Henry and Wheatly, 1992; Henry, 1994; Adamczewska and Morris, 2000), namely among the Anomura (hermit crabs), with the terrestrial coconut crab *Birgus latro*, and the Brachyura (short-tailed crabs), which include the amphibious Ocypodidae (fiddler and ghost crabs). A least five other major lineages of brachyurans can survive prolonged exposure to air, but the most highly terrestrial belong to the Grapsidae, where the halloween crab (*Gecarcinus ruricola*) is completely terrestrial (Bliss, 1968).

However, the problems confronted in terrestrialization are not just anatomical. Semi-terrestrial and terrestrial anomuran and brachyuran crabs continue to use scaphognathite ventilation, the same basic mechanisms as in water-breathing crabs. In some species ventilation seems to be aided by lateral movements of the abdominal mass which acts like a piston to pump air in and out the branchial chamber.

As tempting as it is to go on land where the content of oxygen per litre of ventilated medium is so much higher than in water, particularly in the tropics the high air temperature and sun shining on the carapace creates problems for oxygen transport. As the temperature rises, the solubility of oxygen in the crustacean equivalent of blood plasma decreases dramatically. Accordingly, most tropical or subtropical terrestrial crabs shift from 'plasma' to haemocyanin transport of oxygen. There is no clear relationship between oxygen affinity and terrestriality, but haemocyanin concentration is generally higher in terrestrial crabs. This increases the oxygen-carrying capacity and allows the haemolymph to transport the same amount of oxygen with reduced flow per unit of cardiac activity. The reduced haemolymph flow, in turn, reduces evaporative water loss over the respiratory organs.

But this hypoventilation also brings problems with it. It leads to an accumulation of carbon dioxide, which, in most terrestrial crabs, remains as hydrogencarbonate or is incorporated in the shell as carbonate. But air breathers also have the ability to excrete carbon dioxide directly into the air over the respiratory epithelium thanks to the incorporation of carbonic anhydrase into epithelia of the 'gill-lungs'. This is not only a key factor in the evolution of pulmonary carbon dioxide excretion but also represents an intriguing convergence with craniote lungs. As discussed in Chapter 8, the terrestrial crab genera such as *Birgus*, *Gecarcinus*, and *Gecarcoidea* do not take water with them and carbon dioxide is released via the lungs, whereas bimodal breathers such as the southwestern Atlantic crab, or carangejo granuloso (*Chasmagnathus granulatus*; Grapsidae) extract oxygen through the lungs but release carbon dioxide to transported water through the gills (Halperin et al., 2000).

To better understand the possibilities and limitations of terrestrialization in large crustaceans compared with other animals, let us take a closer look at the completely terrestrial hermit crab *Birgus latro* as a case in point. Specimens of this species, also called the coconut crab or robber crab, can weigh up to 4 kg. Unlike most other hermit crabs, *B. latro* uses its own shell rather than a discarded snail shell, and it is an obligate air breather, entering water only to drink or to lay eggs, and even drowns after some hours in water (Cameron and Mecklenburg, 1973).

So-called branchiostegal lungs extend laterally from the abdomen and account for the entire gas exchange. Experimentally removing the highly reduced gills causes no problems for the animal. So the coconut crab essentially evolved an air-breathing organ without dispensing of its gills until well advanced as an air breather. This development of parallel respiratory organs is also seen in snails and spiders and, among craniotes, in lungfish and many air-breathing teleosts.

The inner lining of the branchiostegal lung is highly vascularized and consists of branching tufts that protrude into the modified air chamber, resulting in a high surface area (Cameron, 1981; Greenaway et al., 1988). The diffusion barrier is thin and haemolymph from the lungs directly enters the pericardial sinus, the crab's equivalent of the left atrium in a craniote (Farrelly and Greenaway, 2005). Blood pressure in this species is the highest reported for any crustacean: 36 kPa. This is about twice as high as human systolic blood pressure.

In spite of this highly derived lung, the respiratory adaptations of *Birgus* leave the impression of being makeshift and making the best of a bad situation. Responses to hypoxia and hypercapnia indicate that *Birgus* occupies an intermediate position in an evolutionary sequence from water to land (Cameron and Mecklenburg, 1973). The lungs are ventilated by the large, anteriorly displaced scaphognathites, which draw air forwards through the lungs. The lung itself has a low extraction efficiency, which is compensated for with a high rate of air flow without counter-current gas exchange. This results in an oxygen extraction of 2–7 per cent, compared with 20–70 per cent in aquatic crabs. Optimization of oxygen delivery is achieved by anatomically constrained adjustments of ventilation and perfusion rather than by modulation of haemocyanin function (McMahon and Burggren, 1979).

So much for a look at the incipient processes of terrestrialization. We have seen many parallels among centipedes, millipedes, arachnids, and crustaceans. Most of these can be characterized as individual adaptive lineages that reflect animals with a relatively low-level metabolic strategy rather than extreme specialization for active life on land.

10.7 Myriapoda

In centipedes, the position of the spiracles and also the fine structure of the tracheae are different in the subgroups Pleurostigmophora and Scutigeromorpha, so it is possible that tracheal systems developed twice within the Chilopoda (Hilken, 1997). In millipedes, on the other hand, the differences are more in degree than in type. In some groups, the tracheae directly supply regions of high oxygen demand such as the CNS and the anterior legs, whereas others possess only tracheal lungs, which end in the open circulatory system. Looking at the myriapods as a whole, their best time probably was back in the oxygen-rich Carboniferous, when massive, millipede-like arthropleurids reached a length of more than 2 m, possibly making them among the largest terrestrial invertebrates ever to walk the earth (Graham et al., 1995; Dudley, 1998). Since then, the group has kept its feet on the ground, as it were, and done a lot of experimenting with tracheae and tracheal lungs.

10.8 Hexapoda

The situation among the non-insectan hexapods is a minefield as well. For instance, some coneheads (Protura) and some springtails (Collembola) have tracheae, whereas others do not. To make matters worse, the location and specific arrangement of these tracheae differs considerably between the groups, making their homology at best questionable. So, right now it is uncertain whether tracheae were already present in stem hexapods and got reduced concomitant with minaturization events that made a dedicated respiratory system obsolete, and subsequently re-evolved using an ancestral genetic template, or if we are facing numerous independent origins of tracheae within hexapods alone. More research on these basally branching hexapods is urgently required in order to shed some light on the course of the evolution of their respiratory system.

Within the Insecta, a tracheal system with two pairs of thoracic stigmata and eight pairs in the abdomen is already present in the Zygentoma (silverfish), a basally branching lineage that never possessed wings. In silverfish, the tracheal supply in the abdomen remains segmentally separate, whereas thoracic tracheae also supply the head. In winged insects (Pterygota), a longitudinal anastomosis connects the abdominal segmental tracheae, and this system remains remarkably unchanged among hemimetabolic insects. Evolutionary directions within the insects primarily involve the pattern of distribution of tracheae and the presence and location of air sacs as well as spiracle structure, including the closing mechanisms. Again, looking at the functional side of the respiratory faculty may be the most promising approach in terms of understanding the morphological diversity, whereas a strictly phylogeny-oriented evaluation remains desirable.

10.8.1 Ontogeny of the tracheal system in a model organism

Diptera (flies) are among the most highly derived insect groups, but we still may be able to learn something about the principles of how the diversity of tracheal structures in hexapods may be generated by looking at the development of the tracheal system in *Drosophila*. There, tracheae originate from respiratory placodes located metamerically along the lateral surface of the embryo. In *Drosophila*, each placode consists of about 80 cells, which divide and migrate inwards to form initial tracheal branches. So far so good. Then, cytoplasmic projections of individual distal tracheal cells ramify to form a series of tubules that terminate in a species-specific fashion at individual internal organs.

At each branching stage, epithelial migration mechanisms and tube formation are distinctive. Fibroblast growth factor (FGF) and a fibroblast growth factor receptor (FGFR) repeatedly activate a general genetic programme to control branch budding and outgrowth, but at each branching stage, the mechanisms that control FGF expression and downstream signal transduction change, depending on the pattern and form of branches (Jarecki et al., 1999; Metzger and Krasnow, 1999; Affolter and Shilo, 2000; Ghabrial et al., 2003). This ontogenetic sequence is not unexpected for a hierarchically branching system, and indeed it is at least analogous to that seen in the mouse lung (Behr, 2010).

More than 30 genes are involved in sequential steps controlling branching morphogenesis. This is caused exclusively by cell migration and cell shape changes rather than by cell division. During the migration phase, cells of each branch are guided to distinct fates, which finally result in the formation of a continuous tracheal network. Each branch forms at key positions a definite number of cell types: fusion cells (responsible for interconnection of tracheae), terminal cells (for gas exchange in tracheoles), and branch cells (portions of the tracheal system that transport oxygen from the spiracles to the tracheoles). We suspect that the fusion cells are an evolutionary innovation of the winged insects. Further research on non-model insects may reveal the mechanisms by which key innovations occur. A recent comparison of *Drosophila* with a beetle, for example, has shown how crosstalk between genes may have resulted in elimination of one spiracle pair and formation of a second longitudinal branch in *Drosophila* and how these innovations may be relevant to survival in these fruit flies (de Miguel et al., 2016).

During terminal branching, FGF expression is regulated by local PO_2 such that tracheal branching matches the cellular oxygen needs, not unlike the mechanisms guiding blood vessel growth during wound healing in mammals. Tracheal branching increases in hypoxia and decreases in hyperoxia (Locke, 1958a; Loudon, 1988). In other words, whereas the major tracheal branches are laid down genetically programmed and controlled by hard-wired developmental cues, branching of the tracheoles is variable and controlled by oxygen demands. This is very similar to the way the circulatory system develops in craniotes.

Now let's take a brief look at air sacs. We see tendencies here that appear to be related both to functional adaptation and to phylogenetic lineages. Large insects, such as dragonflies and locust grasshoppers, have small air sacs distributed along their tracheal system, each flight muscle having its own tracheal supply and associated air sac. Flies and bees—both large and small—have large air sacs (Faucheux, 1974). In addition to their function in distributing air, air sacs help maintain the form and muscle function of the insect irrespective of variable volume of the digestive or reproductive system organs.

10.8.2 Functional morphology and evolution of insect tracheae

The level of complexity of a tracheal system depends in part on the demands placed on it, but also in part on the size of the insect and its phylogenetic position. Small insects or non-volant species with a low metabolic rate tend to have simple tracheal systems that are characterized by small tracheal volumes, a low number of tracheoles, and low diffusing capacities. Volant or fast-running insects, on the other hand, have highly complex tracheal systems with large volumes, volume-variable air sacs, tracheoles penetrating the organs, and extremely high diffusing capacities. Tracheal volumes are between 1 and 50 per cent of the total volume of the insect, depending on the species and developmental stage. Examples are the larvae of the silkworm *Cossus* with a tracheal volume of 1.5–3 per cent of the body volume, 1.3 per cent in instars of stick insects, 5–10 per cent in pupae of moths, 6–10 per cent for the water beetle *Dystiscus*, and for some other adult insects 30–50 per cent (Krogh, 1920b; Buck, 1962; Schmitz and Perry, 1999).

How well does the tracheal system match the respiratory needs in small and large insects and is the oxygen delivery more challenging in large insects? This hypothesis has been tested in grasshoppers many times. In *Schistocerca americana*, the tracheal volume scales with mass to the 1.3 power, this is larger than the exponent for metabolism (0.8), suggesting a greater respiratory capacity in larger grasshoppers, if tracheal diffusing capacity is proportional to volume (Lease et al., 2006). The tracheal diffusing capacity in the jumping legs in adults have about four times as much diffusing capacity as second instars. Differences are mainly due to larger surface-to-volume ratios and thinner tracheal walls in adults. In addition, adults' leg muscles have more intracellular tracheae and mitochondria than those of instars (Kirkton et al., 2005). But large grasshoppers can also compensate for their increase in body size by quadrupling convective gas exchange due to increasing abdominal pumping frequency and the tidal volume thanks to their air sacs (Greenlee and Harrison, 2004).

Further evidence for matching of structure and function in the tracheal system was provided by comparison of adults of species of different sizes, ranging from 0.07 to 6.4 g. The maximal tracheal system conductance scales with mass to the 0.7 power: similar to the scaling factor for metabolic rate in general. But ventilation rate ($f \cdot V_T$) in hypoxia scales directly with body mass, suggesting that convection is a key factor of gas exchange in all species, and body size does not affect the safety margin for oxygen delivery (Jarecki et al., 1999).

However, diffusing capacity is expressed in terms of oxygen flux per unit driving pressure, and since it is to be expected—particularly in an actively flying insect—that the driving pressure in the musculature would be very much greater than in the haemolymph, virtually all of the gas exchange would take place in the tracheoles. So in the end, for insects—but not necessarily for other tracheated arthropods—both the lateral and terminal gas-exchange models lead to the same conclusion.

True tracheal lungs also occur in insects, particularly in larvae. Caterpillars of the Brazilian skipper butterfly *Calpodes ethlius* have been studied because of their economic importance as leaf-rolling defoliators of canna plants. The eighth (last) pair of abdominal spiracles have thin-walled tracheal tufts not unlike the tracheal lungs of centipedes and onychophorans, surrounded by a large number of haemocytes (Wigglesworth, 1953, 1972; Wigglesworth, 1983a).

10.8.3 Spiracle closing mechanisms

Most insects have some mechanism such as sieve plates, hairs, bristles, or the like that keep foreign objects from entering the tracheae. As explained in Chapter 7, there are two types of spiracle closing mechanisms: the internal and external ones (Mill, 1985). But this does not mean that they evolved only twice. Completely unrelated groups such as grasshoppers, cockroaches, and lepidopterans have an external closing apparatus, whereas an internal one is found in bees and lepidopterans (Beckel and Schneiderman, 1957; Nikam and Khole, 1989; Schmitz and Wasserthal, 1999). Yes, lepidopterans have both types!

10.9 The origin and evolution of discontinuous gas exchange cycles in tracheated arthropods

Many arthropods that have tracheae also breathe discontinuously, particularly during rest or metamorphosis in insects. A prerequisite for discontinuous respiration is closable spiracles. This is the case in some arachnids, centipedes, millipedes, and as well as in insects. The plesiomorphic condition is undoubtedly continuous gas exchange since we must assume that the structures that close the stigmata had not yet evolved. Also, in addition to DGC, cyclic respiratory patterns are known, whereby the spiracles never fully close but a rhythmic pattern of carbon dioxide release is observed. In some insects and possibly in other tracheated arthropods, continuous, discontinuous and cyclic respiration are a continuum influenced by the PO_2 in the surrounding medium and the metabolic rate (Bradley, 2006, 2007; Contreras and Bradley, 2009). The closing phase lowers the PO_2 in the insect, but when metabolic rate increases, this phase disappears, leading to a cyclic respiratory pattern. A further increase in metabolism shortens the flutter phase and when it is completely eliminated, continuous respiration is the result.

Looking at the evolution of breathing patterns in arthropods, the most closely related taxon, the velvet worms have relatively high respiratory rates and breathe continuously with low safety margins. This observation is consistent with the hypothesis that discontinuous respiration and very low metabolic rates constitute a derived trait and have appeared independently in several groups such as centipedes and arachnids, in addition to insects. Some examples are given in the following paragraphs.

In centipedes we see incipient evolution of DGC. Representatives of the Scutigeromorpha show continuous respiration, while some Lithobiomorpha show weak periodic patterns as does *Cormocephalus elegans*, a representative of the crown group Scolopendromorpha. But two other scolopendromorphan species in the same genus, *C. morsitans* and *C. brevicornis*, show identical DGC patterns to those found in insects (Klok et al., 2002).

Among arachnids, the pseudoscorpions, solifugids, and mites/ticks have discontinuous ventilation but it differs not only from what we see in insects but also among groups and—as we have seen in centipedes—also among species within the same group. But unfortunately, as the high-level systematics of arachnids is uncertain, it is not possible to draw any conclusions regarding the evolution of DGC in this taxon. But one thing that turns up is the unusual role of oxygen.

In the pseudoscorpion *Garypus californicus* (Lighton and Joos, 2002), for example, internal hypoxia induces a decrease in interburst phase length. This suggests that the opening phase is actually triggered by hypoxia: just the opposite to what we see in insects. We remember: hypoxia as a trigger is common in water-breathing animals but is rare in terrestrial ones. But the pseudoscorpion is not alone, since a hypoxia trigger is also suspected for the African tortoise tick *Amblyomma marmoreum*. In addition, this tick differs from insects in that it has no clear flutter phase.

Solifugans, as fossorial animals, may often be exposed to hypoxic or hypercapnic environments in deep, sealed underground burrows (Lighton, 1998).

They do perform DGC (Lighton and Fielden, 1996), but internal hypoxia triggers the flutter phase rather than the opening phase, and increases interburst phase length as in insects.

Now looking at mites and ticks: the giant red velvet mite *Dinothrombium magnificum*, which is also fossorial, is assumed to perform DGC during diapause-like phases (Lighton and Duncan, 1995). And the American dog tick *Dermacantor variabilis* breaths discontinuously before a blood meal. During the blood meal it stays discontinuous in males but becomes continuous in females. This can be explained by the nearly 100-fold increase in weight and also in metabolic rate, particularly in females (Fielden et al., 1994; Fielden et al., 1999).

The best-studied group regarding DGC is insects (see Chapter 8). During activity the majority of insect groups display a plesiomorphic continuous respiration pattern combined with various cyclic patterns during other life phases. DCG is undoubtedly a derived state and probably developed independently at least five times in the following groups: Dictyoptera (among roaches and mantids), Orthoptera (grasshoppers and crickets), Coleoptera (beetles), Lepidoptera (butterflies and moths), Hymenoptera (wasps, bees and ants), and Diptera (flies) (Marais et al., 2005). Having said that, it is possible that the technical improvement may reveal that DGC is much more common in small insects than previously thought (Schneiderman, 1960; Lighton, 1996; Bradley, 2006; Chown et al., 2006; White et al., 2007; Chown, 2011).

So, is there a single origin of DGC or perhaps several, all leading to the same result in diverse lineages? To date, there are six different hypotheses about how DGC may have arisen, whereby the hypotheses are not necessarily mutually exclusive and certainly are not animal-group specific. These are (1) the hygric hypothesis—DGC reduces respiratory water loss (Lighton, 1996); (2) the chthonic hypothesis—DGC optimizes subterranean breathing in hypoxic and/or hypercapnic environments by the generation of adequate partial pressure gradients (Lighton and Berrigan, 1995; Lighton, 1998); (3) the chthonic-hygric hypothesis which is a combination of (1) and (2); (4) the oxidative damage hypothesis—DGC protects against the toxic properties of oxygen (Hetz and Bradley, 2005; Bradley, 2006); (5) the emergent property hypothesis—DGC is a non-adaptive process, which arises as an epiphenomenon of the interaction of oxygen and carbon dioxide control, both regulating the spiracular opening; and (6) the strolling arthropod hypothesis—DGC arose as protection against parasitic invaders.

Many authors now favour the hygric hypothesis (Chown et al., 2006; White et al., 2007), but there are also arguments against it (Lighton, 2007). Two such arguments are the fact that many inhabitants of dry habitats, such as some desert ants, are in fact continuous breathers, and also that the elimination of DGC does not necessarily result in increased water loss (Lighton and Turner, 2008). As implied earlier, it is not unlikely that the evolutionary *raisons d'être* behind DGC may be different in different groups, and that it arose independently several times. One strong argument for the repeated, independent origin of DGC, as pointed out previously, is its presence even in certain non-insect arthropods, such as centipedes and arachnids.

CHAPTER 11

Respiratory faculties of aquatic craniotes

Before you can do what you want to do, you always have to do something else. So let us first take a look at the non-craniote chordates to better understand the respiratory faculties in craniotes. One of the structures that characterize chordates in general is the perforated branchial basket in the anterior part of their gut. As the name 'branchial' implies, this structure, which functions in filter feeding, is also assumed to have a respiratory function. But let's have another look.

In the amphioxus *Branchiostoma lanceolatum*, the branchial basket is supplied with haemolymph from the ventral aorta (Figure 11.1). As mentioned in Chapter 4, this fluid has a vanadium-containing substance for which a respiratory function has yet to be demonstrated, but no red blood cells. A series of tiny, ventrally located contractile elements called bulbilli pump the haemolymph through spaces in the branchial bars and it is collected in dorsal aortic roots just as in craniotes. But unlike craniotes, there are no thin-walled gas exchange surfaces in the 'gills' in these cephalochordates, but parallel vessels do supply the nephridial organs. More importantly: the branchial basket in *Branchiostoma* makes up only 1–2 per cent of the total morphometrically measured oxygen diffusing capacity of the animal (Schmitz et al., 2000).

So if the 'gills' really just supply the nephridial organs and/or serve in filter feeding, how does *Branchiostoma* breathe? To understand this we need to take a look at the late developmental stages of the animal. *Branchiostoma* has an asymmetric larval stage in which a small number of gill openings exit on the right-hand side: the mouth is on the left side. During metamorphosis, the gill openings on the right migrate across the ventral midline to the left side, new ones form on the right, and the mouth takes its final, anterior position. The new branchial basket becomes enveloped in an atrium. During this process, the relatively straightforward larval coelom develops folds that permeate most of the body including the wall of the atrium, providing it with a very thin water–coelom diffusion barrier. This surface is ventilated by the same water current that serves in filter feeding. When the animal swims using muscle-powered undulation, the coelomic fluid is pushed around, presumably resulting in a circulation of dissolved metabolic gases (Schmitz et al., 2000). The segmental muscles also are spongy and have a fluid-filled cavity (myocoel) facing the outer surface of the animal. So the muscles can exchange respiratory gases directly through the skin and movement of the myocoel fluid distributes them. When the animal is not moving and the metabolic rate falls, convection presumably ceases. In other words, the very craniote-like circulatory system of our non-craniote relative probably has no significant gas exchange function. The ciliated epithelium of the branchial filter ventilates the atrial gas exchange surface of unique coelomic spaces and passive movement of coelomic fluid functionally takes on the gas transport function provided by the circulatory system in craniotes. The

Respiratory Biology of Animals: Evolutionary and Functional Morphology. Steven F. Perry, Markus Lambertz, Anke Schmitz, Oxford University Press (2019).
DOI: 10.1093/oso/9780199238460.001.0001

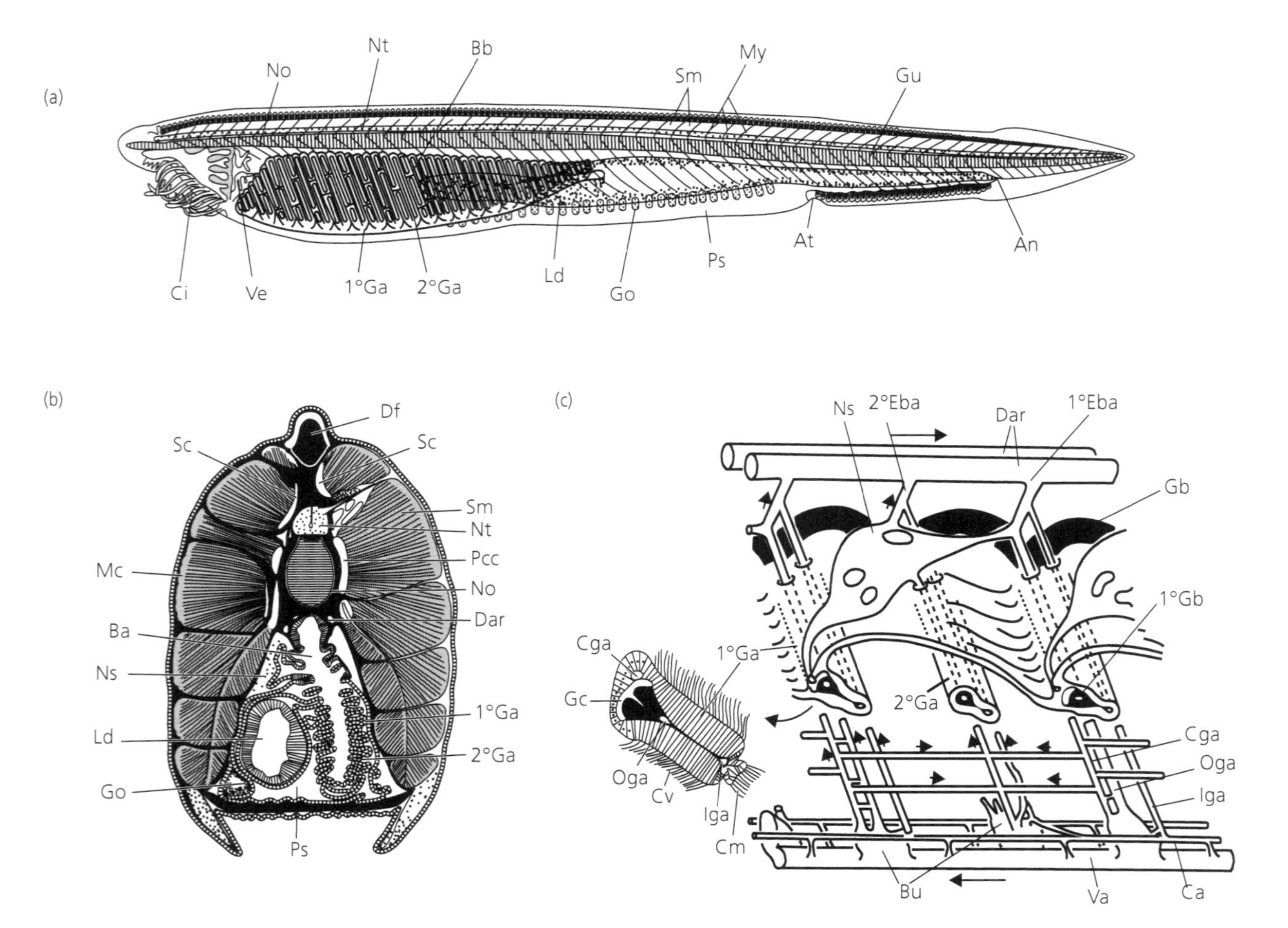

Figure 11.1 Structure of the branchial apparatus in the amphioxus, *Branchiostoma*. (a). The branchial apparatus consists of at least 20 gill arches on each side, the number becoming doubled by the formation of secondary arches. Part (b) represents a cross section through the middle of the gill apparatus. The gut, including the branchial apparatus—originally located in the unsegmented body cavity (coelom)—becomes secondarily surrounded by a peribranchial space (Ps), while the coelom invades various spaces: for example, sclerocoel (Sc), parachordal coelom (Pc), gill arch coelom (Gc). The myocoel (Mc) is a fluid-filled, muscle fibre-free region on the lateral surface of the segmental muscles, which serves in the direct ventilation of the musculature. Part (c) is a detail of the branchial apparatus. Ventilated water is drawn by action of ventilatory branchial cilia (Cv, in inset) in through the mouth, moved across the gills, collected in the peribranchial space (atrium) (Ps) and exits through the atriopore (At in part (a)). A mucus net, formed in the endostyle (not labeled) at the base of the branchial apparatus is carried dorsally by mucous-transporting cilia (Cm) and then moved by further cilia to the gut (Gu, in part (a)). Arrows show direction of haemolymph flow. All labels in alphabetical order: An, anus; At, atriopore; Ba, lumen of branchial basket; Bb, branchial basket; Bu, bulbilli; Ci, oral cirri; Ca, collateral artery; Cga, coelomic gill artery; Cm, mucus-transporting cilia; Cv, ventilatory cilia; Dar, dorsal aotic roots; Df, dorsal fin; Go, gonads; Gu, gut; Ld, liver diverticulum; Iga, Inner gill artery; Mc, myocoel; No, notochord; Ns, nephridial sinus; Nt, neural tube; Oga, outer gill artery; Pcc; Parachordal coelom; Ps, Peripranchial space (Atrium); Sm, Segmental musculature; Ve, velum; Va, ventral aorta; 1°Ga, primary gill arch; 1°Gb, primary gill bar; 1°Eba, efferent branchial artery of primary gill arch; 2°Ga, secondary gill arch; 2°Gb, secondary gill bar; 2°Eba, secondary efferent branchial artery. After Westheide and Rieger, 2007; Perry, 1989.

convergence with bivalve molluscs is astounding, but more about this later in Chapter 14.

Of the three tunicate groups—Thaliacea, Apendicularia, and Ascidia—the first two are pelagic and planktonic, and together make up fewer than 150 species. The Ascidia (ascidians, sea squirts), with some 4000 species, have a pelagic, tadpole-shaped larval stage but as adults they are sessile filter feeders. The branchial basket of large individuals contains several thousand gill slits (Goldschmid, 2007). Like amphioxus, these ascidians draw in water by ciliary action through a broad inflow orifice and move it through the filter apparatus of the branchial basket into to a peribranchial space, whence it flows out through a relatively narrow outflow orifice (Figure 11.2). The ciliary beat is under central, neurological control.

It is assumed that gas exchange takes place in the gill slits themselves, which are well supplied with circulatory vessels. Unlike amphioxus, a central heart, which can reverse the direction of pumping, is present and extensive coelomic spaces are lacking. In the haemolymph, in addition to vanadocytes, in which a sulphur–protein–vanadium complex results in an intracellular vanadium concentration some 100,000 times higher than the surrounding sea water, cells containing high concentrations of iron, chromium, niobium, tantalum, titanium, and manganese are present (Goldschmid, 2007). The possible respiratory function of these metal-containing cells is unknown but on exposure to air the haemolymph turns black.

A little bit of terminological clarification will now be helpful. The plesiomorphic structure of gills in craniotes is undoubtedly a series of gill pouches that connect the central lumen of the pharynx with the outside pharyngeal surface. The principal anatomy is best illustrated using a shark as a model, although their discussion follows later (but see Figure 11.6c later in this chapter) These pouches have an array of respiratory structures, described in more detail later, on their anterior and posterior surfaces. This respiratory structure array plus half of the wall that separates one pouch from another is called a 'hemibranch' (half-gill), and the whole wall (branchial septum) plus its two hemibranchs is called a 'holobranch'. On the other hand, each pouch plus the two hemibranchs that border it is called a 'respiratory unit'. To avoid confusion and missing the forest because of all the trees, so to speak, we shall not go into any detail about the pattern of branchial blood vessels supplying and draining gills. For this there are a large number of good comparative craniote anatomy books which you are encouraged to consult.

11.1 Myxinoida

Hagfish completely lack a cartilaginous gill skeleton in the holobranch septum, and an external cartilaginous net, such as that seen in lampreys, is incomplete or lacking. The array of gas exchange structures on the gill pouch walls consists of a series of primary and secondary respiratory folds, which lie more or less parallel to the direction of water flow (Figure 11.3). The blood spaces within these folds contain pillar cells that closely resemble those of sharks and bony fish (see section 11.3). Blood flows through the spaces between the pillar cells in the opposite direction of water flow: the counter-current model already exists in the most basally branching lineage of extant craniotes.

In hagfish, the ciliated ventilatory pump of the cephalochordates is functionally replaced by a muscle-driven pharyngeal velum (Marinelli and Strenger, 1959; Mallatt and Paulsen, 1986).This strange device is a curtain (the true meaning of the Latin word, *velum*) of tissue that is attached to the pharynx and opens forward onto an inverted T-shaped plunger. When the plunger moves down it traps pharyngeal water behind the velum. This water is then pushed caudally into the gills when the plunger is raised, the free lateral edges of the velum acting as a valve. So the velum is a positive pressure pump with a fixed unit volume, but the minute volume can be varied by changing the frequency.

11.2 Petromyzontida

Unlike hagfish, lampreys migrate into or permanently inhabit bodies of fresh water, where they lay their eggs in sandy stream bottoms. There they develop a filter-feeding (ammocoete, meaning 'sand-bed') larval stage with an amphioxus-like lifestyle. Although they do have ciliated epithelium on their gills, it is just used for transport of the alimentary mucus sheet and not for water propulsion. The gills, which resemble miniature versions of the adult ones, are ventilated with a muscle-powered velum (Rovainen, 1996).

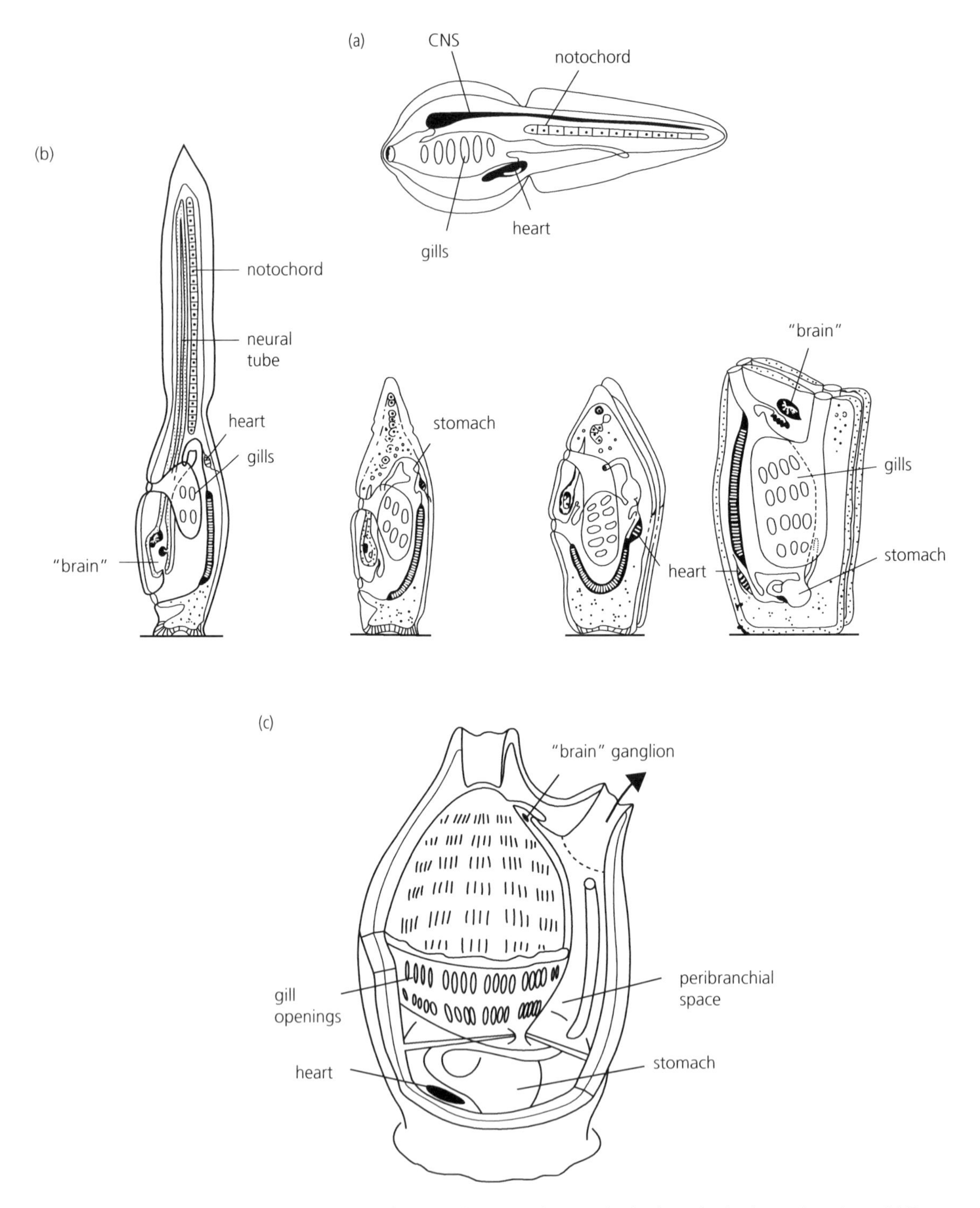

Figure 11.2 The Ascidia, the most species-rich group of tunicates, have a muscle-powered tail and notochord as larvae, shown in part (a). The larvae attach to the substratum, resorb the tail and most of the central nervous system and become sessile filter feeders (sequence (b)), whereby the locomotor system is reduced, most of the central nervous system (CNS) is eliminated, the brain becomes reduced to a ganglion and the branchial basket is greatly enlarged. Large adult organisms, (c), may have several thousand gill openings to the peribranchial space. After Westheide and Rieger, 2007.

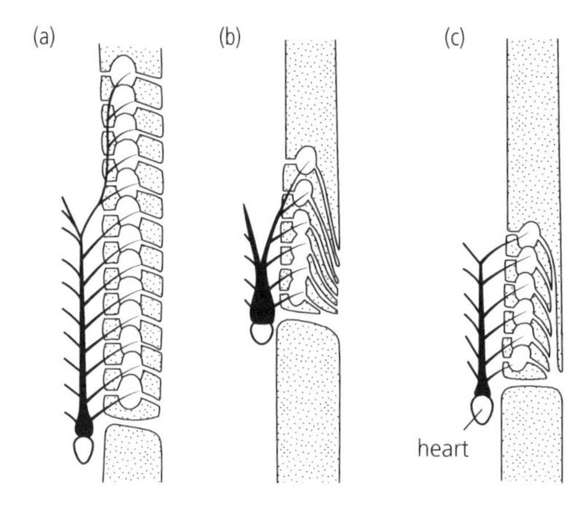

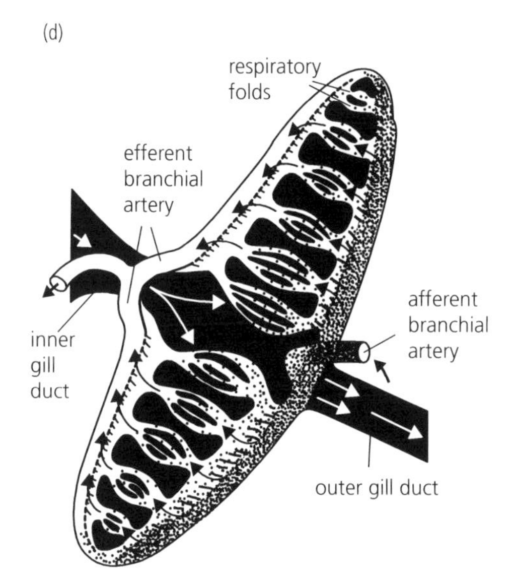

Figure 11.3 Gill structure in hagfish. The branchial region and the blood supply on the right-hand side of the body, viewed dorsally in three different hagfish genera ((a), *Epatretus*; (b), *Paramyxine*; (c), *Mxyine*) are illustrated. The respiratory unit, or gill pouch, exemplified for *Myxine* in (d) consists of respiratory folds that are oriented parallel to the direction of water flow. Since the blood enters from the afferent branchial arteries distally and flows proximally it the efferent branchial artery (black arrows) and the water (black with white arrows) flows in the opposite direction, the counter-current model is established. After Westheide and Rieger, 2015.

The gill structure of the adult lamprey is very different from that of hagfish and similar to sharks (Marinelli and Strenger, 1959; Kempton, 1969; Mallatt, 1996). It is composed of raised ridges, called 'filaments', that are oriented on the gill pouch wall parallel to the water flow (Figure 11.4). The filaments used to be called primary lamellae. Flat, gas exchange structures jut more or less at right angles to the filament, and are accordingly called 'secondary lamellae' but we shall henceforth just refer to them as lamellae. They extend on both sides, from near the free edge of each filament to its base, where it attaches to the branchial septum. This is fundamentally different from hagfish, in which the folds on the filaments run parallel to the filamentar axis. In lampreys, near the attachment of the filament to the septum, the lamellae taper, forming a free space, or water channel, along the base of the filament.

During metamorphosis the velum changes its function, in the adult serving as a filter that prevents food from entering the gills, and adult lampreys have superficial branchial constrictor muscles that they use in gill ventilation (Rovainen, 1996). Since the mouth in these ectoparasites is usually blocked by attachment to the host, the branchial cavity is filled passively by non-ventilatory backflow through the gill openings, the water being moved by recoil of the cartilaginous branchial basket that surrounds the gills, but cartilaginous gill arches are lacking. Due to the relatively narrow external opening of the gills, the flow velocity decreases as water fills the pouches and the lamellae are not ventilated during inspiration.

During ventilatory outflow, water is forced at high velocity through a relatively narrow opening at the end of the gill pouch, thereby creating a negative pressure—much like those vacuum pumps in the chemistry lab—and pulling water in between the lamellae, then into the water channel, and out of the gill. The lamellae do not collapse because the pressure on both sides is the same. The direction of blood flow in the lamellae is from the base near the water channel toward the free end of the filament: countercurrent to the flow of water during expiration. So in spite of the dramatically different anatomy of the gas exchange structures, the counter-current model appears conserved.

11.3 Chondrichthyes

In jawed craniotes (Gnathostomata), the structure of the gills (filaments plus lamellae with their pillar cell systems) is the same as in adult lampreys, but the

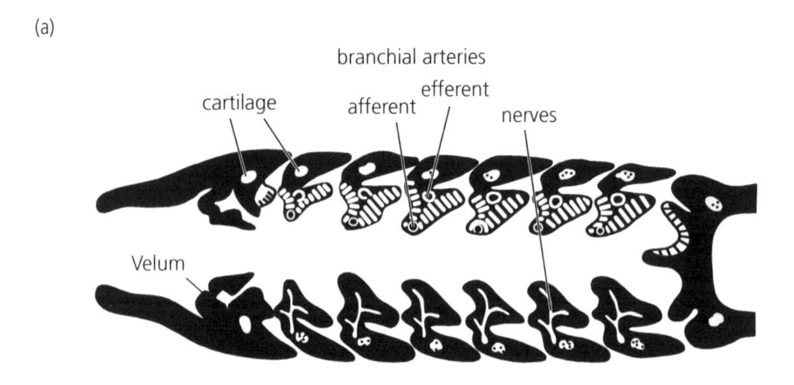

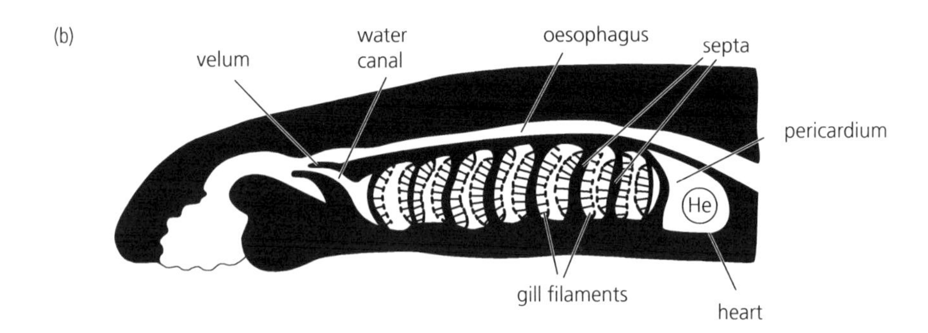

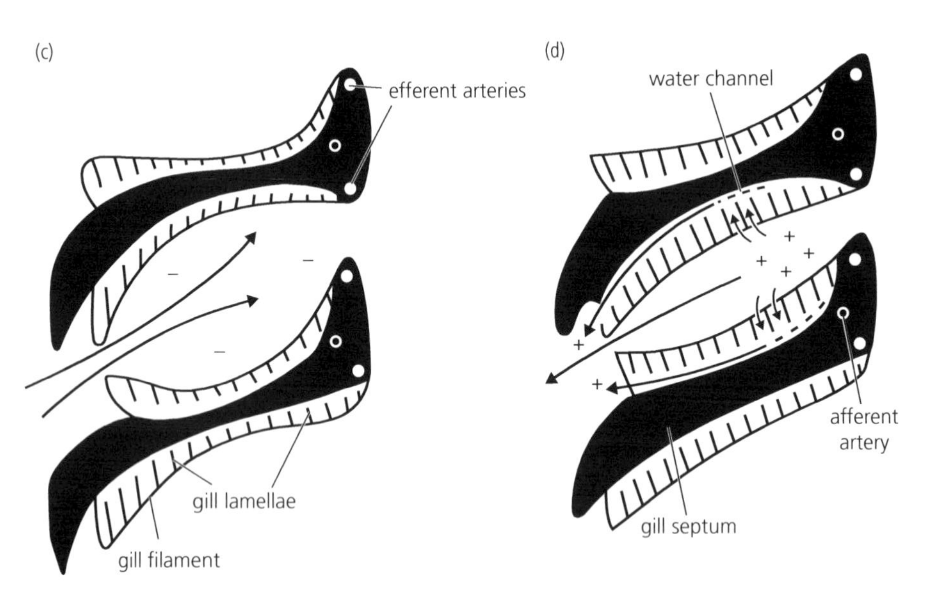

Figure 11.4 The branchial region of the lamprey, *Petromyzon*, in schematic frontal (part (a)) and parasagittal section (part (b)). Parts (c) and (d) illustrate gill ventilation in schematic frontal section of a respiratory unit (gill pouch) on the left-hand side viewed dorsally, lateral to the left. During inspiration (part (c)) when the animal is attached by its mouth to an object or host, water is drawn in through the gill openings, by relaxation of the constrictor musculature and dilation of the branchial region as the elastic cartilage skeleton (part (a)) springs back to its resting state. The pressure conditions during filling of the gill pouches differ dramatically from the flow-through ones during inspiration (part (c)) and are not simply reversed: the lamellae are not ventilated. During expiration (part (d)), as the constrictor musculature (not shown) contracts, pressure rises in the gill pouches (+ signs). Water escapes by flowing between the gill lamellae and out through the water channels at their base. Compare also Fig. 11-6(d). The water flowing rapidly over the tips of the gill filaments at the outlet lowers the relative pressure, drawing more water between the lamellae. Deoxygenated blood enters the septa through paired afferent arteries proximally and leaves, oxygenated, from the same septum through unpaired efferent arteries, which may lie proximally (parts (c) and (d)) or distally (part (a)) in the septum, depending on the ventral or dorsal location. Modified after Westheide and Rieger, 2015. and Perry, 1989.

septa between gill pouches now contain a branchial support system that consists of internal (medial) gill arches (Mallatt, 1996) and also have cartilaginous rays extending into the relatively thin holobranchial septa. Each gill arch is made up of a series of five parts—termed pharyngo-, epi-, cerato-, hypo-, and basibranchial elements—jointed to each other (Figure 11.5). Of these, only the epi- and ceratobranchial elements support gills. The elements are connected to the skull and to each other by sophisticated branchial musculature that is directly innervated by the branchiomeric cranial nerves V, VII, IX, and X (see Chapter 13). Unlike in lampreys, in which the cartilaginous basket lies peripheral to the gills, the branchial skeleton of gnathostomes lies medial to the gills and is of neural crest origin. Recent studies, however, revealed that the gill epithelium of chondrichthyans, like that of hagfish, is endodermal (Gillis and Tidswell, 2017).

In gnathostomes, inspiration is not passive as in lampreys, but is actively brought about by levator muscles of the gill arches. Dorsally, the levator muscles raise the pharyngo- and epibranchial cartilages, while ventrally it is the hypobranchial muscles which pull the gills caudally and open the lower jaw. Expiration is caused by contraction of the dorsoventrally oriented external constrictor muscles, which squeeze the entire branchial region. The deeper interarcual muscles and the branchial adductors attach directly to the branchial cartilages, constricting each gill arch separately by decreasing the angle between the arch elements and at the same time raising the floor of the gill region. Some parts of the complicated hypobranchial muscles also can cause constriction of the gill region. In contrast to the intrinsic gill arch muscles, the hypobranchial group consists of modified segmental muscles of the trunk.

During inspiration, the mouth and the spiracles open and the branchial dilator muscles contract. This draws water into the mouth and from the oral pharynx into the gill pouches, whereby external flaps of the branchial septa act as valves to prevent water from entering the gill region from behind. The mouth is then closed, forcing water into the gill pouches (Figure 11.6). The pharynx is constricted by a contraction wave of external and internal constrictor muscles, progressing from the front to the back of the pharynx. This forces trapped water into the gill pouches and over the respiratory surfaces. This push–pull mechanism causes a nearly continuous unidirectional flow of water across the gills.

Deoxygenated blood from the ventral aorta is pumped into the afferent branchial arteries, which are located on the branchial skeleton, and from

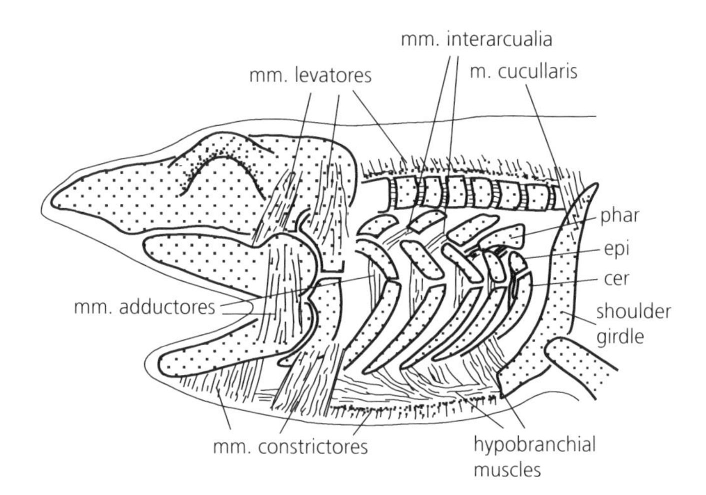

Figure 11.5 Basic structure of the gill skeleton and the most important musculature in Gnathostomata. Of the five skeletal elements (Pharyngo-, Epi-, Cerato- Hypo- and Basibranchial cartilages or bones) only the first three (phar, epi, cer) are illustrated here and only the Epi- and Ceratobranchial parts support gill filaments. The *mm. levatores* extend above all gill arches as *levator arcuum branchialis* and aid in inspiration while the external constrictor muscles (*mm. constrictores*, here onls partially shown) cover the entire gill region in sharks and effect expiration. The mm. *interarcualia* and *adductor branchialis* are internal constrictor muscles that aid in forcing water out of the pharynx. After Westheide and Rieger, 2015.

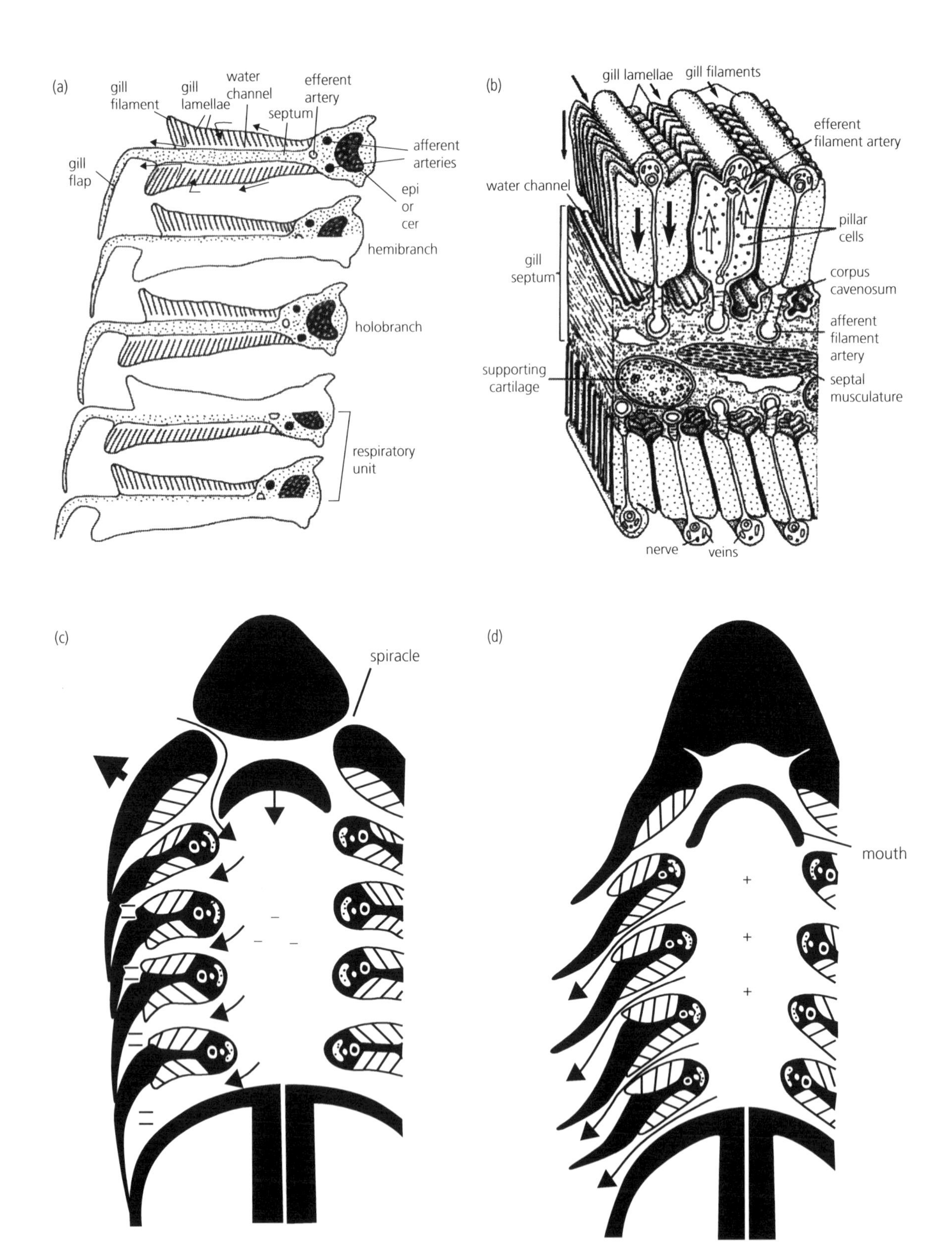

Figure 11.6 Parts (a) and (b) show the anatomy and terminology for shark gills. Note that the gill septum is much longer than the gill filament and that a water channel is present at the base of each filament as in the lamprey. Also note that the direction of blood flow through the gill lamellae (white arrows in part b) is opposite that of the water flow (black arrows) between them. Branchial region in a shark. (c), Expiration. The mouth closes and the branchial region constricts, increasing the pressure in the pharyngeal cavity and forcing water over the gills. (d), Inspiration. The mouth opens and the branchial region expands, closing the gill flaps and preventing water from entering by any other path than through the mouth or the spiracles. After Westheide and Rieger, 2015.

there into the afferent filament arteries. Following counter-current gas exchange in the lamellae, the oxygenated blood leaves the gills via the efferent filamentar and the efferent branchial arch arteries, arriving in the dorsal aorta for distribution to the body.

11.4 The operculum

We have left out one important element of the structure of the branchial region of many gnathostomes: the gill covers or opercula (sing., operculum). In various extinct (†) and extant jawed fish groups (placoderms†, chimaeras, acanthodians†, paleoniscids†, sarcopterygians, actinopterygians), opercula appeared and the gill septa became at least partially reduced. These anatomical modifications have functional consequences. If funnel-shaped gill pouches are no longer present, we would expect that the high water velocity at the distal end of the filaments would be reduced and ventilation of the lamellae would no longer be effective. But muscular adduction of the operculum can also increase the velocity of water leaving the gills (Figure 11.7). Reduction of the holobranch septum frees the filaments and—as long as each filament is separately supported by a cartilaginous ray—provides the advantage that water can flow directly between the lamellae and out of the gill, and the energy-wasting deflection into narrow water channels on the septal surface is no longer necessary.

11.5 Basic structure and function of the gills in Osteognathostomata, particularly Actinopterygii

In bony, jawed craniotes (Osteognathostomata), the skull and the operculum make the superficial constrictor muscles seen in sharks and rays redundant. Instead osteognathostomatans have highly differentiated opercular and gill arch muscles. But the structure of the gills in the most basally branching actinopterygian (ray-finned fish) radiations such as Cladistia (bichir and reed fish) and Chondrostei (sturgeons and paddle fish) are very shark-like, with nearly complete holobranch septa. In more crownward groups, this septum becomes reduced and is completely gone in the most species-rich teleost lineage, the Percomorpha (perch-like fish). Each filament is supported by a cartilaginous rod (Figure 11.7). A pair of filament adductor muscles, which may represent a rudiment of the old superficial constrictor muscles, adjusts their position and optimizes water flow between the lamellae. When the adductor muscles are relaxed, the filaments spread apart such that the tips even touch and all of the respiratory water is forced over the lamellae: the so-called branchial curtain. Under conditions where less gas exchange efficiency is required, the adductor muscles contract, allowing water to escape without contacting the respiratory surfaces. In addition, blood flow through the filamentar and lamellar vasculature is not homogeneous and favours the more basal regions of both structures during periods of reduced activity. This active adjustment of teleost gills results in an optimal ventilation/perfusion relationship in the lamellae. In addition to the principal branchial vascular architecture with afferent and efferent arteries as previously described for sharks, osteognathostomatans have a so-called central venous sinus that receives blood from the efferent filamentary arteries. It is assumed to have mechanical functions, but also nutritive ones.

However, in spite of these dramatic changes in the anatomy of the respiratory apparatus, the basic process of gill breathing does not really change much. As in sharks, the process begins when the mouth opens and water enters (Figure. 11.7). As the mouth closes, water is forced into the pharynx. At the same time the operculum swings outward (abduction), forming an opercular valve prohibiting the water from entering from behind. Then, with the mouth closed the operculum swings inward (adduction), forcing water out over the gills and the process starts over again.

Most ray-finned fish have four functional holobranchs, namely gill arches III, IV, V, and VI. Lungfish can have up to five gill arches, but their respiratory function is very different among the lungfish species. In the Australian lungfish *Neoceradotus forsteri*, for example, the gills are the main respiratory organ, whereas in the South American lungfish *Lepidosiren paradoxa*, their contribution is virtually non-existent. In African lungfish, *Protopterus*, only three of the five gill arches have gas exchange tissue.

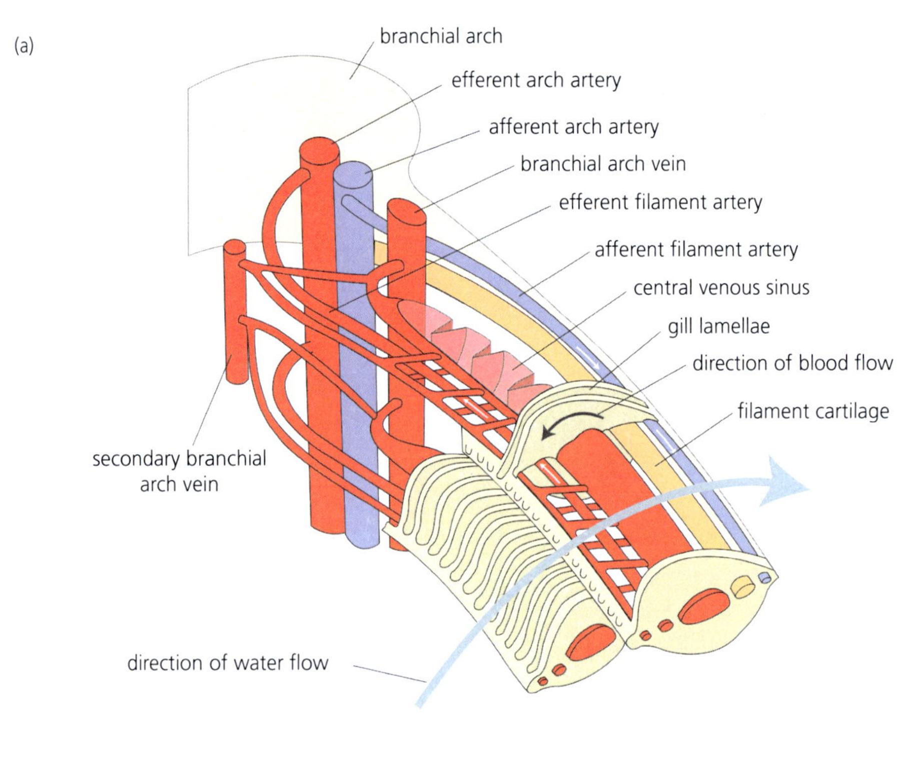

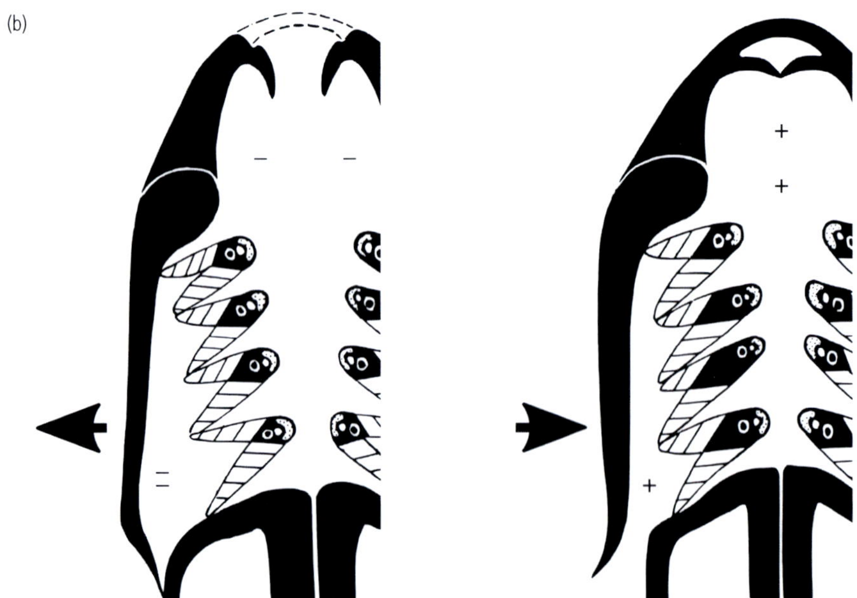

Figure 11.7 Structure of a typical teleost (percomorph) gill with a completely reduced septum (a). Note the complicated vasculature including the central venous sinus (part of the secondary circulatory system), by which the stiffness of the filament can be adjusted. The oxygenated blood it contains may also serve as a reserve. Water flow (large blue arrow) and blood flow (small black arrow) through the gill lamellae are in opposite directions. Branchial region of a teleost fish. (b), left, Inspiration. The mouth opens and the operculum abducts, reducing pressure in the pharynx and drawing water in, whereby the gills are ventilated. (b), right, Expiration. The mouth closes and the alar valve prevents water from escaping anywhere except over the gills. The operculum then constricts, forcing water out of the branchial cavity. Note that when the gill arch septa are reduced, the tips of the gill filaments can be adjusted to touch, forming a branchial curtain. Water exits directly to the supopercular space and not through a water channel as in sharks and lampreys. After Westheide and Rieger, 2015; Perry 2008.

Numerous morphometric estimates of surface area and diffusion barrier thickness conducted over the past four decades have revealed that within the cartilaginous and the ray-finned fish, the diffusing capacity correlates well with the lifestyle (e.g. sluggish bottom dweller vs constant rapid swimmer) rather than with the systematic position. (For references, see Hughes, 1972; Hughes and Morgan, 1973; Hughes, 1984; Fernandes et al., 2007.) But in addition, recent studies have corroborated earlier ones that show that killifish or trout, respectively, can quickly change morphological parameters that result in an increase in diffusing capacity when placed in hypoxic water (Soivio and Tuurala, 1981; Ong et al., 2007).

11.5.1 Optimizing gill function

As discussed earlier, water flow is normally unidirectional in craniotes with internal gills, the exceptions being lampreys and sturgeons. In sturgeons, water flow is normally unidirectional but may become tidal when the animal searches for food with the mouth submerged in sediment.

With an eye on the biophysical constraints we saw in Chapter 3, it is not surprising that respiratory faculties show morphological mechanisms that optimize their efficiency, but because of the limited solubility of oxygen in water and wildly fluctuating diurnal/nocturnal oxygen partial pressures and the metabolic demands of activity, one would expect not only anatomical but also neurophysiological mechanisms. These will be discussed in more detail in Chapter 13, but one surprising and perhaps particularly revealing phenomenon should be mentioned here.

Since the pericardial cavity in sharks is surrounded by the branchial spaces, it makes biomechanical sense that ventricular contraction (systole) should coincide with branchial constriction, and that filling of the ventricle (diastole) and the sinus venosus should occur simultaneously with dilatation of the gill arches. And what one actually finds is not only a coupling of ventilation and perfusion *rates* but also of *events*. A dogfish shark (*Scyliorhinus canicula*) shows a one-to-one coupling of expiratory events and ventricular systoles at rest, and following exercise, the rates change but whole-number ratios may be conserved (Taylor et al., 2006). In bony fish, this neurological coupling persists in spite of the fact that the heart may lie caudal to the gills rather than being surrounded by them and that the rigid skeleton supports the sinus venosus, reducing the biophysical aspect of event coupling.

This arsenal of highly versatile and effective ventilatory mechanisms was certainly instrumental in allowing the evolution of the enormous diversity in size and aerobic activity seen among present-day jawed fish. Estimates of extraction of dissolved oxygen from inspired water range from 60 per cent in sharks to more than 80 per cent in teleosts, and as mentioned in Chapter 5, the water in the wake of a fish school may contain no measurable oxygen at all (Domenici et al., 2007).

11.5.2 Ram ventilators

Since fish have a flow-through system from front to back beginning at the mouth, it is possible for a fish just to swim with its mouth and operculum open and save the energy of using its breathing muscles. Breathing frequency falls to zero (or becomes continuous, depending on your point of view), also releasing heartbeat from any respiratory constraint. Independently, many lines of fishes that inhabit flowing water evolved the practice of this so-called ram ventilation to some extent, and several lines of pelagic neoselachians and bony fish such as tunas and billfishes (e.g. swordfish and marlins) as adults are incapable of ventilating their gills at rest. These sharks, for instance, actually would drown if they stop swimming, but contrary to some popular anecdotes, this is not true for sharks in general, most of which have perfectly functional respiratory musculature. Also the remora suckerfish (Perciformes: Echeneidae), which attach themselves to other fish, boats, sea turtles, and even SCUBA divers using a sucker (actually a modified dorsal fin) on the top of the head, employ ram ventilation when their host is swimming. Comparing constant-swimming neoselachians and bony fish such as tunas and billfish, we see amazing similarity.

The shortfin mako shark (*Isurus oxyrinchus*), weighing roughly 5 kg, ram ventilates while cruising at 1.4 km h^{-1}, while the much smaller skipjack tuna (*Katsuwonus pelamis*) with one-third of this weight, swims at almost twice the speed (Wegner et al., 2012).

Both species of ram ventilators meet their metabolic demands, attaining more than 50 per cent extraction of oxygen from the inspired water, but through different strategies (Stevens, 1972; Bushnell and Brill, 1991, 1992). The slower-swimming shark has a lower metabolic rate and moves less water through the gills than the tuna, but because the interlamellar spaces are larger (56 μm as opposed to only 20 μm in the tuna), the resistance to water flow is lower in the shark and the flow rates over the lamellae are similar in the two species. In addition, the shark lamellae are larger, resulting in a longer contact time between respiratory water and the lamellar surface than in the tuna. This is compensated by the thinner water–blood diffusion distance in the tuna, which is comparable with the air–blood diffusion distance in a mammalian lung! The tunas can achieve such thin diffusion barriers because their filaments are virtually immobile and fused together by crossbeams. This results in a constant-volume gill, analogous to the constant-volume avian lung, and with the same result: the respiratory organs can have extremely thin cell coverings because they do not need to take volume changes and bending into account. Incidentally, the labyrinth ABO of anabantid fishes are also constant volume and also display air–blood barriers as thin as those in mammalian lungs (Munshi, 1985; Graham, 1997).

11.6 Lissamphibia

In general, larval lissamphibians are aquatic gill breathers whereas amniotes don't have a larval stage at all and do not perform aquatic gas exchange. Let's have a closer look at this.

In present-day Anura (frogs and toads), both internal and external gills can be present, the external gills replaced by internal ones at a later developmental stage. The internal gills are derived from an anterior row of filaments on the gill arches, and the external ones, from the posterior row. Interestingly, the same pattern is seen in neoselachians and also among highly derived teleosts such as the loaches (Cobitidae). The gills are resorbed during metamorphosis in anurans. External gills are also found in unrelated groups of air-breathing fishes such as lungfish and cladistia (Figure 11.8). Among Caudata

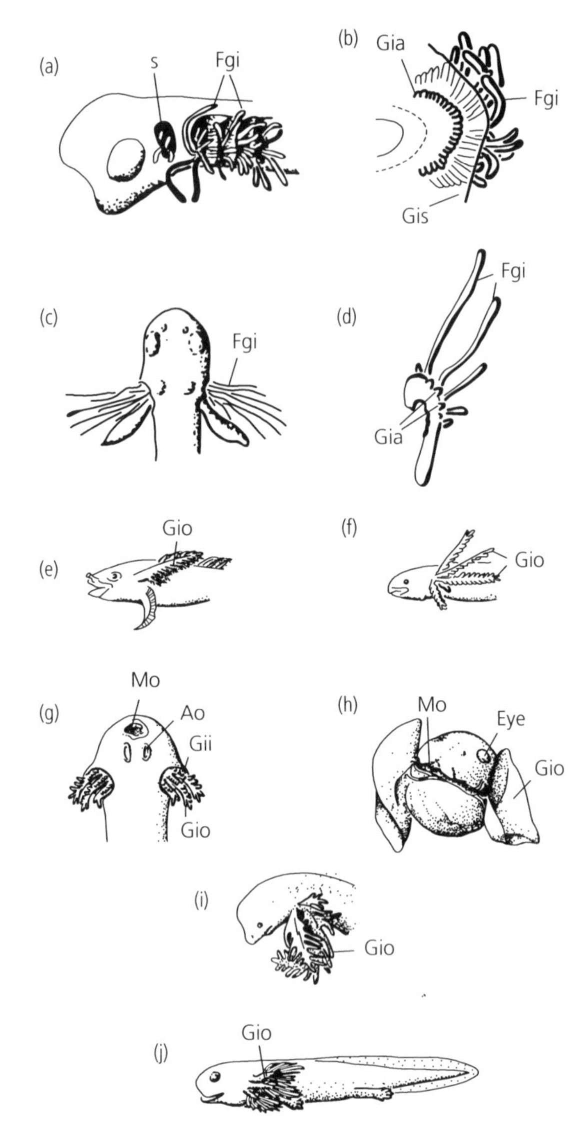

Figure 11.8 Primary ((a) – (d)) and secondary ((e) – (j)) larval gills in fish and amphibians. Similar filamentous gills (Fgi) extending from the posterior side of the gill septum (Gis) in fetuses of two shark genera ((a), *Acanthias*; (b), *Mustelus*) and in the former also from the spiracle (S), which has no gas-exchange function in the adult. The larval teleost fish *Cubitis* shows a very similar gill development. Both in the shark and the teleost the anterior side of the gill arch (Gia) shows no filamentous gill development. Secondary, external gills can be superficially similar as in lungfish ((e), *Polypterus*; (f), *Lepidosiren*) and anuran amphibians ((j), *Salamandra*) or highly diverse as is anuran amphibians ((g), *Rana*, (h), *Gastrotheca*) and gymnophionans ((i), *Hypogeophis*). Ao, adhesive organ; Gii, internal gill; Gio, external gill; Mo, mouth. Perry, 1989.

(salamanders and newts) they are common in such distantly related groups as Proteidae, Ambystomidae, and Necturidae, where external gills are maintained throughout life, whereas others (e.g. Hynobiidae, Cryptobranchidae, Sirenidae, and Amphiumidae) maintain only internal gills. In the Gymnophiona (caecilians), larvae have three pairs of feathery external gills, two of which are resorbed and the other is shed upon metamorphosis (Himstedt, 1996). In fact, the identification of a gill opening in a juvenile caecilian was the first unequivocal anatomical character that allowed their inclusion within amphibians rather than snakes (Müller, 1831).

In addition to gills, most lissamphibians also use the skin for aquatic gas exchange (for relevant references, see Feder and Burggren, 1992), and skin breathing may not only be important for carbon dioxide release in aquatic lissamphibians. Anecdotal reports claim that terrestrial toads accomplish the same by sitting in puddles of dilute urine.

And then there are the often highly terrestrial Plethodontidae (lungless salamanders) that entirely lack gills and lungs, and are completely reliant on the skin for gas exchange. Plethodontids show many adaptations that enhance the efficiency of cutaneous gas exchange, which functions equally well in water and on land. Fine capillaries penetrate into the epithelium, reducing the diffusion barrier to less than 10 µm. As in other salamanders and some anurans (Noble, 1931), the erythrocytes fragment into tiny haemoglobin-filled packets that can easily enter capillaries, making lissamphibians the only craniotes besides mammals that can have nucleus-free red blood cells. And since the lungs are lacking, separation of blood in the heart is not necessary, and the interatrial septum has a large foramen, creating a functional analogue of a fish heart.

Specializations for cutaneous respiration are also seen in anurans, probably the best known being the permanent skin folds of the Lake Titicaca frog *Telmatobius culeus*, reminiscent of those in the giant salamander *Cryptobranchus alleghenienisis* and the hair-like papillae displayed in the flanks and thighs of the male African 'hairy frog' *Trichobatrachus robustus* during the breeding season. A recent study indicates that at least late in their developmental stage these 'hairs' cannot serve in gas exchange, because there is no way for the blood to return to the skin (Barej et al., 2010). In earlier developmental stages, however, a venous return probably exists, but this remains to be verified based on suitable specimens.

There are, though, also lungless frogs, such as the recently discovered *Barbourula kalimantanensis* from Borneo (Bickford et al., 2008). However, to date very little is known about respiratory adaptations or mating habits of this species. And last but not least, there are also lungless caecilians such as *Microcaecilia iwokramae* (Wake and Donnelly, 2010) and *Atretochoana eiselti*, the latter actually being the largest known species of tetrapod without lungs (Wilkinson and Nussbaum, 1997).

11.7 Cutaneous gas exchange in amniotes

The group Amniota is characterized by a large number of traits that make the animals well suited for life on dry land. These include a waterproof skin, internal fertilization, and the lack of a larval stage and functional gills. Most Amniota are not capable of aquatic gas exchange and even in groups that are completely marine it tends to account for no more than a quarter of the total oxygen consumption (King and Heatwole, 1994). In the sea snake genera *Acalyptophis* and *Lapemis* during voluntary diving, oxygen is continuously removed from the lung until about 50 per cent remains, at which time the animal usually surfaces (Seymour and Webster, 1975). During diving, blood tends to bypass the lung (right–left shunt), partially recirculating in the deoxygenated state. This is viewed as a mechanism to avoid losing oxygen to the water via the cutaneous circulation (Seymour and Webster, 1975). During long, forced dives, the oxygen reserves are depleted completely and the blood pH drops due to anaerobic metabolism, indicating that cutaneous gas exchange alone is not capable of supporting an active sea snake.

The Nile softshell turtle *Trionyx triunguis* is capable of limited buccopharyngeal and cutaneous gas exchange (Girgis, 1961). Numerous papillae extend into the oral cavity and are supplied by branches of the carotid artery. Because of their thick diffusion barrier (approximately 100 μm), these structures probably play a role only during extreme hypoxia. Girgis (1961) reports that approximately 30 per cent of the aquatic respiration is buccopharyngeal, the remaining 70 per cent being cutaneous. In another soft-shelled turtle species, *Apalone spinifera*, cutaneous gas exchange may account for as much as 37.5 per cent of the total oxygen uptake. But as in the sea snake, all sources of non-pulmonary gas exchange combined—while helpful in allowing survival of a resting animal—are not sufficient to support an active turtle.

In two other aquatic turtles, *Sternotherus odoratus* and *S. minor*, the ratio of buccopharyngeal to cutaneous gas exchange is closer to 70 per cent versus 30 per cent (Belkin, 1968). Based on the published histological sections (Heiss et al., 2010), the exchange barrier in the oral papillae in *Sternotherus* appears to be pretty thick, as in *Trionyx*, making the degree of its contribution to oxygen uptake questionable. The primary function of the papillae probably lies in carbon dioxide release, since the KCO_2 for tissue is much greater than the KO_2. Perhaps during prolonged submerged periods, carbon dioxide can be released via these structures, thereby reducing its systemic partial pressure and delaying the breathing reflex. This would be advantageous because it would increase the time an animal can spend under water.

In studies designed to simulate underwater hibernation in the painted turtle *Chrysemys picta*, animals maintained at 3°C and denied access to air were able to survive in normoxic but not in anoxic water (Ultsch et al., 1984). The rate of oxygen uptake in normoxic water was calculated to be sufficient to sustain life at 3°C, but not to oxidize the lactate accumulated during metabolism under anaerobic conditions. Repeating the experiments at 10°C and sequentially blocking the possible sites for aquatic gas exchange showed the skin, rather than the buccopharyngeal cavity or the extensive urinary bladder, to be the source of extrapulmonary gas exchange (for references, see Jackson, 2013).

11.8 Do turtles have a cloacal water lung?

In his excellent review of the role of the urinary and cloacal bladder in turtles, Jørgensen (1998) states: 'The view that the cloacal bladders function in respiration became widely adopted, despite the fact that experimental evidence does not support such a role'. In fact, no cryptodiran turtle has been shown to engage in gas exchange with the urinary or accessory (cloacal) bladders, in spite of their often villous internal structure.

But having said that: several species of pleurodires do remain voluntarily submerged for prolonged periods up to several days (King and Heatwole, 1994), and the cloaca is actively ventilated. Physiological studies on *Myuchelys latisternum* showed that about one-quarter (a bit more at 30°C than at 20°C) of the total oxygen uptake was extrapulmonary. Of this, about half took place in the buccopharyngeal cavity and the other half occurred in the cloaca and through the skin, with a tendency for the cloaca to be more important. A similar situation was reported for the closely related *Elseya albagula* (FitzGibbon and Franklin, 2010). So certain turtles, together with sea cucumbers and dragonfly larvae, appear to be the only animals to have a functional water lung.

CHAPTER 12

Respiratory faculties of amphibious and terrestrial craniotes

Lungfish, amphibians, and amniotes breathe air. This should not be a surprise to anyone, and the main part of this chapter will deal with them. But it is probably less well known that all of the major groups of Actinopterygii (ray-finned fishes) except the Chondrostei (sturgeons) and Clupeomorpha (herrings and allies) contain at least one air-breathing group. So we shall start off on the left foot as it were and deal first with them before returning to the mainstream of terrestrial air-breathing craniotes.

12.1 Air-breathing fish

In general, fish that breathe air have three problems: first, as fish, they spend most of their time in the water and they ventilate the branchial region with water, if for no other reason than to maintain the proper blood pH with the help of the ionocytes (aka 'chloride cells') located there. Second, the gills, skin, and anything else contacting the water will release carbon dioxide, resulting in a high blood pH. The same high-affinity haemoglobin that works well at a high pH to extract oxygen from water may not be suitable for air breathing at a lower pH. And third and finally, they have to find some way of ventilating their ABO that is not mutually exclusive with water breathing.

The most basally branching extant lineage of ray-finned fish—the Cladistia (aka Polypteriformes, bichirs and reed fish)—have paired, lung-like, air-filled organs that originate ventrally from the posterior pharynx and are capable of some gas exchange. In other ray-finned fishes, the swim bladder is dorsal and unpaired. In many cases it is used as an ABO. In the Ginglymodi (aka Lepisosteiformes, gars) and Halecomorphi (aka Amiiformes, bowfins) the respiratory swim bladder is supplied bilaterally from a branch of the sixth branchial arch, just like the lungs of tetrapods. Also the structure of the swim bladder is strongly reminiscent of a frog lung, so that the designation 'pulmonoid swim bladder' is justified. Respiratory swim bladders in other teleost groups receive their blood supply from an unpaired swim bladder artery, which branches off the dorsal aorta and under 'normoxic' conditions would be expected to contain oxygenated blood that has just passed through the gills. Under hypoxic conditions the swim bladder would receive deoxygenated blood. Blood that has been oxygenated by the respiratory swim bladder is returned to the general venous circulation and not directly to the heart as in lungfish and tetrapods.

Since the swim bladder is often used for gas exchange, perhaps a few words on this organ are in place although it is dealt with in more detail elsewhere. Except in the Cladistia, it is a dorsal derivative of the pharynx, posterior to the gills. In the plesiomorphic condition it remains connected to the gut, either directly or by a pneumatic duct. This is known as the physostome condition. The swim bladder is filled either by pushing air in from the

Respiratory Biology of Animals: Evolutionary and Functional Morphology. Steven F. Perry, Markus Lambertz, Anke Schmitz, Oxford University Press (2019).
DOI: 10.1093/oso/9780199238460.001.0001

pharynx or by oxygen liberation from the blood (see section 4.6.5 in Chapter 4). Among the more highly derived teleosts, the pneumatic duct is only present in developmental stages and in adults oxygen secretion is the only mechanism for filling the swim bladder. This is known as the physoclist state.

Graham (1997) describes air breathing in fish as a 'mosaic': it has arisen and developed independently (i.e. from non-air breathing ancestors) perhaps more than 60 times. The factors leading to air breathing in fish are multiple and the prerequisites are few. Those groups that have lungs, or a pulmonoid/respiratory swim bladder tend to develop only the skin as an accessory aerial gas exchange organ, whereas those with a non-respiratory or secretory swim bladder also modify the gills, the opercular or branchial cavities, the pharynx, the pneumatic duct, or the stomach or intestine: virtually any surface that can contact air.

Of the two options, gas secretion tends to present more possibilities for the exploitation of ecological licences than does gas uptake. On example illustrates this tendency. The basally branching teleost group Osteoglossomorpha has numerous groups in which the swim bladder is respiratory. One air-breathing subgroup, the Notopteroidea, has eight species whereas the gas-secreting sister taxon, Mormyroidea, has 198 (Berenbrink et al., 2005). This could either mean that the Notopteroidea got it right the first time or that the Mormyroidea are much more capable of further specialization, possibly with the help of their gas-secreting swim bladder. Air breathing can be developed in other parts of the body, but oxygen secretion is limited to the swim bladder and the *rete mirabile* of the eye.

One particularly interesting group is the Otophysi. This group is characterized by contact between the swim bladder and the otic region via an osseous chain, the Weberian apparatus, named after the German anatomist and physiologist Ernst Heinrich Weber (1795–1878). With a tersely filled swim bladder, the fish can sense vibrations in the water and hear approaching danger or prey. One group within the Otophysi, the Characiformes, have a two-chambered swim bladder, whereby the posterior chamber is respiratory and the anterior one contacts the Weberian apparatus. Also within this Otophysi, catfish (Siluriformes) are primarily sluggish creatures and really professional bottom-dwellers, many of which also live in poorly oxygenated habitats. They use virtually every part of the body, sometimes even the swim bladder, for air breathing. In the walking catfish *Clarias*, branched structures extend dorsally from the branchial chambers and serve as ABOs. In the closely related *Heteropneustus*, the chambers are unbranched but penetrate deep into the musculature, forming long sacs (Moussa, 1956; Munshi, 1961; Hughes and Munshi, 1973; Maina and Maloiy, 1986).

Other highly successful air-breathing fishes include percomorph species that are either physoclist or have completely reduced the swim bladder. Channiformes (snake head fish), Blennioidei (blennies), Gobioidei (gobies), and Anabantoidei (climbing perches) are some examples. In all of these groups and also in the swamp eel *Synbranchus*, the pharynx and/or gill chambers are the preferred location for air breathing.

The fine structure of the gas exchange surfaces in the widely phylogenetically diverged genera *Channa*, *Clarias*, and *Anabas*, are strikingly similar although of completely different origin. In all cases, the superficial blood channels are extremely tortuous, forcing blood cells close to the exchange surface and resulting in diffusion distances of less than 0.5 μm: similar to a mammalian lung (Munshi, 1985). Among the Anabantoidei and the Channiformes, so-called labyrinth organs have developed independently. These are highly specialized, more or less constant-volume organs that are located above the gills and can be filled by pushing air in from the mouth, expelling water. Alternatively, air can be drawn in when water is emptied to the mouth by forcing it into the labyrinth organs from behind by opercular contraction (Peters, 1978). Some anabantoid species, such as the blue gourami (*Trichopodus trichopterus*), are obligate air breathers and cannot cover their entire oxygen demand by breathing water, while others such as the Siamese fighting fish (*Betta splendens*) are facultative air breathers. A recent study has shown that this state is so firmly established that *T. trichopterus* cannot enhance larval gill growth when raised under hypoxia, whereas the facultative air breather *B. splendens* can (Mendez-Sanchez and Burggren, 2017).

From the point of view of breathing mechanics, irrespective of phylogenetic lines, there are basically two kinds of ABOs: constant-volume ones and inflatable ones. In the constant-volume ones such as labyrinth organs, the fish have to see that they remove water from the ABO and fill the space with air. This process, common to many species that have the ABO in the head or pharynx, has been described as a four-cycle mechanism (Brainerd, 1994a). Ray-finned fish such as *Polypterus*, which have lung-like organs, draw air into the ABO using elastic energy stored between the ganoid scales during forced expiration (Brainerd, 1994b), and breaths can be 'topped up' using the buccal pump (Graham et al., 2014). But *Polypterus* exhibits a special inspiratory pathway through the spiracle. Although these fish can inhale through the mouth, the spiracles are actually their preferred route (Graham et al., 2014).

The question of how the swim bladder is initially filled in larvae needs more study. Busse et al. (2006) point out that although the larvae of physoclist species are at first physostome, they may never have access to the surface (e.g. in deep sea forms) and/or are so small that they risk adhering to the surface film once they break it. In addition, some larvae lack haemoglobin for regulation of the swim bladder filling. And the surface tension within a larval swim bladder of only 100 μm in diameter must be extremely high. Although surfactant is present in all adult fish species studied (Daniels et al., 2004), tiny larvae have not yet been investigated.

To sum up, there are numerous examples of air breathing, ray-finned fishes but unlike arthropods and lobe-finned fish they show little or no tendency to become permanently terrestrial.

12.2 The respiratory faculty of the Lissamphibia

The basic structure of amphibian lungs is superficially similar to that of pulmonoid and other respiratory swim bladders, and even the lungs of snails. They are sacs that contain a waffle-like, edicular parenchyma (see also section 12.3) that is supported and pulled up by smooth muscle and elastic tissue-containing trabeculae (diminutive of the Latin *trabs*, meaning a support beam) that extend around the curvature of the lung in a net-like fashion (Figure 12.1). When the lungs are filled with air and the glottis is closed, contraction of the smooth muscle pulls on the respiratory

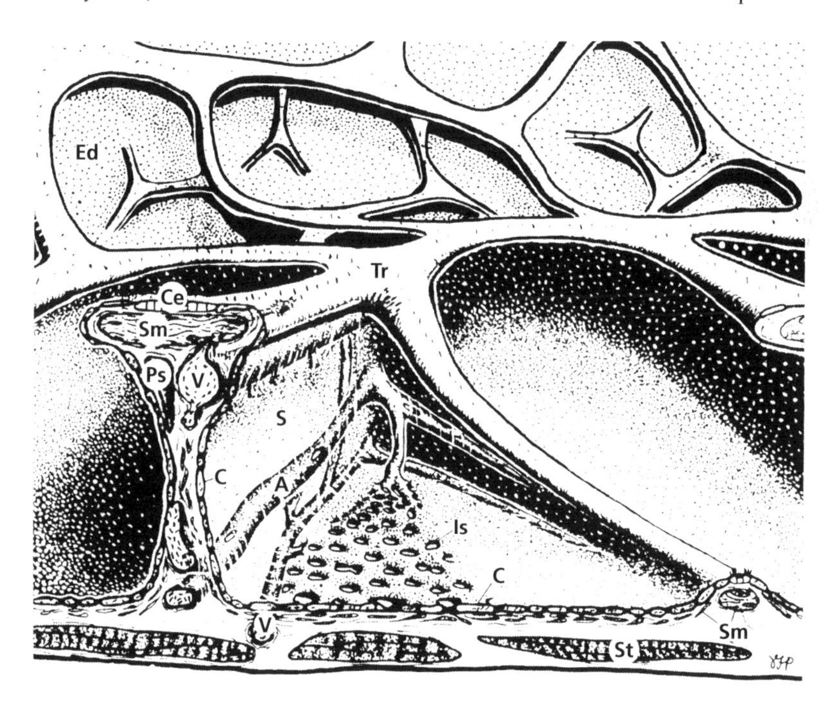

Figure 12.1 Artistic representation of an edicular parenchyma, as found in the lungs of amphibians and in many reptiles. A superficially similar parenchymal structure is also seen in lungfish and in the respiratory swim bladders of basal radiations of actinopterygian fishes. A, intrapulmonary branch of the pulmonary artery; S, Central leaflet of the inter-edicular septum; Ed, edicula; Ps, perivascular space; Tr, trabecula with its ciliated epithelium (Ce); Sm, smooth musculature and elastic tissue of trabeculae; St, striated muscle present in some species; V, intrapulmonary branch of the pulmonary vein, which carries oxygen-enriched blood out of the lung. Note the complete capillary (C) net on both sides of the inter-edicular septum. The central leaflet can contain in addition to connective tissue, blood vessels, nerves and smooth muscle. Perry 1998.

septa that connect the trabeculae with the inner lung wall and holds them upright: sort of analogous to a suspension bridge. These trabeculae are covered by a clearance-type epithelium, consisting of ciliated cells and secretory cells: serous secretory cells or mucus-secreting goblet cells. In addition, neuroepithelial bodies are found in the trabecular epithelium, and these structures are believed to help regulate the smooth muscle tension on the respiratory septa.

The overall shape of the lungs in lissamphibians roughly corresponds to the overall body shape of the animal: usually being roundish or ellipsoid in anurans, sausage-shaped in caudates, and very elongated in gymnophionans. The lungs usually are paired organs (or totally lacking, see Chapter 11), but gymnophionans have a unilateral (left-sided), at least partial pulmonary reduction. In caudates, the lungs are stretched along the body wall and suspended by pulmonary ligaments (mesopneumonia), allowing buoyancy regulation at different states of filling when the animals are in the water. In anurans, the mesopneumonia are reduced, and the lungs collapse toward the glottis during expiration. Aside from this overall simplicity of lissamphibian lungs, there is some structural diversity found among them (Marcus, 1937), and a broad-scale revision of their pulmonary morphology would be a worthwhile endeavour.

At the histological level, amphibian lungs also closely resemble those of most of their piscine relatives. A single type of epithelial cell assumes the dual function of lining the gas exchange surface and producing surfactant (see Chapter 3).

As opposed to the four-cycle mechanism seen in air-breathing ray-finned fish, lungfish and tetrapods have been described as having a two-cycle breathing mechanism (Simons et al., 2000). We shall come back to this later and modify it, but for now it suffices to know that (phase 1) air is gulped or drawn into the mouth and then (phase 2) forced into the lungs with the aid of jaw and hypobranchial muscles. Air is also expelled from the lungs, aided by contraction of the transverse trunk musculature, a muscle group lacking in ray-finned fish. To understand this so-called buccal pump breathing in amphibians, it helps to first have a closer look at the entire cardiorespiratory faculty.

Amphibians, like fish, have a single aorta leaving the heart. The aortic trunk gives rise, sequentially and bilaterally, first to the pulmocutaneous arteries, which supply oxygen-poor blood to the respiratory organs (lungs/gills and skin). These arteries derive from the bilateral sixth branchial arch. Since the fifth branchial arch degenerates in amphibians that lack external gills, the next more anterior paired vessels that branch off in postmetamorphic frogs and toads are the bilaterally aortic vessels, which, as in amniotes, are derived from the fourth aortic arch. These vessels carry at least partially oxygenated blood to the body with exception of the head, which is supplied by the carotid arteries, derivatives of the third branchial arch. This is the last blood to leave the heart and contains the highest oxygen saturation. Separation of the blood masses occurs in the heart due to the location of the opening of the right atrium near the aortic trunk and the left atrium more peripherally. Secondary blood separation occurs in the aortic trunk due to the action of a spiral valve located there. This mechanism works reasonably well in a resting toad with some recirculation of deoxygenated blood (right-to-left shunt), but when the animal is metabolically challenged by hypoxia the efficiency actually increases, that is, the right–left shunt decreases (Gamperl et al., 1999).

Now back to breathing mechanisms (Figure 12.2). Having supplied the lungs with oxygen-poor blood, the amphibian must now ventilate them and oxygenate the blood as it flows through the pulmonary capillary bed. The positive-pressure buccal pump has been best studied in frogs and toads (Gans et al., 1969; Macintyre and Toews, 1976). There, the two-step sequence of events is (1a) the mouth closes, opening the nostrils by lever action on the premaxillary bone, and the floor of the mouth is lowered, drawing fresh air into the buccal cavity, (1b) the glottis opens, allowing air to escape from the lungs. The flank muscles, in particular the *m. transversus* aids in expelling air, which jet-streams over the stored air in the buccal cavity and exits through the nostrils. (2) With the glottis still open, the nostrils are closed and the buccal floor is raised, forcing the stored air into the lungs and finally (3a) the glottis closes. Neurobiological studies have revealed, particularly in amniotes, that what is indicated here as 3a is actually an independent third, post-inspiratory phase, known in mammals and turtles (Richter et al., 1987; Douse and Mitchell, 1990). Since the lungs can have a larger volume than the buccal cavity, this process can be repeated several times to fill them.

Buccal ventilation in frogs and toads involves an unusual mechanism mentioned previously, which is

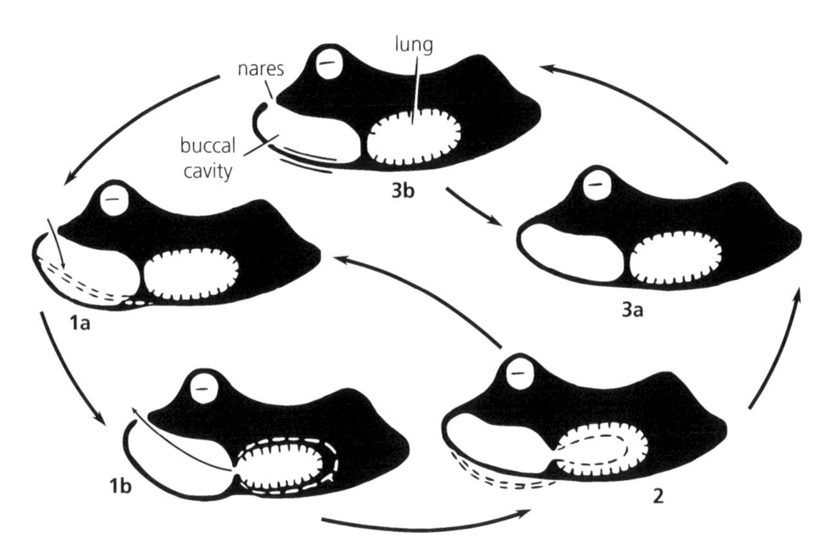

Figure 12.2 Schematic representation of the breathing cycle in many anuran amphibians, anterior to the left. Dashed outlines indicate the condition existing before the one shown with solid lines. The breathing cycle alternates between a ventilatory (1 and 2) and non-ventilatory 3 periods. In the ventilatory period, following a preparatory phase in which the nares are opened and the floor of the buccal cavity is lowered, the glottis is opened (1) and the lung collapses, allowing air to jet stream out of the nares. In 2 the nares close and the glottis remains open while the floor of the buccal cavity is raised, forcing air into the lungs. This procedure may be repeated several times until the lungs are filled. In 3 the glottis is closed and the lungs remain filled (3a) while oscillatory activity may continue in the buccal cavity (3b). After Westheide and Rieger, 2015.

lacking in caudates and gymnophionans, whereby pressing the mouth closed rotates the premaxillary bone and opens the nares. In fact, there is really quite a variety of respiratory sequences seen among amphibians, which should encourage us not to lean too heavily on 'the frog' as a model organism, particularly when it comes to understanding evolution (see Chapter 15). The cane toad (*Rhinella marina*) seems even be able to ventilate its lungs in the experimental absence of the buccopharyngeal pump (Bentley and Shield, 1973). In addition, exclusively aquatic genera such as *Xenopus* or *Pipa* only surface very briefly in order to breathe and also have a highly modified and abbreviated buccal pumping mechanism, whereby inspiration occurs only after air has been expired from the lung and buccal cavity, in comparison to the ranids and bufonids, which fill the buccal cavity first (see earlier-mentioned phase 1a) (Brett and Shelton, 1979).

12.3 The respiratory faculty of the Amniota

In general, the structure of amniote lungs differs fundamentally from that of amphibians in that they are primarily multichambered and contain branching airways that supply independent chambers, which tend to be taxon-specific in their number and location. They are generally suspended by mesopneumonia but in many taxa are attached partially or completely to the body wall and/or the viscera that surround them. Only in most mammals are the lungs completely free to glide within closed pleural cavities, and the mesopneumonia usually are accordingly reduced.

The parenchyma of amniote lungs is, in its plesiomorphic condition, superficially similar to that that of amphibians, but shows more variation in its distribution and specific type (Figure 12.3). In particular, in most reptilian lungs it is heterogeneously distributed, with deep parenchyma near the mouths of the chambers and sac-like, sometimes even parenchyma-free regions in the periphery. The deep regions with tubular, honeycomb-like air spaces are called 'faveolar' parenchyma (from Latin *favus*, meaning honeycomb). If the air spaces are not much deeper than they are wide, the parenchyma is 'edicular' (from Latin *aediculum*, meaning wall niche), and if the trabeculae lie directly on the surface of the lung, it is 'trabecular'. In addition, the trabeculae tend to be

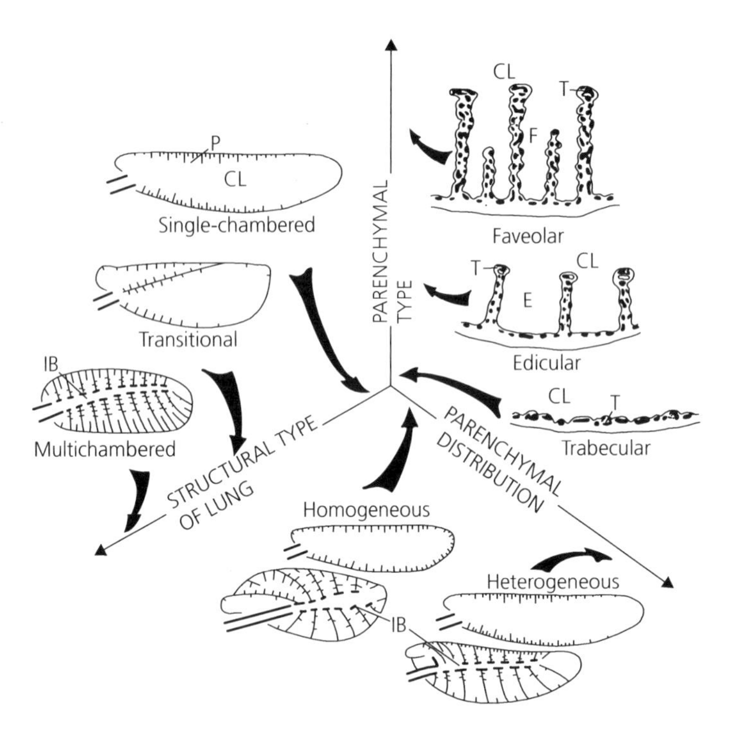

Figure 12.3 Three independent structural variables in reptilian lungs: structural type, parenchymal type and parenchymal distribution. In the single-chambered structural type, and intrapulmonary bronchus is lacking and the lungs consist of one chamber. The transitional lung type has a small number of chambers emanating from a short intrapulmonary bronchus. The multichambered type has a long intrapulmonary bronchus and a large number of chambers. The faveolar parenchymal type consists of tubular air spaces that are deeper than they are wide. In edicular parenchyma the spaces are about as wide as they are deep, and in trabecular parenchyma the septa are so short that the trabeculae lie directly on the inner surface of the lung. Types of parenchymal distribution range from homogeneous to heterogeneous and may vary even within a single species. CL, central lumen; E, edicula; F, faveola; IB, intrapulmonary bronchus; T, trabecula. After Perry, 1998.

more branched than in amphibians, and are classified as first-, second-, and third-order trabeculae, the latter often reaching the floor of the lung. Mammalian and avian gas exchange tissue is highly specialized and does not fall under these classifications of parenchymal types.

A fundamental difference to amphibian lungs, however, is found at the histological level in amniotes. There, two discrete types of epithelial cells are present and assume different functions. One, called 'type I' cells, spread out and constitute the gas exchange surface. The other, accordingly called 'type II' cells, is responsible for producing surfactant (see Chapter 3).

Also opposed to extant amphibians, the plesiomorphic condition in amniotes is negative-pressure, aspiration breathing, in which the ribs are swung downward, forward, and outward, reducing the pressure in the body cavity and expanding it, filling the lungs with air if the glottis is open. Amniotes probably did not 'invent' aspiration breathing. It has, for instance, been observed in one species of aestivating lungfish (*Protopterus amphibius*), but not in others (Lomholt et al., 1975; Delaney and Fishman, 1977; Perry et al., 2008), and body wall muscle contraction associated with breathing has even been seen in ammocoete larvae of lampreys (Hoffman et al., 2016). The reverse seems more likely: lissamphibians eliminated the possibility of costal breathing when they eliminated their ribs.

12.3.1 Lepidosauria

Lepidosauria (Greek: 'scaly reptiles') is the most species-rich group of extant reptiles and is made up of the Squamata (Latin: also refers to scaly reptiles, but includes only lizards and snakes) and the Rhyncocephalia, the latter group consisting of a single extant species, the tuatara (*Sphenodon punctatus*).

Just as there is no typical lepidosaur, there is also no typical lepidosaurian lung. The morphological pulmonary diversity spans almost the entire range seen in amniotes, ranging from single-chambered sacs to multichambered organs and everything in between. Instead of a boring description of the lungs in each component subgroup, we shall just defer you to Perry (1998) for that and move on to some more interesting stuff.

In spite of their huge structural diversity, there are some overriding features of all lepidosaurian lungs, the most important of these being a branched pulmonary bauplan (Lambertz et al., 2015). Well, at least developmentally (compare also Chapter 15). If we begin with lizard-like lepidosaurs (including the tuatara), we usually see rather sac-like, single-chambered lungs in the adults. The originally branched nature is betrayed only by the subapical entrance of the extrapulmonary airways. This is the result of an early ontogenetic branching event, resulting in the anterior-most region of the lung literally overgrowing the extrapulmonary bronchus. Internally, the lungs of several species show additional remnants of the initial branching in the form of septa extending into the central pulmonary lumen (Figure 12.4). These septa result in a series of lobes or niches, and persist as surfaces that enlarge the potentially respiratory surface area, as can be seen in many geckos and lacertid lizards, for example. The parenchyma frequently is of the edicular type, in which the honeycomb-like respiratory structures are not deeper than they are wide. The parenchyma in most cases is also heterogeneously distributed, becoming flatter distally, with the smooth-muscular trabeculae resting directly on the lung surface. Others, such as tegu lizards (*Salvator* and *Tupinambis*), have no lobes and

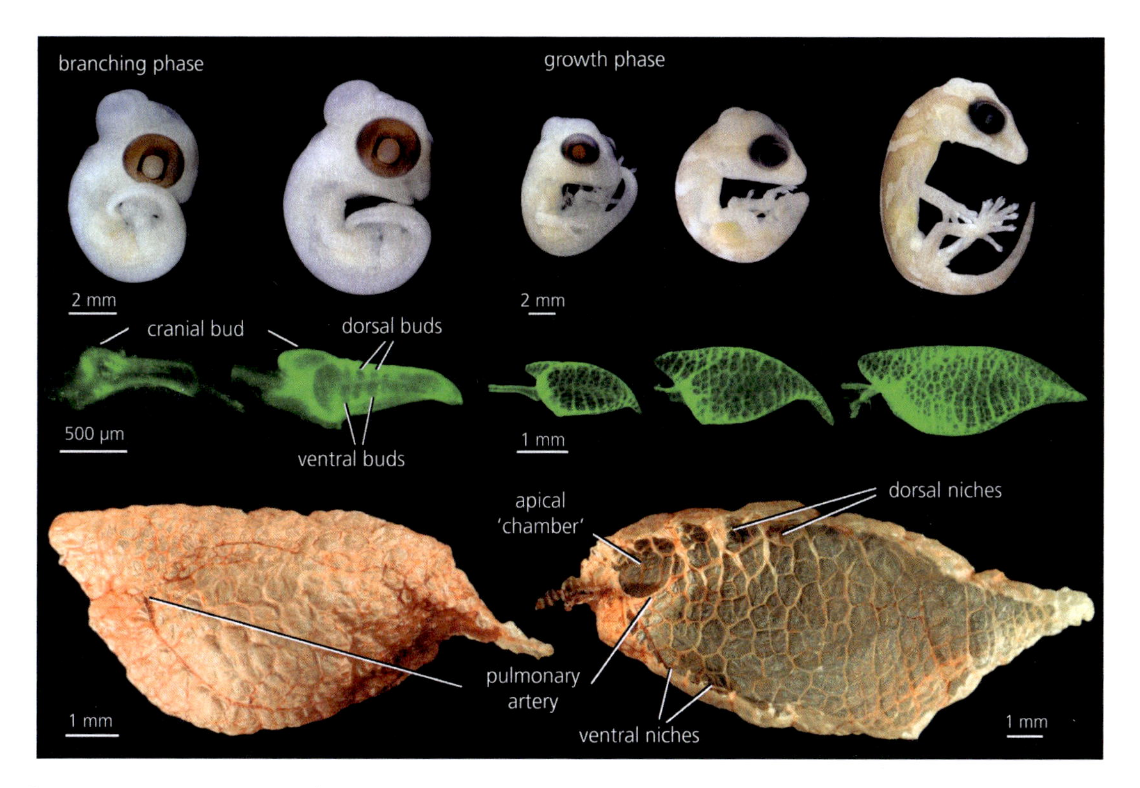

Figure 12.4 Ontogenetic sequence of the single-chambered lung of the gecko, *Paroedura picta*. The upper row of full profiles of the foetal stages corresponds with the second row, showing corresponding stages of lung development. Note the cranial, dorsal and ventral buds form a transient multichambered lung. Traces of this developmental stage are still present in the form of dorsal and ventral niches and in the pathway of the pulmonary artery in the adult lung (dried preparations in lower row: left-hand illustration, intact left lung viewed laterally; right-hand illustration, right lung, medial side removed, viewed medially showing lateral wall with niches. After Lambertz (2016b).

niches, but rather a homogeneously distributed faveolar parenchyma (Perry, 1983). The structural type and distribution of the parenchyma has consequences for breathing mechanics and gas exchange (Perry and Duncker, 1980, see also later in this section).

Between the single-chambered and the multichambered types there is what we call the transitional lung type (Figure 12.5), found mainly among members of the Iguania (i.e. iguanas, chameleons, agamids, and close relatives). There, the apical branch, whose further growth causes the subapical entrance of the extrapulmonary bronchi as described earlier, is bordered by a large septum that extends far into the lung. This results in the division of the lung into two discrete chambers: a small anterior one and a larger posterior one. A short intrapulmonary bronchus supplies these two chambers individually, although in many species it is hardly recognizable. Additional smaller septa, being the remnants of early branching events as well, can be present to various degrees in both of these chambers.

In what appears to be a remarkable case of re-evolution (compare also Chapter 15), monitor lizards (*Varanus*) and the closely related gila lizards (*Heloderma*) use the branched pulmonary bauplan, which persists during early ontogeny in all lepidosaurs, as a template to develop a true multichambered lung as adults, in which an intrapulmonary

Figure 12.5 The transitional lung of an *Iguana iguana*: dried right lung in lateral view, lateral wall removed, anterior to the right. Part (a) shows the two main chambers, the apical and posterior chamber. Parts (b) and (c) show details of the short intrapulmonary bronchus. Part (d) is a schematic representation of the communication of the intrapulmonary bronchus with the two chambers, the location marked with an asterisk in part (a). The orifice to the posterior chamber is indicated by the left-hand white disc and the associated arrow,b, indicates the view shown in (b). The orifice to the apical chamber is indicated by the upper, white ovoid disc and the associated arrow,c, indicates the view shown in (c). Original M.L.

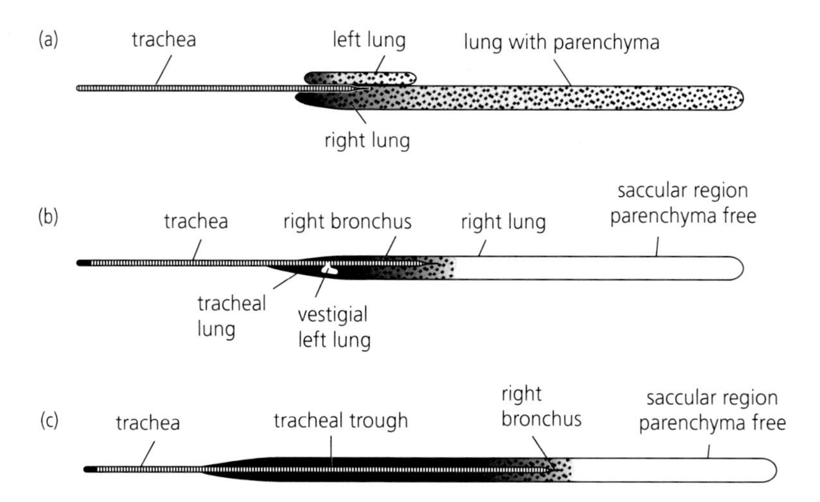

Figure 12.6 Three snake lungs in ventral view, showing different forms of development. (a), the rubber boa, *Charina*; (b) the colubrid rat snake, *Pantherophis*; (c), a rattlesnake, *Crotalus*. In part (a), since two lungs is the plesiomorphic state for squamates, a well-developed right lung and a reduced left one as in *Charina* is considered to reflect the plesiomorphic state rather than a derived one. Also characteristic of squamate lungs in general, parenchyma extends to the posterior end of both lungs, whereas in the lungs illustrated in parts (b) and (c) the posterior part of the lung is saccular and lacks parenchyma and pulmonary vasculature. Note also the subterminal bronchial entrance as in other reptilian groups. In part (b), the tiny vestigial left lung of *Pantherophis* also shows a subterminal bronchial entrance, whereas the right lung has an intrapulmonary bronchus and a short tracheal lung anterior to the bronchial bifurcation. The rattlesnake in part (c), on the other hand, has no vestige of a left lung, but embedded in the faveolar parenchyma is a long, trough-like tracheal extension typical of viperid snakes and the right bronchus is vestigial. After Wallach, 1998.

bronchus supplies several discrete chambers (Becker et al., 1989). The walls of these chambers, at least in certain monitor lizards, can also be perforated, making extrabronchial intercameral connections. These may be important for air distribution (see later in this section).

Probably one of the most intriguing evolutionary patterns evident among lepidosaurs is the repeated and convergent appearance of a serpentiform (snake-like) body shape. True snakes (Serpentes) are the most species-rich radiation of scaled reptiles, but almost all other squamate lineages also exhibit serpentiform species. Most of these species share not only a similar external appearance, but also show similar adaptations in the anatomy and location of their internal organs, including the lungs. One lung usually is strongly reduced or even absent, whereas the other one is greatly elongated. But the earlier-mentioned subapical bronchial entrance is still recognizable. Which lung becomes reduced in general is taxon specific and highly conserved. In snakes, for instance, the left lung is reduced, whereas in amphisbaenians it is the right one. The degree of one-sided reduction varies: snakes, for example, as adults may have no left lung at all (many colubrids and vipers), a tiny vestigial left lung (many cobras), or a shorter but properly developed one (pythons) (Figure 12.6). Several unrelated species of snakes also have a remarkable additional site for gas exchange in the trachea: a so-called tracheal lung (Wallach 1998). This structure of course has nothing to do with the tracheal lungs of arthropods and is just an unfortunate synonym.

In squamates in general, a left-sided at least partial reduction is more common, perhaps due to the presence of the stomach on the left side and the habit of many species to swallow large prey whole. But as seen with amphisbaenians and their right-sided reduction, both conditions occur, but are usually in a conservative, taxon-specific fashion. But there is no rule without an exception. The European glass lizard *Pseudopus apodus* shows a so-called fluctuating asymmetry: some specimens have a much shorter left lung, some, a much shorter right lung, and in some, both lungs are about the same length (Lambertz et al., 2018). The developmental causes of this unusual phenomenon remain obscure, but are of great interest from an evolutionary perspective.

And there are more exceptions. One radiation of gymnophthalmid lizards (*Calyptommatus* and relatives) is composed of sand swimmers, with a serpentiform body and strongly or completely reduced legs. Despite their body elongation, the lungs are short, stout, and of approximately equal length. In fact, the entire arrangement of their viscera is not asymmetrical and 'snake-like', but the organs lie in sequence along the body axis (Klein et al., 2005).

Intrapulmonary structures such as pronounced septa, lobes or niches rarely occur in serpentiform squamates and the lungs in general are pretty much sac-like. Here, the multichambered lungs of blind snakes (Typhlopidae), represent an exception. In fact, membranous regions (particularly in the caudal part) can be completely devoid of parenchyma and pulmonary blood vessels in many species. These sac-like regions may serve in air storage and/or passive ventilation or simply as a space holder to allow effective serpentine locomotion, but they certainly have no direct gas exchange function.

Gas exchange takes place in the parenchymous regions (Figure 12.7). Given the highly variable structural types of lungs in lepidosaurs, the body mass-specific respiratory surface area also differs considerably, being largest in those species with multichambered lungs. A similar variability exists in the barrier thickness. In general, the diffusion distance in lepidosaurian lungs is much greater than in birds and mammals. Stereology-based studies designed to estimate these factors are few and far between (see summary in Perry, 1998). Typical barrier thicknesses tend to fall between about 0.5 and 1 µm, being thinnest in animals with single-chambered lungs and faveolar parenchyma. In such lungs, pronounced capillary bulging is also common, resulting in exposure of a large proportion of the capillary surface.

Like all other amniotes, lepidosaurs are aspiration breathers and movements of the ribs together with the associated volume and pressure changes of the trunk are responsible for the ventilatory air movements (Figure 12.8). The musculature involved are hypaxial body wall muscles, which include, in addition to the intercostals, also the oblique, transverse, and rectus muscles. Due to the 'lizard-like' sprawling gait of many lepidosaurs and the lateral undulation during walking, the trunk muscles wear

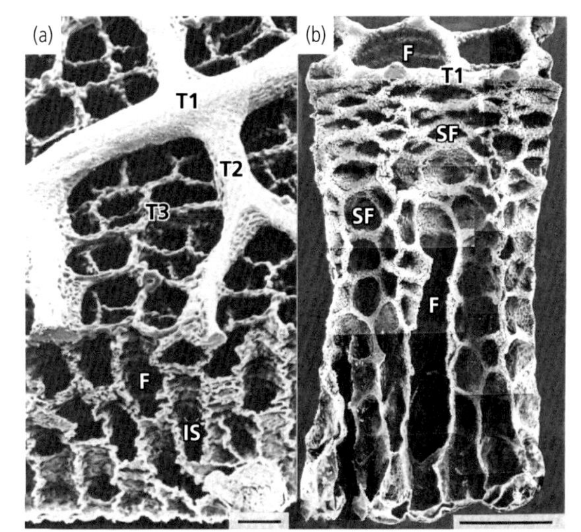

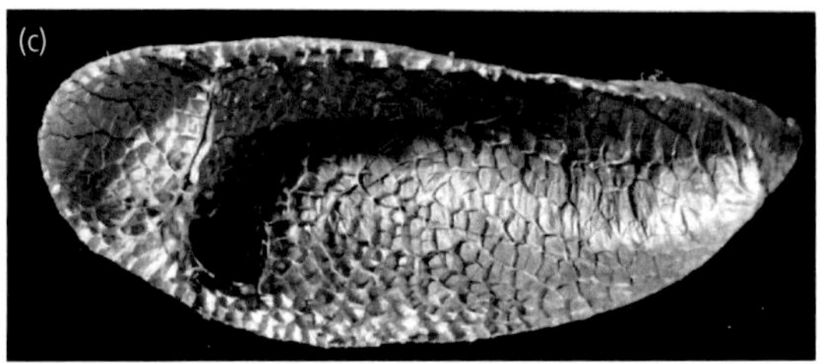

Figure 12.7 Examples of parenchymal types in reptilian lungs. Parts (a) and (b) show faveolar parenchyma. In (a), the prenchyma of the lizard *Podarcis* consists of niches (see also Fig. 12-1) bounded by trabeculae of 1st (T1) and 2nd (T2) order and within the niches, trabeculae of 3rd (T3) and even 4th and 5th order (not labeled) can be seen. Faveoli (F) are separated from each other by interfaveolar septa (IS). In part (b), the highly specialised faveolar parenchyma of the colubrid snake, *Nerodia*, only shows 1st and 2nd order trabeculae, but the faveolar walls possess secondary faveoli (SF) in their upper portions. Scale bars indicate 0.5 cm. Part (c) shows the typical edicular parenchyma of the giant gecko, *Rhacodactylus*. Note the heterogeneous distribution, with trabecular parenchyma in the posterior region (to the right in the picture). The lung is 10 cm long. After Perry, 1998.

two hats, as it were. They are postural, both during standing and locomotion, and also are respiratory. So particularly during locomotion, lizards have a problem breathing and walking at the same time, just when increased ventilation would be necessary. This paradox has been termed 'Carrier's constraint' in honour of David R. Carrier who first described it (Carrier, 1987).

Theoretically, the animals can do only one of these things at a time, but at least the savannah monitor *Varanus exanthematicus* has developed a way to circumvent this constraint. When walking, an auxiliary

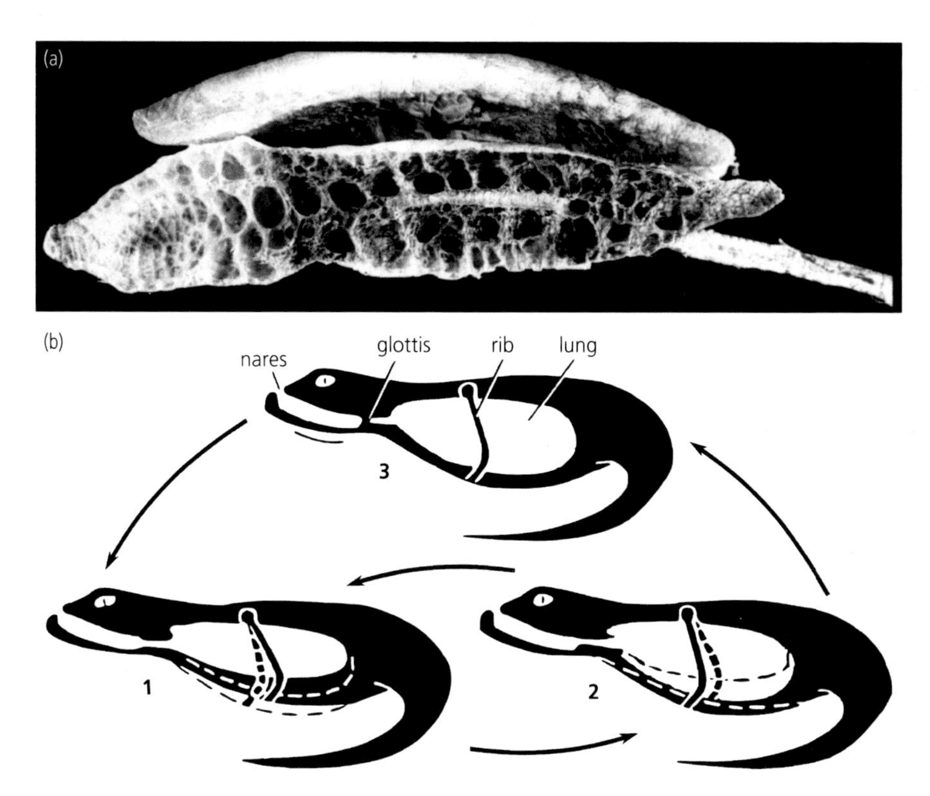

Figure 12.8 (a) Typical multichambered lung of the gila lizard, *Heloderma*, with numerous separate chambers, each communicating with the long intrapulmonary bronchus, clearly visible extending along the centre of the lung. The opened, right lung is 15 cm long. The shorter left lung can be seen in the background. (b), A typical respiratory cycle or a squamate reptile. Dashed outlines indicate the condition existing before the one shown with solid lines. The breathing cycle alternates between ventilatory (1 and 2) and non-ventilatory (3) periods. The ventilatory period begins (1) with expiration followed by inspiration (2). This cycle can be repeated several times before ending in the post-inspiration (3), when the glottis is closed and the non-ventilatory period begins. The plesiomorphic condition is aspiration breathing using the ribs and intercostal muscles is illustrated here. Some reptiles such turtles (Fig. 12-11) and crocodilians (Fig. 12-15) have non-intercostal accessory breathing mechanisms including trunk muscles while some lizards employ buccal/gular pumping, not unlike that illustrated for amphibians in Fig. 12-2. After Perry, 1998; After Westheide and Rieger, 2015.

ventilatory mechanism is used: gular pumping (Owerkowicz et al., 1999). Gular pumping is superficially similar to buccal pumping, already described for amphibians, but involves not only the floor of the mouth, but also the extensive hyoid apparatus of the throat, making the volume of air 'swallowed' independent of the size and shape of the jaws. While gular pumping as a way around Carrier's constraint has been only experimentally demonstrated for one species, given the similar hyoid anatomy in other monitor lizards, it is likely there as well and has also been demonstrated in the agamid *Uromastyx* (Al-Ghamdi et al., 2001). But gular pumping can also serve another function in reptiles. Saxicolous lizards such as *Uromastyx* wedge themselves into rock crevices and use gular pumping to overinflate their bodies when threatened, making them difficult to pull out (Smith, 1946).

While this is the only known active auxiliary ventilatory mechanism demonstrated to date in lepidosaurs we shouldn't count our lizards before they hatch. In other lineages, certain passive mechanisms and structural arrangements exist that significantly influence ventilatory performance. These involve septation of the body cavity. In squamates, especially modification of the mesentery results in coelomic subdivision, which reduces the motility of the viscera during breathing, thereby increasing ventilatory efficiency. While several lineages of squamates, particularly among those that used to be united as

'Scincomorpha', exhibit at least partial, mesentery-derived intracoelomic compartmentalization (Klein et al., 2005), it has been most extensively studied in tegu lizards (Klein et al., 2003a; Klein et al., 2003b; Klein et al., 2003c). There, the mesentery forms a so-called post-hepatic septum, which is instrumental in maintaining 'order' within the coelomic cavity during inspiration. This faculty appears to be intimately connected with the homogeneous, faveolar lungs of these animals, which show the lowest compliance of any reptilian lungs studied to date (compare to earlier discussion).

If we now come back to the monitor lizards, we see another intriguing aspect that is unique among lepidosaurs. We already mentioned pores connecting adjacent chambers of varanid lungs, allowing air to travel a route alternative to the intrapulmonay bronchus. This is reminiscent of the condition encountered in crocodiles (see section 12.3.4.1). Now, in general air flow in lepidosaurs is tidal: air moves in and out of the lung by the same path. Experimental and simulation-based analysis of air movements in the lungs of the green iguana (*Iguana iguana*) showed that the inspired air reflects off the posterior pulmonary wall and flows back along the edges in an anterior direction. During expiration, flow in the anterior direction continues. As attractive as it may sound, just as there is no justification to consider the saccular regions of snake lungs to be related to avian air sacs, we do not consider this peripheral unidirectional air flow to be avian-like (see section 12.3.4.2) (Farmer, 2015; Portugal, 2015). The jet-streaming of air flowing through the iguana lung just reaches the posterior end of both chambers and bounces back along the edges. You basically see the same thing happening when you fill a beer glass (Figure 12.9).

Analogous to that of birds, the savannah monitor on the other hand, does show a special form of unidirectional intrapulmonary airflow since it has a multichambered lung, a long intrapulmonary bronchus, and the intercameral perforations described previously. It has been shown experimentally that

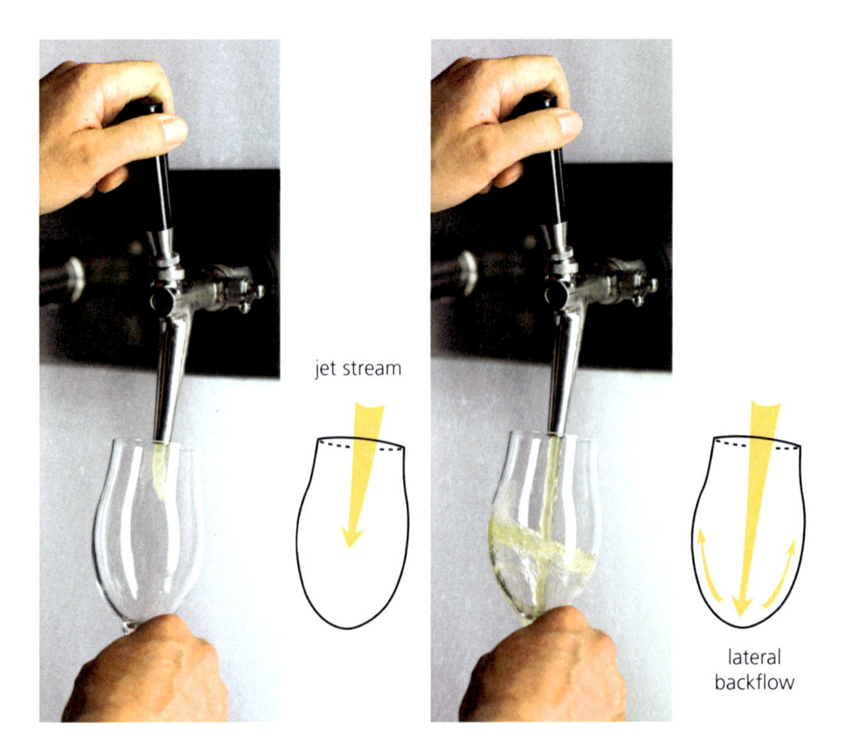

Figure 12.9 Filling a beer glass demonstrates that inflating lung chambers by jet-streaming results in peripheral flow in the opposite direction already during 'inspiration'. Original M.L.

air masses move extrabronchially through the intercameral connections from back to front during inspiration (Schachner et al., 2013). In many varanids the lungs literally wrap around the heart forming a pronounced cardiac lobe (Becker et al., 1989). Such cardiac lobes are also seen independently in crocodiles and mammals. We hypothesize that the primary physiological advantage of this anatomical configuration lies in the intrapulmonary circulation of air masses whenever the heartbeat frequency exceeds respiratory frequency, such as during non-ventilatory periods (see also section 12.3.4.1 on crocodilians). This is really a win–win situation: the mechanical work of the heart is reduced because it is surrounded by air, and the lung gets auto-ventilated by the heart 'for free' in form of a mechanical event-coupling between gas circulation and perfusion.

12.3.2 Testudines

Turtle lungs can be considered as representing sort of the archetype of the branched, multichambered lungs of amniotes. The trachea branches into two extrapulmonary bronchi, which give rise to the two lungs, respectively. An intrapulmonary or first-order bronchus continues within each lung from which a lateral and a medial row of secondary branches arise (Figure 12.10). These generally assume a chamber-like appearance in the adult. In marine (Chelonioidea) and softshell turtles (Trionychidae), these chambers exhibit a rather narrow lumen, enhancing their bronchial appearance. The posterior region of the lung usually ends in a more or less sac-like terminal chamber. The second-order branches or chambers can have additional subdivisions, logically characterized as third-order in the hierarchical system. Different types of third-order branches can occur. The first lateral chamber is known to give rise to one or two discrete subchambers and the proximal region of the lateral chambers contains a distinct medial lobe. The terminal regions of both the lateral and the medial chambers can exhibit additional, usually dichotomous branching events, which persist in the adult as discrete intracameral septa bordering additional lobes. This basic pulmonary bauplan is shared by all turtle species, but certain taxon-specific expressions of these branching patterns also exist (see also Chapter 15).

The parenchymal type that lines the second- and third-order branches in chelonian lungs is primarily edicular and can range from a rather homogeneous distribution (e.g. softshell turtles) up to a highly heterogeneous one in which the distal regions of the chambers are characterized by trabecular parenchyma, resulting in an almost sac-like appearance (e.g. pond turtles). Adaptive responses and a certain degree of plasticity are also evident (Schachner et al., 2017), but apparently restricted to these most terminal regions of developing lungs, which are directly involved in gas exchange.

Histologically, the lungs of turtles in general exhibit the typical reptilian architecture. Early morphometric studies aimed at estimating the morphological diffusing capacity are scarce and do not strictly follow most recent stereological methods (Perry, 1976, 1978). But the pulmonary gas exchange barrier thickness of about 0.5 μm falls within the range seen in other reptiles.

Turtles are unique among extant craniotes because their ribs are incorporated into the characteristic shell. This construction comes at the cost of a complete loss of costal motility and turtles are the only extant amniote lineage in which costal breathing is structurally impossible. But the shift towards this encapsulated bauplan was associated with a fundamental reorganization of the body wall musculature as well, whereby especially the hypaxial group is relevant to the respiratory mechanism (see also Chapter 15). Intercostal muscles are useless in an animal with akinetic ribs: they never fully develop and are lacking in adult turtles. But also the oblique body wall muscle layers have become highly modified and are represented by a single *m. obliquus abdominis* that extends along the posterior flanks. The deepest layer of the hypaxial group, the transverse musculature, which plesiomorphically forms a continuous layer enveloping the body cavity, on the other hand, splits into an anterior *m. transversus thoracis* (which regrettably was named *m. diaphragmaticus* by Ludwig Heinrich Bojanus (1776–1827) in his otherwise unsurpassed descriptive anatomy of the European pond turtle from 1819–1821) and a posterior *m. transversus abdominis*

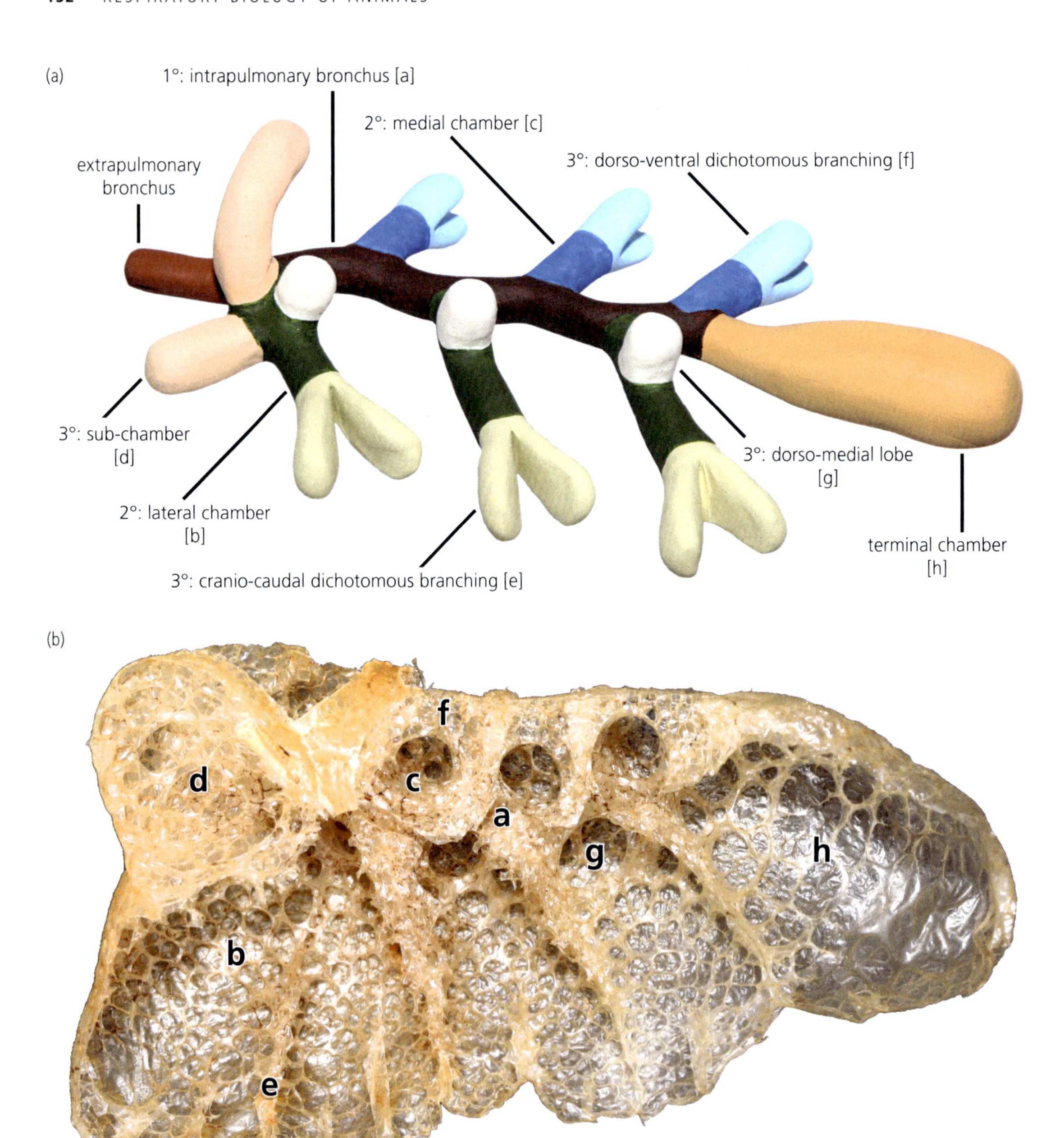

Figure 12.10 Dorsal view of an artistic model (a) and dried left lung (b), opened ventrally, of a slider turtle (*Trachemys scripta*). From a primarily unbranched (1°) intrapulmonary bronchus [a], a secondary sequence (2°) of branching occurs, forming lateral and medial chambers (only 1st chambers, [b] and [c], labelled here), which ends in a terminal chamber [h], may have been present in the multichambered lungs of the first amniotes (see text). Further (3°) monopodial and dichotomous branching events [e] and [f] and formation of unpaired subchambers [d] and lobes [g], respectively, may follow. Original M.L.

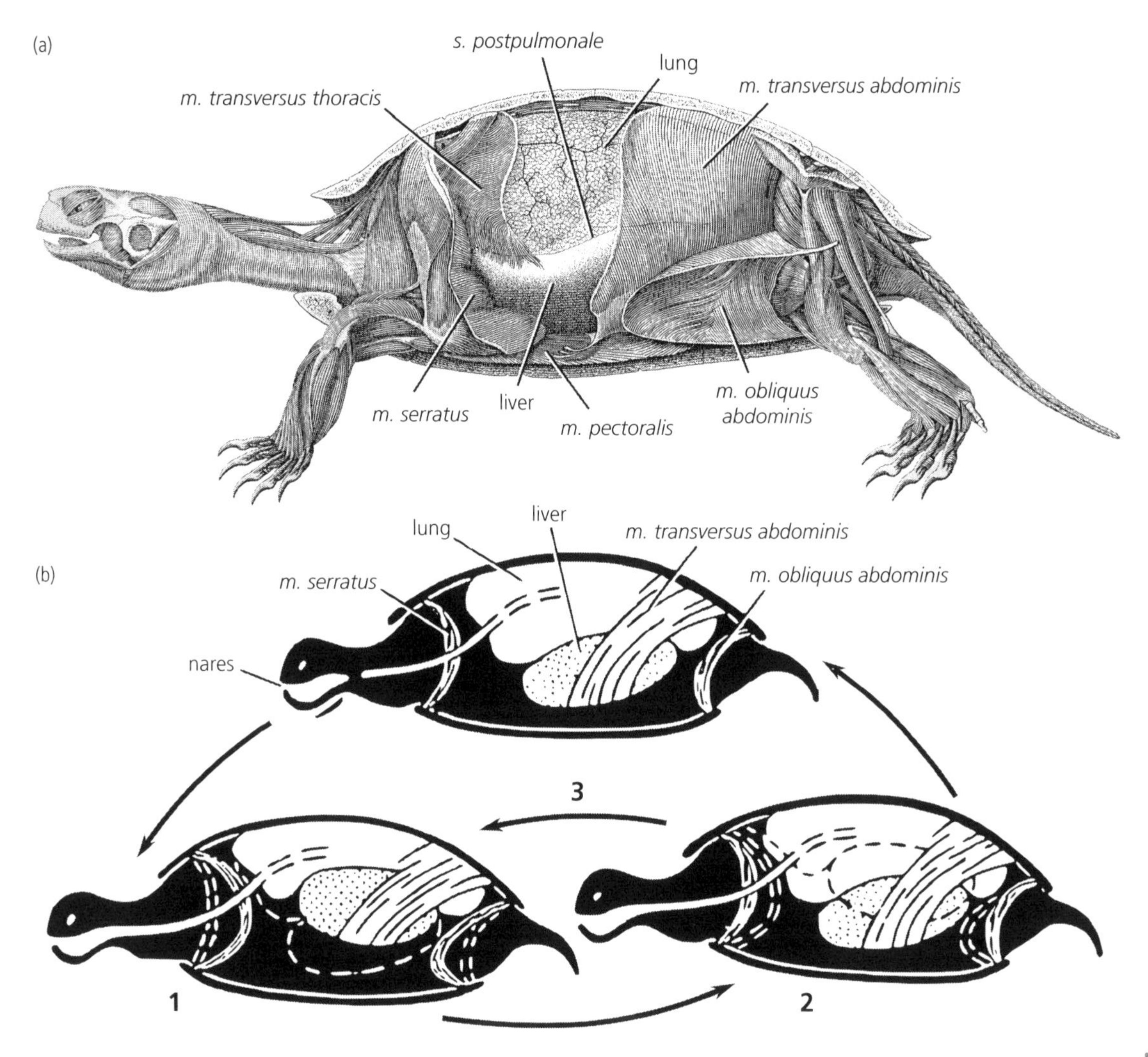

Figure 12.11 (a), Respiratory musculature in the European pond turtle, *Emys* and (b), chelonian respiratory cycle. The main inspiratory (*mm. obliquus abdominis* and *serratus*) and expiratory (*m. transversus abdominis*) muscles as well as accessory breathing muscles (*mm. pectoralis* and *transversus thoracis*) were known to the early investigators, but their exact function, shown in part (b) as well as in Fig. 12.12, was not demonstrated until much later. Part (b), represents a typical respiratory cycle of a turtle. Dashed outlines indicate the condition existing before the one shown with solid lines. The breathing cycle alternates between a ventilatory (1 and 2) and non-ventilatory (3) periods. The ventilatory period begins with expiration (1), effected by contraction of the *m. transversus abdominis*. This is followed by contraction of the *mm. obliquus abdominis* and *serratus*, causing inspiration (2). This cycle can be repeated several times before ending in post-inspiration (3), when the glottis is closed and the non-ventilatory period begins. Oscillatory movement of the buccal floor probably serves olfaction. (a) after Bojanus, 1819-21; (b) After Westheide and Rieger, 2015.

(Figure 12.11a). Of these two transverse muscles, only the latter is found in all species. So now, how do turtles breathe (Figure 12.11b)?

In fact, it took nearly two centuries before the functional morphological insights of Robert Townson on their ventilatory mechanism became universally recognized. In 1799, Townson proposed that the motor of the ventilatory process was exactly the two earlier-mentioned abdominal body wall muscle groups. The oblique abdominal muscle was identified as being responsible for inspiration and the transverse abdominal, for expiration. But it was not until the seminal studies of Carl Gans in the 1960s that Townson's observations, which were based on simple vivisections, could be corroborated electromyographically and the 'belief' that turtles breathe using amphibian-like buccal pumping was proven wrong. Two studies (Gans and Hughes, 1967; Gaunt and Gans, 1969a, 1969b) correlated the activity of several muscles with simultaneous recordings of changes of the intrapulmonary pressure and removed any lasting doubts about this (Figure 12.12).

When the cup-shaped oblique abdominal muscle contracts, it becomes flattened. This decreases the pressure within the shell, including the lung. When the glottis is open, this causes inspiratory air flow in a special chelonian form of aspiration breathing.

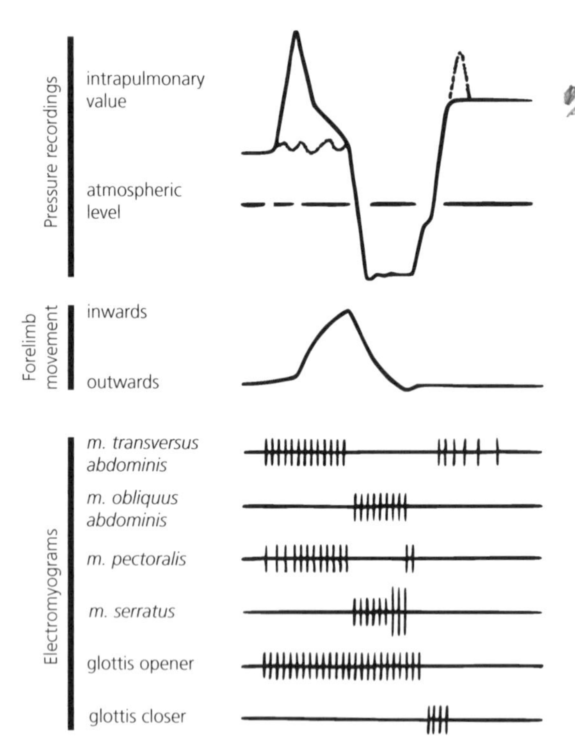

Figure 12.12 Simultaneous electromyographic activity and intrapulmonary pressure recordings in the terrestrial turtle *Testudo graeca*. Increased intrapulmonary pressure clearly correlates with *m. transversus abdominis* activity supported by forelimb movement effected by *m. pectoralis*, and decreased intrapulmonary pressure correlates with *mm. obliquus abdominis* and *serratus* activity. Outward movement of the forelimb also correlates with periods of decreased intrapulmonary pressure. Since the glottis opener is active during this entire period, increased intrapulmonary pressure results in expiration and decreased pressure, in inspiration. Gans & Hughes, 1967.

This process is further assisted by activity of shoulder girdle muscles, particularly the *m. serratus*. In principle, this mechanism can be seen as a functional analogue to the diaphragm-assisted ventilatory mechanism of mammals (see section 12.3.3), whereby the main anatomical difference is that the muscles involved are outside the body cavity. Expiration on the other hand is just like in many other tetrapods: contraction of the transverse muscles presses the viscera up against the dorsally situated lungs, increasing intrapulmonary pressure and—when the glottis is open—resulting in expiration. Since intrapulmonary pressure changes are the result of general changes in pressure within the shell, it follows that retraction of the neck (Mans et al., 2013) and extremities will also effect breathing. In fact, at least during rest, movement of the pectoral girdle assists the ventilatory process.

Summarizing, the general chelonian reorganization resulted in a new functional division of the trunk. The ribs, now being integrated into the rigid shell, are responsible for torsional control during locomotion and are devoid of any respiratory function, whereas the modified body wall muscles have taken over the ventilatory role (see also Chapter 15). But, as we have already seen regarding the influence of retraction of the extremities on intrapulmonary pressure, it is not surprising that locomotion can have a direct influence on ventilatory performance. Any overarching generalizations are difficult to justify though, partly because experimental studies are limited to a small number of species. Marine turtles, for instance, do not breathe at all while moving on land. They have to stop their cumbersome crawling to breathe (Jackson and Prange, 1979). On the other hand, the ventilatory pattern of the fully terrestrial box turtle *Terrapene carolina* seems not to be affected by locomotion at all (Landberg et al., 2003), while semi-aquatic species such as the red-eared slider *Trachemys scripta elegans* appear to lie somewhere between these two extremes: they continue to ventilate their lungs during terrestrial locomotion, but with a reduced tidal volume compared with resting conditions (Landberg et al., 2009).

Finally, it should be mentioned that turtles—although they routinely do not do so for respiratory purposes—can employ buccal pumping. Similar to the situation previously described for the saxicolous lizards, also the pancake tortoise *Malacochersus tornieri* is for instance able to inflate itself using this strategy when hiding in rock crevices (Ireland and Gans, 1972).

So much for breathing on land, but most extant turtle species are at least semi-aquatic. To date, no study has examined the ventilation in swimming turtles in any detail, comparing taxa with different locomotor strategies. All that is known is that turtles can adjust their level of pulmonary filling to control buoyancy (Jacobs, 1939; Jackson, 1969; Milsom and Johansen, 1975; Peterson and Gomez, 2008). Electromyographical recordings are lacking, but it is highly probable that the two groups of transverse muscles are responsible for moving air from anterior to posterior regions of the lung as well as from the left to the right side, possibly in combination with another largely neglected and poorly understood lung-associated muscle: the *m. pulmonalis*, aka *m. striatum-pulmonale* (Hansemann, 1915; George and Shah, 1954).

12.3.3 Mammalia

Probably everyone reading this book already knows that mammals have bronchoalveolar lungs, where branching cartilage-reinforced bronchi and cartilage-free bronchioles lead to alveolar sacs that are composed of clusters of tiny gas exchange units called alveoli (Figure 12.13a, b). In positive casts they resemble bunches of grapes, hence the Latin name 'alveolus', which actually means 'tiny bunch of grapes'. But unlike in reptilian lungs, the airways in mammals are purely gas-conducting structures, covered with 'clearance' epithelium, made up mostly of ciliated cells and mucous-secreting goblet cells. Particularly in terminal regions of the airways, as in reptilian and amphibian lungs, we find embedded in this epithelium, autonomically innervated groups of endocrine-like cells, termed neuroepithelial bodies. The endocrine-like cells have dense-core vesicles, which, in turn, contain serotonin and neuroendocrine active peptides, which are secreted basally in response to local hypoxia but not to hypoxaemia (for references, see Scheuermann, 1987). Connecting adjacent alveoli are interalveolar pores of Kohn, named after the German physician Hans Nathan Kohn (1866–1935). Their function probably is in effectively spreading out the surfactant and they may also play a certain role for intrapulmonary air distribution.

Most mammalian lungs are asymmetrical, the left lung tending to have a more simple branching structure then the right one. The right lung often has a separate lobe, called the cardiac lobe, which wraps around the heart and may participate in cardiogenic air distribution. As already hypothesized for monitor lizards earlier, surrounding the heart by air rather than by soft tissue reduces the mechanical work of heartbeat. Many species have up to 20 generations of bronchial branching before giving rise to respiratory bronchioles, which have some gas exchange function and finally ending blindly in alveolar ducts and sacs. The number of lung lobes is independent of the branching pattern. Bronchial branches located within the lobes are held open by cartilaginous rings. The bronchioles, on the other hand, enter the lobes, lack cartilage, and are held open by their attachment to the parenchyma. During inspiration, air is drawn into the bronchioles, the attachment to the surrounding lung tissue keeping them open. During expiration, the parenchyma surrounding the bronchioles becomes smaller as air leaves it, pulling on the bronchiolar walls and keeping them from collapsing. In people suffering from emphysema, the parenchyma does not get smaller and the bronchioles collapse, trapping the inhaled air and destroying more and more lung tissue. Bad news.

Branches of the pulmonary artery carry deoxygenated blood from the right ventricle of the heart into the lung, closely following the branching bronchi and bronchioles. Upon reaching the alveoli, the deoxygenated blood enters a sheet of capillaries in the alveolar walls. During lung development, a double capillary net is present, with a layer of capillaries facing each side of an interalveolar septum that separates each alveolus from its neighbour. This condition reminiscent of that of most reptiles is maintained in monotremes: echidnas and the platypus. In addition, it is also found in many marine mammals such as whales, dugongs, and manatees, where it is presumed to be useful in storing a large blood volume and providing a large oxygen-carrying capacity (see also Chapter 5). It may also give diving animals some advantage in rapid gas exchange upon surfacing.

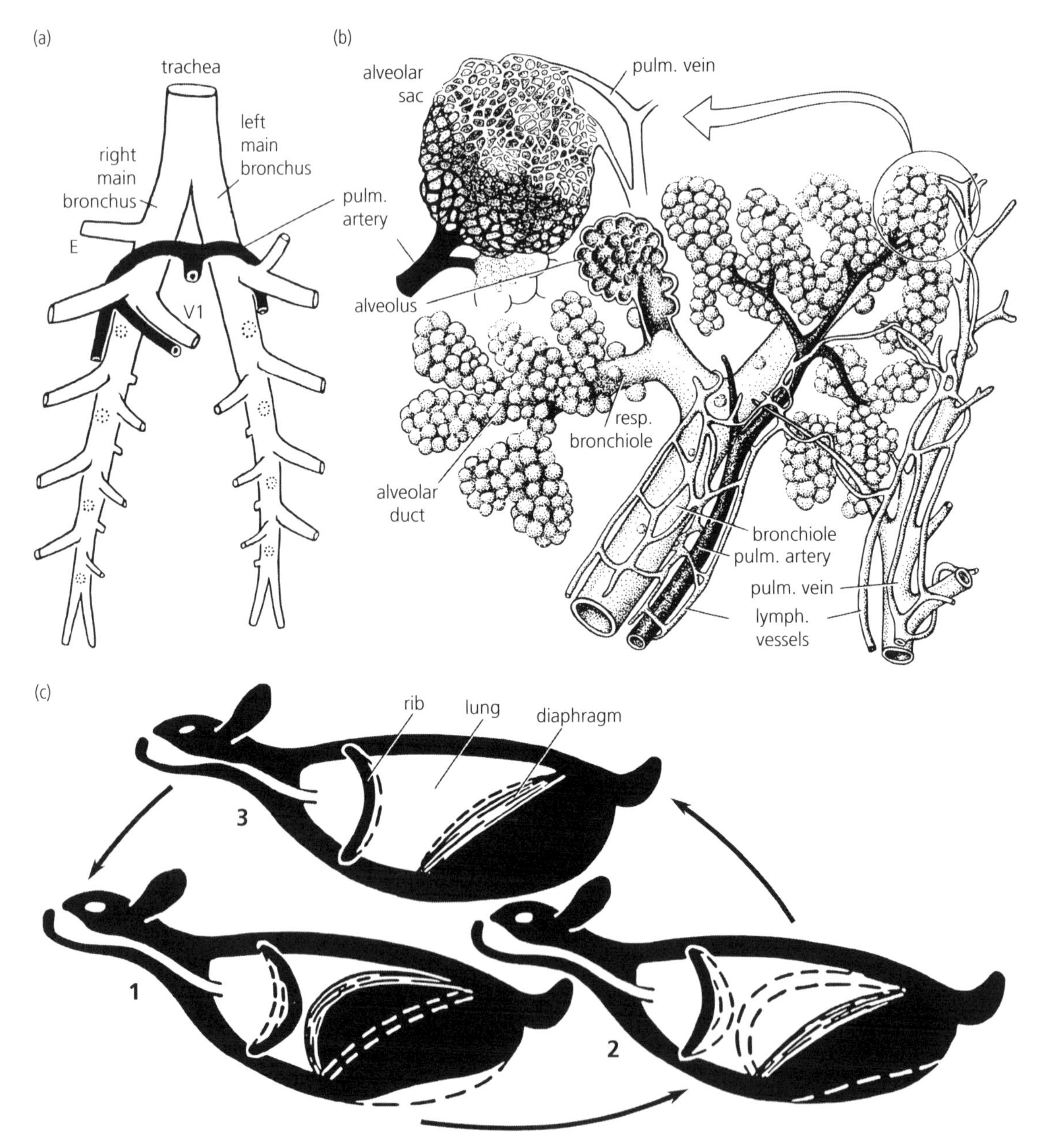

Figure 12.13 Mammalian lung structure ((a) and (b)) and (c) breathing pattern. Part (a) shows in ventral view, the bronchial branching pattern in a rodent. Branch V1 goes to the cardiac lobe, mentioned in the text. The location of the eparterial branch, E, is important for establishing the homology of the first branching generations. Compare the left lung (right-hand side in this ventral view) with the turtle lung in Fig. 12-10. Part (b) is an artist's rendition of the terminal branches of a mammalian lung, showing the alveoli (alveolus) located in alveolar sacs, which connect through alveolar ducts and respiratory bronchioles that have a few alveoli, to the bronchioles. Note that the pulmonary artery lies close to the airway, whereas the pulmonary vein, which contains oxygenated blood, exits distally and lies in the connective tissue separating lung lobes. In contrast to reptiles (Figs. 12.5, 12.11, 12.15), in which the glottis completely closes during the non-ventilatory period, the breathing cycle (part (c)), in most mammals is continuous. Dashed outlines indicate the condition existing before the one given with solid lines. In analogy to reptiles, the breathing cycle begins (1) with expiration, in which the diaphragm is passively pulled forward or during forced expiration, pushed by contraction of abdominal and intercostal muscles. This is followed (2) by an inspiration effected by contraction of the diaphragm and a forward movement and lateral splaying of the ribs. Only a single rib represents the rib cage in this simplified representation. 3 represents the transitory post-inspiratory phase in which the larynx may be partially closed. After Westheide and Rieger, 2015.

The pulmonary veins, as they leave the alveoli, do not follow the same path as the arteries, but rather they go to the periphery of the lobes and enter the interlobular connective tissue. This arrangement prevents contact of small arterial and venous vessels, and makes counter-current gas exchange between blood entering and leaving the lungs unlikely. The larger, lobular vessels run parallel to each other and leave and enter the lung at the point where the airways also enter. This is called the root, or the hilus, of the lung. But even there, the artery and the vein tend to be on opposite sides of the bronchus.

Minimizing contact between afferent and efferent pulmonary vessels appears to be a general structural principle of pulmonary circulation, and is not limited to mammalian lungs. In the peripheral, saclike regions of the lungs of the red-eared slider turtle, for example, small branches of pulmonary arteries and veins tend to cross at angles approaching 90°.

The lungs are the largest surface of the animal in contact with the environment, with all its pollutants and pathogens. The human lung has a surface area about the size of a tennis court, so probably larger than your apartment. So it is reasonable that lungs have an important immune function in addition to being gas exchange organs. Lodged in the pulmonary epithelium are dendritic cells: members of the monocyte cell line that send cellular extensions to the lung surface and capillaries, where they check out air-borne and blood-borne material. They are in contact with alveolar epithelial cells, in particular the type II cells and can influence the surfactant immune response. Mammalian lungs furthermore are supplied by an extensive system of lymphatic vessels and bronchus-associated lymphatic tissue.

The alveoli of mammals—with diameters in the order of 100 μm and less—must be inflated and deflated with each breath. Taking into account what we know about the high surface tension phenomena associated with small radii of curvature (see Chapter 3), without even measuring it one can predict a low compliance and high work of breathing in mammals compared with birds and reptiles, where only the most flexible parts of the respiratory apparatus need to be moved. But mammals have two tricks up their sleeve that allow them to be high-performance animals in spite of the difficult biomechanical hand that they have been dealt. First of all, they have extremely effective pulmonary surfactant. But even taking this into account, the compliance of mammalian lungs still lies one to two orders of magnitude below the values for reptiles and birds. The second trick is the diaphragm. All mammals have one. The functional unit of diaphragm together with intercostal muscles and the completely closed pleural cavities easily allows generation of transpulmonary negative pressures exceeding the colloid osmotic pressure and vapour pressure of blood, which set the limits of meaningful aspiration breathing. All this without any negative effect on the rest of the viscera.

The diaphragm undoubtedly is a key structure for the breathing mechanism in mammals (Figure 12.13c). There are entire volumes dedicated to its functional integration in the respiratory performance of the thorax (e.g. Roussos and Macklem, 1985), so we will restrict ourselves here to a brief summary of its anatomy and principal actions. This dome-shaped muscle, which upon contraction flattens and expands the pleural cavity, is composed of two different structures. Well, actually three. During ontogeny, first a membranous structure derives from the posterior wall of the pericardial cavity and merges with those of the cranial nephric fold to form a so-called postpulmonary septum, which in turn separates the lungs from the remaining viscera. This postpulmonary septum becomes subsequently invaded by musculature, thus forming an active structure involved in breathing. However, it is important to recognize that there are in fact two different muscular units, namely the costosternal and the crural part, respectively. The combination of these two groups of muscles with their different fibre orientation, which in principle can be seen in all mammals ranging from monotremes to us humans, eventually allows for the effective flattening. The diaphragm acts in concert with the remaining musculature of the thorax, in particular the intercostal muscles, in decreasing intrapulmonary pressure and thus generating inspiratory air flow. Expiration on the other hand is largely brought about by elastic recoil of the lungs themselves, but may be supplemented by muscular activity.

12.3.3.1 Gas exchange and aerobic scope in the Mammalia

Mammalian lungs demonstrate the ventilated pool gas exchange model (see Chapter 3), according to which an equilibrium is reached between inspired air and incoming blood. Fully oxygen-saturated blood can never reach atmospheric PO_2 levels and does not exceed that of the end-expired air. Nevertheless, a flying bat reaches oxygen consumption rates similar to those of flying birds in the same weight range (about 35 ml O_2 min^{-1} 100 g^{-1}). Also looking at aerobic scope (difference between resting and maximal $\dot{V}O_2$), mammals measure up pretty well, most of them coming in at around 20, with racehorses and pronghorn antelopes (*Antilocapra americana*) exceeding 60 (Bishop, 1999). Now even this may seem modest compared with 100 in a bee or 300 in a moth, with their tracheal systems. But insects the size of racehorses? Tracheae also have their downside.

Small bats such as *Pipistrellus pipistrellus* also have a large aerobic scope, nearly double that of non-volant mammals in their size range. Given the mechanical disadvantages of the mammalian lung, if bats did not exist one would probably predict that they never could. One of us (S.F.P) suspected that small bats might have well-developed respiratory bronchioles analogous to the avian neopulmo and/or large interalveolar pores of Kohn. However, detailed airway casts and scanning and transmission electron micrographs showed nothing unusual: bats just have huge lungs. Well, that is not all. They are regular flying machines with a very large heart and large blood volumes as well (Jürgens et al., 1981).

12.3.4 Archosauria: what crocodilian reptiles and birds have in common, and what not

As archosaurs, crocodiles and birds have several structural lung characteristics in common (Figure 12.14). The anterior part of the multichambered

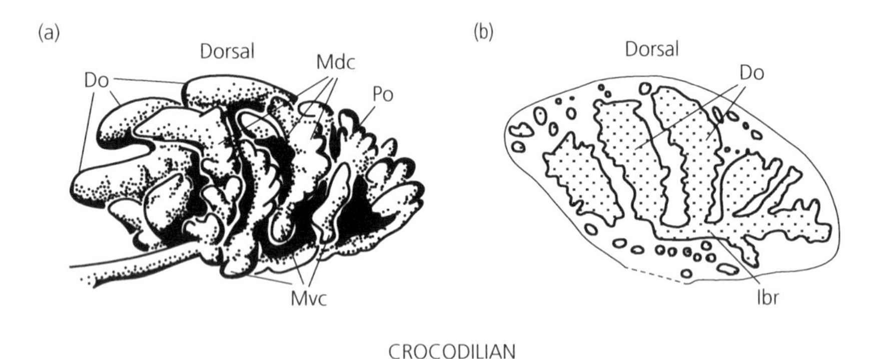

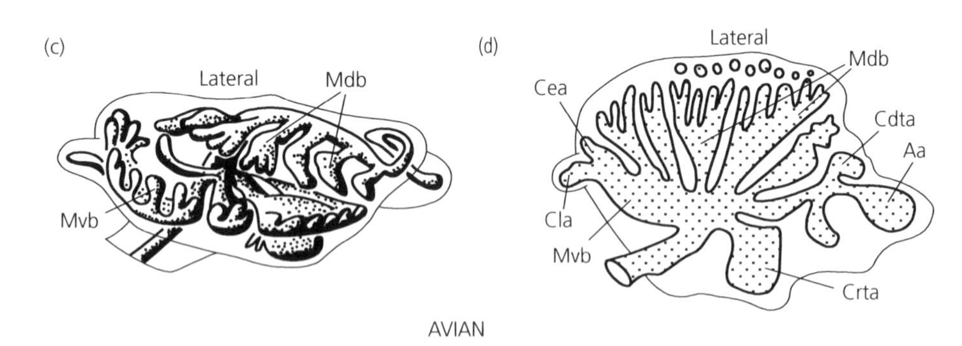

Figure 12.14 Comparison of developmental stages of crocodilian (upper) and avian (lower) lungs. Note the similarity between crocodilian (parts (a) and (b)) and avian ((c) and (d)) lungs in three-dimensional reconstructions or injection preparations ((a) and (d)) and representative sections ((b) and (d)). Air sac development (Aa, abdominal air sac; Cdta, caudal thoracic air sac; Cea, cervical air sac; Cla, (inter)clavicular air sac, Crta, cranial thoracic air sac) takes place only in the avian lung. The left-hand three-dimensional reconstructions show similar monopdial branching patterns in developing crocodilian and avian lungs and division into two regions: an anterior region with a small number of chambres (Do, Mdc, Mvc) in the crocodilian lung or secondary bronchi (Mvb) in the avian lung, and a posterior region with a large number of chambres (Po) or secondary bronchi (Mdv) in crocodian and avian lungs, respectively. Crocodian lungs: Do, dorsal chambres; Mdc, mediodorsal chambres; Mvc, medioventral chambres. Avian lungs: Mdv, mediodorsal bronchi; medioventral bronchi). After Perry, 1989.

crocodilian lung contains a cartilage-reinforced, more or less straight intrapulmonary bronchus that gives rise to rows of secondary bronchi. Because these secondary bronchi diverge sharply without any change in the direction of the main bronchus and are much smaller in cross section than the latter, we speak of a monopodial branching pattern. The openings to the major secondary bronchi tend to spiral around the main bronchus in a counterclockwise fashion in the left lung, clockwise in the right lung. Bird lungs go through developmental stages that are very reminiscent of crocodilian lungs (Locy and Larsell, 1916a, 1916b; Goodrich, 1930; Sanders and Farmer, 2012).

In crocodilians, the secondary bronchi form long, arching chambers that contain the gas exchange surfaces (Figure 12.15a). These surfaces are often perforated, and even the surfaces that make up the walls between adjacent chambers are perforated. The latter perforations allow cardiogenic air flow during apnoeic rest periods (Farmer, 2010). Extrabronchial, intercameral air flow has even been demonstrated

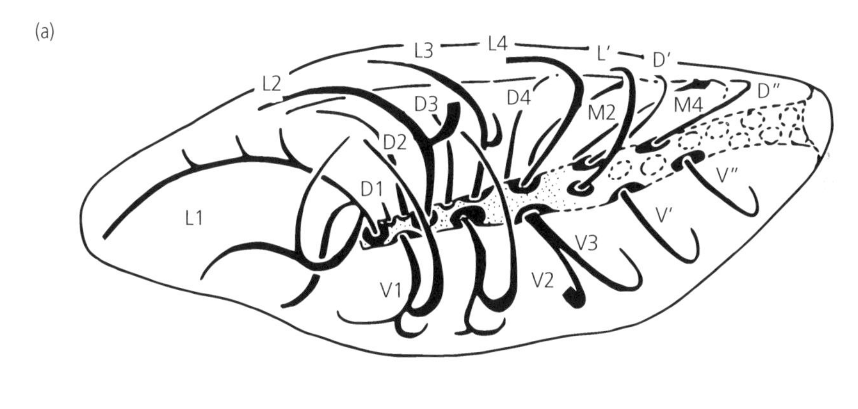

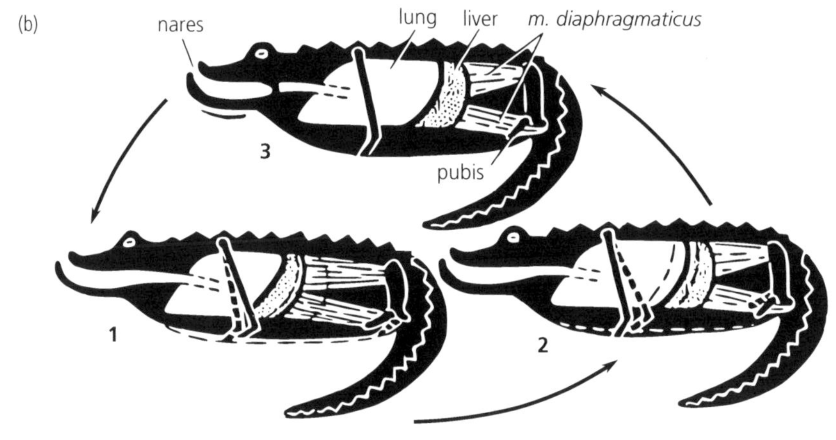

Figure 12.15 Basic lung structure and breathing mechanism in a crocodilian reptile. A, Sketch of the left lung, anterior to the left, in a Nile crocodile, *Crocodylus niloticus*, showing the trajectory of chambres. Thick lines approach the observer. The anterior part (stippled) of the intrapulmonary bronchus is cartilage-reinforced and gives rise to four sets of chambers: dorsal (D1-4), lateral (L1-4), and ventral (V1-3). Medial chambers are lacking in the most anterior part, and only in the second and fourth sets of chambers (M2, M4) are illustrated here. The posterior part of the intrapulmonary bronchus is not cartilage reinforced and gives rise to an irregular array of numerous chambers indicated by hashmarks, ending in a small terminal chamber. Part (b) shows a typical respiratory cycle for a crocodilian reptile. Dashed outlines indicate the condition existing before the one shown with solid lines. The breathing cycle alternates between ventilatory (1 and 2) and non-ventilatory (3) periods. The ventilatory period begins (1) with expiration followed by inspiration (2). This cycle can be repeated several times before ending in the post-inspiratory (3), when the glottis is closed and the non-ventilatory period begins. Costal inspiration may be aided by contraction of the *m. diaphragmaticus* and retraction of the liver, whereby the kinetic pubis is swung back by contraction of the *mm. ischiopubis* and *ischiotruncus* (not labled). After Perry, 1998 and Westheide and Rieger, 2015.

during breathing episodes in several crocodilian species (Farmer and Sanders, 2010; Schachner et al., 2013; Farmer, 2015).

In birds (Figure 12.16), the secondary bronchi do not contain gas exchange surfaces but give rise to tubes called parabronchi, which are surrounded by three-dimensional meshworks called the 'blood–air–capillary net', where gas exchange takes place (Figure 12.17). In the fetal avian lung, parabronchi from posterior secondary bronchi (called the medioventral bronchi) grow toward their counterparts from the anterior secondary bronchi (called mediodorsal bronchi) and merge with them in a so-called plane of anastomosis. Also, the avian respiratory system develops a series of five pairs of air sacs, all of which are lacking in crocodilians. Incidentally, not only birds but also all saurischian dinosaurs are also hypothesized to have had lungs with separate gas exchange regions and air sacs. There is more about this in Chapter 15.

12.3.4.1 Crocodylia

Crocodilians have a unique (but see also pterosaurs, Chapter 15) breathing mechanism that allows them to circumvent Carrier's constraint and also to adjust their centre of buoyancy while in the water (Figure 12.15b). The liver lies posterior to the lungs and is firmly encapsulated by a post-hepatic septum. The posterior margin of this septum attaches to the so-called *diaphragmaticus* muscles (not homologous to the mammalian diaphragm), which extend to the pelvis, and in particular to the mobile pubic bones. During inspiration, the *diaphragmaticus* muscle contracts, pulling the liver with it and causing a negative pressure in the closed pleural spaces, which contain the lungs. When the glottis is open the animal breathes in: if the glottis is closed and the animal is in the water, the lungs are stretched in a posterior direction, changing the centre of buoyancy. At the same time, the contraction of the *ischiopubis* and *ischiotruncus* muscles swing the pubis bone down and back, making room in the abdomen for the liver (Farmer and Carrier, 2000). This so-called hepatic piston mechanism (Gans and Clark, 1976) is unique to extant crocodilians and there is no hint of anything similar in any other group with the possible exception of extinct pterosaurs (see Chapter 15). But having said this, at least in the estuarine crocodile *Crocodylus porosus*, costal breathing is used to such an extent that the hepatic piston can appropriately be considered an accessory breathing muscle rather than the main breathing mechanism. For further details see Munns et al. (2012), also for references to earlier studies.

A recent study on American alligators (*Alligator mississippiensis*) furthermore demonstrated an accessor expiratory function of the *m. iliocostalis* (an epaxial muscle!) mediated through the uncinate processes on the vetrebral ribs when breathing under certain conditions on land (Codd et al., 2019).

12.3.4.2 Aves

The proper designation of a bird's respiratory system is 'lung–air sac system', since it is composed of two anatomically discrete parts: the virtually constant-volume lung and extremely flexible air sacs, which lie outside the lung proper. The gas exchanger consists of a flow-through system of tubes called parabronchi (Figure 12.17). Their walls open to air capillaries, which form a meshwork with blood capillaries. The air capillaries are less than 10 μm in diameter and the air–blood barrier is commonly less than 0.1 μm thick. The air capillaries, because of their small radius of curvature, would certainly fill with fluid from the blood were it not for surfactant produced at the entrance to the blood–air–capillary net and spreading onto the gas exchange surface. And this extremely thin diffusion barrier can only remain stable because the lung is firmly fixed to the curvature of the dorsal body wall and has a virtually constant volume. For more information the reader is referred to Duncker (1971) and Maina (2005, 2017).

Five pairs of extremely flexible air sacs are attached to the lung, four of the pairs being involved in moving air through the lungs (Figure 12.16a). Of these, one pair, the interclavicular air sacs, fuse in most species to form a single sac, so that seven air sacs (three anterior and four posterior) remain to propel air through the lung. During both inspiration and expiration, air moves in the same posterior-to-anterior direction in those parabronchi that connect posterior and anterior regions of the lung: the part called the 'palaeopulmo'. But in most birds, a second group of parabronchi develop (the 'neopulmo') connecting the posterior thoracic air sacs and the abdominal air sacs with the intrapulmonary

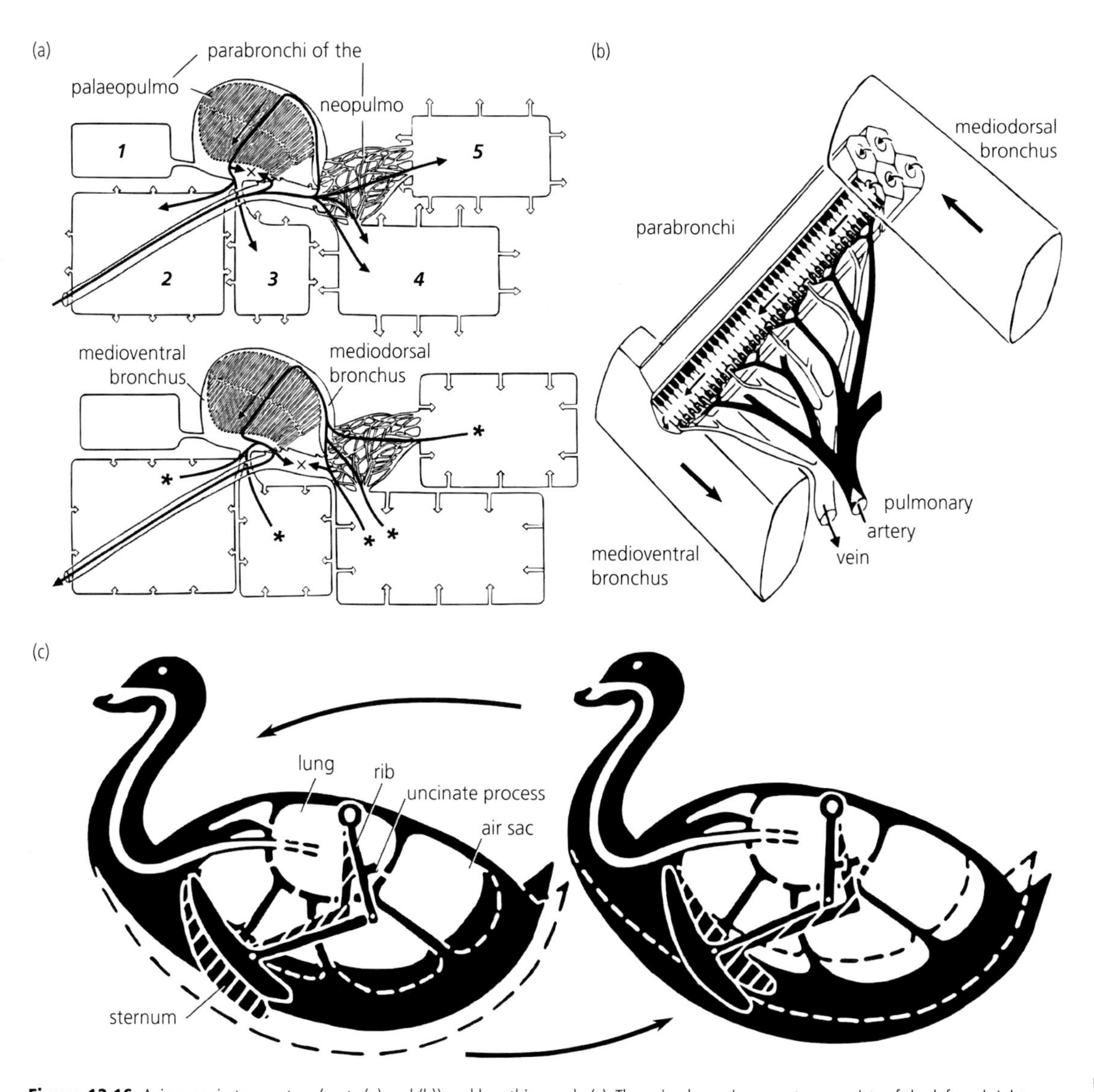

Figure 12.16 Avian respiratory system (parts (a) and (b)) and breathing cycle (c). The avian lung air-sac system consists of the left and right lungs and air sacs. The lungs usually consists of two parts: the paleopulmo and the neopulmo. Each of these parts is composed of flow-through tubes called parabronchi, which are surrounded by a blood-air-capillary network, in which gas exchange takes place (b). For details of the parabronchial structure see Fig. 12.17. The air sacs are paired structures (with the exception of the interclavicular air sac, which is fused to form a single sac in most species) and form two groups: the anterior group, consisting of (2) the interclavicular and (3) the cranial thoracic air sacs, and the posterior group, consisting of (4) the caudal thoracic and (5) the abdominal air sacs. The cervical air sac (1) is not relevant in ventilation. During inspiration, all ventilatory air sacs increase in volume (white arrows in part (a) upper drawing, see also part (c)) and decrease in volume during expiration (white arrows in part (a) lower drawing, see also part (c)). During both inspiration and expiration the parabronchi of the paleopulmo are unidirectionally ventilated whereas those of the neopulmo are ventilated bidirectionally. Black arrows indicate the direction of airflow in (a) and (b). A sheath containing the blood-air-capillary net (part (b)) is supplied with deoxygenated blood by branches of the pulmonary artery. The breathing cycle (part (c)) in most birds is continuous. Dashed outlines in part (c) indicate the condition existing before the one given with solid lines. In analogy to reptiles, the breathing cycle begins (left) with expiration, whereby the ribs are moved forward by lever action of the appendico-costal muscles (not shown) on the uncinate processes, rocking the sternum forward. Only a single rib is shown in these simplified drawings. Expiration is caused primarily by action of the external oblique muscle sheet (not shown), pulling the ribs back. Or during forced expiration, pushed by contraction of abdominal and intercostal muscles. Inspiratory and post inspiratory (right) phases are shown here as combined pending further research. After Westheide and Rieger, 2015.

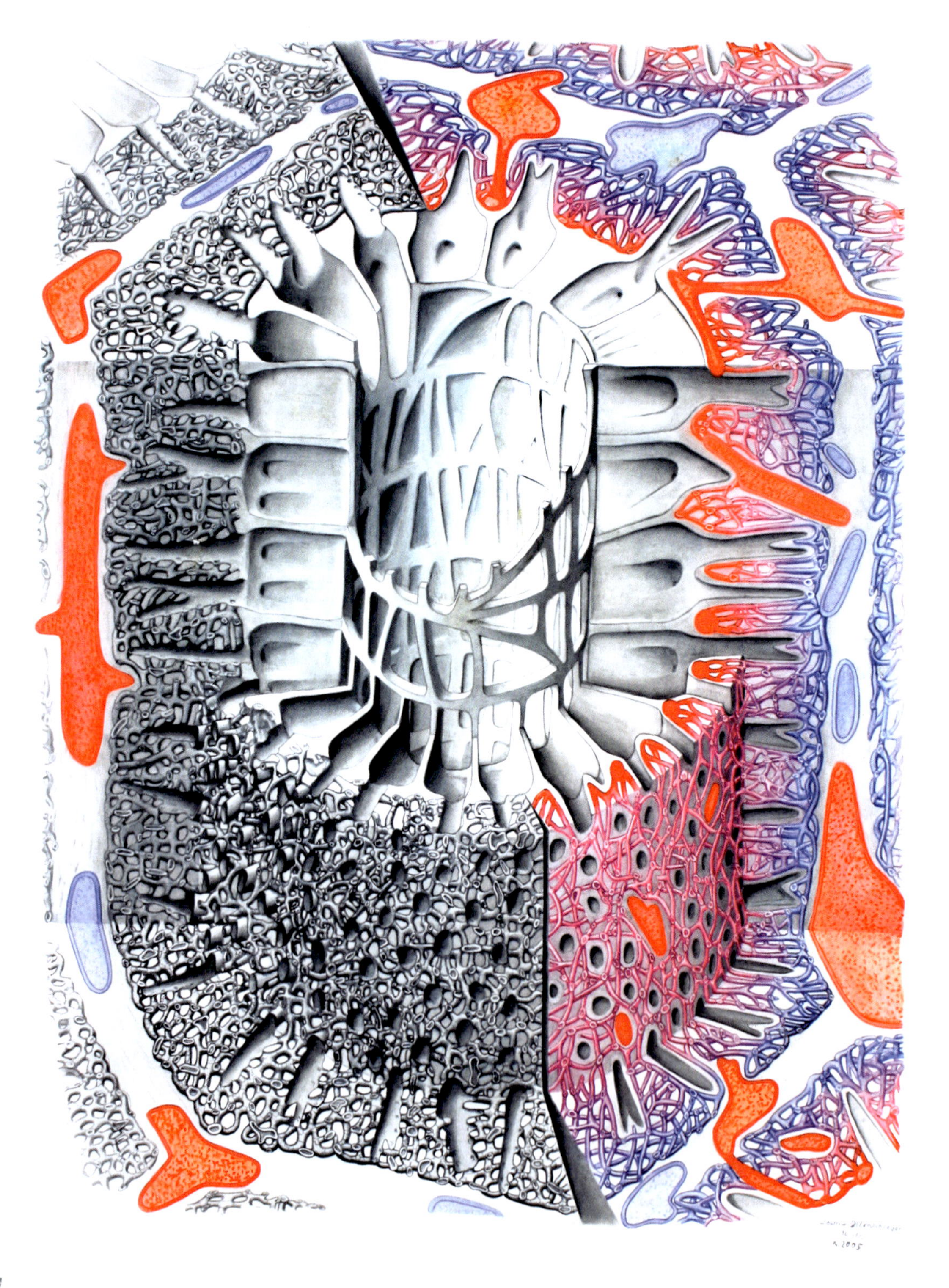

Figure 12.17 Detailed illustration of a parabronchus. Gas exchange takes place in the parabronchial mantle, in which atria give rise to smaller infundibula, which in turn lead to a network of fine air capillaries (illustrated on the left half). These air capillaries are interwoven with blood capillaries (illustrated on the right half). Although avian gas exchange in general is best characterised by the cross-current model (see text and Fig. 12.16), deoxygenated blood (blue) enters these capillaries and becomes oxygenated (red) in a local counter-current-like fashion as it moves toward air of higher oxygen partial pressure. Oxygenated blood is collected near the parabronchial lumen and directly removed by branches of the pulmonary vein, thereby minimizing the chance of oxygen loss by counter-current exchange. Drawing by Beatrice Ollenschläger (Bonn, 1995, modified 2005) after Duncker (1971).

bronchus. In the neopulmo, air moves bidirectionally from the intrapulmonary bronchus into the air sacs during inspiration and in the opposite direction during expiration. But the important thing is that the air flow is directed and the cross-current model is maintained (see also Chapter 3).

Now, let's look at the breathing process in a little more detail (Figure 12.16c). During inspiration, about half of the air traverses the neopulmo and is stored in the parabronchi of the neopulmo and in the posterior air sac group at end inspiration. The other half of the inspired air bypasses the posterior air sac group, traverses the palaeopulmo from back to front, and is stored in the parabronchi of the palaeopulmo and in the anterior air sac group. During expiration, the air from the posterior air sac group is pushed back through the neopulmo in the opposite direction and then displaces the air that is in the palaeopulmo, moving in the same direction as during inspiration, and is at least in part exhaled. The air that was in the paleopulmo and in the anterior air sac group is simply exhaled. And all of this is accomplished without any mechanical valves!

Unfortunately, it is falsely stated in many textbooks that during inspiration air fills the posterior air sac group (without mentioning the neopulmo), then moves to the 'lung' (palaeopulmo?) during expiration, to the anterior air sacs during the next inspiration, and is exhaled in the second expiration. This process conveniently leaves out the bidirectional flow in the neopulmo, thereby emphasizing the 'necessity' of unidirectional air flow and also requires that the palaeopulmo have the same volume as the posterior air sacs, leaving no flexibility for variability in tidal volume. So forget it!

The motor for inspiration is provided by the external oblique muscle and a special division of the external intercostal muscles, called the appendicocostal muscles. The latter muscles attach to lever arms, called uncinate processes, on the posterior surface of the ribs and to the anterior margin of the rib behind it. Their contraction causes the ribs to swing forward and out, depending on the articulation angle. A ventral, jointed extension attaches the free end of the ribs to the sternum, rocking the latter down and forward with each inspiration. The extremely flexible air sacs are carried along with this motion and air is drawn into and through the lungs as described earlier. Because the air sacs are so flexible, birds can't passively breathe out like we can. The forced expiration is done using a complex array of hypaxial muscles (Codd et al., 2005).

Unlike mammals and most reptiles, directed air flow is crucial for breathing in birds and is the key to the cross-current model. But it is not true that the air flow must be unidirectional: remember, about half of the parabronchi, namely those of the neopulmo, are bidirectionally ventilated.

CHAPTER 13

Control of breathing in craniotes

Many convergent faculties exist in the control of breathing not only among various invertebrate groups, but also between invertebrates and craniotes. This convergence is probably dictated by similar requisites of animals in general and of those with elevated levels of aerobic metabolism in particular. Let us assume for the time being that the principle of Hetz and Bradley (2005) applies in general and that craniotes also maintain tissue oxygen levels at the evolutionary Pasteur point (see Chapter 1). This is a reasonable assumption, because high levels of molecular oxygen could result in the formation of ROS, which are detrimental to life but may also have an important messenger function. Some oxygen is necessary for aerobiosis, though. Seen from this perspective, a respiratory system must not just supply oxygen. It must supply the right amount of oxygen. And there is no reason to assume that craniotes are any different from invertebrates in this respect.

13.1 The craniote brain and innervation of respiratory muscles

To explain the central nervous control of breathing, we first need to provide a general description of the craniote brain. Traditionally, the brain is made up of three parts: the forebrain, or prosencephalon; the midbrain, or mesencephalon; and the hindbrain, or rhombencephalon. The forebrain, in turn, consists of the paired cerebrum (telencephalon)—which is the seat of reception and processing of chemical signals in all craniotes and cognition in most—and the diencephalon, which in addition to being the connection between the cerebrum and the hindbrain, is also the origin of light-sensitive and neuroendocrine organs such as the retina, epithalamus/pineal body, hypothalamus, and posterior pituitary gland (neurohypophysis). The prosencephalon is separated from the hindbrain, or rhombencephalon, by the midbrain (mesencephalon) and the metencephalon. The latter is technically the most anterior part of the rhombencephalon and is composed dorsally of the cerebellum and ventrally of the pons. Although prosencephalon-derived centres and optical and acoustic information processed in the mesencephalon can participate in hormonally regulated and descending control of the respiratory system, such as calling in frogs and birds or in human speech, respectively, the main centres for breathing are located in the so-called brainstem. The brainstem is defined as the hindbrain, with the exception of the cerebellum.

For our present purposes, it is enough to know that the brainstem gives rise to between 7 and 9 of the 10–12 cranial nerves that arise from the brain, including those innervating the jaws (V: trigeminal), the gills (VII: facial; IX: glossopharyngeal; X: vagal; XI: vagal accessory), and the hypobranchial musculature (XII: hypoglossal). These parts of the brain also contain the corresponding motor centres (nuclei) for these cranial nerves, which are directly involved in breathing movements. In the brainstem, we are dealing with the most 'ancient', and one of the most poorly understood parts of the brain: the so-called reticular formation. It will therefore only be possible to paint a broad picture of the neuronal control of breathing.

Respiratory Biology of Animals: Evolutionary and Functional Morphology. Steven F. Perry, Markus Lambertz, Anke Schmitz, Oxford University Press (2019).
DOI: 10.1093/oso/9780199238460.001.0001

The control of breathing, be it in ventilation of the gills or of lungs, resides in the brainstem. But as we shall later see, it remains to be clarified if the initiation of breathing is a property of pacemaker neurons in the brainstem or if it relies on external stimuli. In any case, the neuronal pathways that originally controlled gill breathing still appear to be active in controlling ventilation in terrestrial tetrapods, including mammals and birds.

13.2 Respiratory organs

Although a more detailed account of the structural diversity of gills and lungs appears in Chapters 11 and 12, a brief summary of those points relevant to the control of breathing will be given here. To keep things simple for the present purposes, we use the word 'fish' for any craniote that is not a tetrapod. Fish are primarily water breathers, and although many have secondarily evolved air breathing, none have completely eliminated the gills. Fish gills are derivatives of the pharynx and except in some specialized groups, such as adult lampreys, they are unidirectionally ventilated by water flowing in the mouth (or the nose or spiracles as in the case of hagfish or chondrichthyans and sturgeons, respectively) and out through gill slits or opercular opening. Except for the most basally branching lineages of craniotes such as hagfish and larval lampreys, water is moved through the gills by jaw muscles and specialized gill muscles, which are innervated by cranial nerves V, VII, and IX–XII.

Lungs are blind-ending sacs that are derivatives of the posterior pharynx. In buccal pumping of amphibians, many of the muscles involved in lung inflation and deflation are homologous to those used by fish in ventilating their gills. Since amniotes are plesiomorphically rib breathers, the activation of the intercostal and related hypaxial muscles as well as the mammalian diaphragm are coordinated with the activity of the branchiomeric and hypoglossal cranial nerves V, VII, IX, X, and XII, which are primarily involved in regulating airflow at the glottis or larynx.

13.3 Sensing of respiratory gases

Like invertebrates, craniotes must first detect the presence of oxygen and carbon dioxide before the cyclic activity of respiratory muscles can be initiated or modified. Also as in invertebrates, water breathing species tend to use ambient oxygen for this purpose, whereas in air-breathing species the signal comes from the circulating pH or PCO_2. The detection of and reaction to ambient and systemic PO_2 and PCO_2 is complex to say the least and many pieces of the puzzle are the object of ongoing research.

In addition to these primary triggers for breathing, there are a host of other substances ranging from simple divalent ions such as Ca^{2+} and Mg^{2+} to neurotransmitters such as glutamate and gamma-aminobutyric acid (GABA), indole and catecholamines. Many of these substances probably have been around before and throughout the evolutionary history of craniotes and have been incorporated into the complex control of breathing. These substances are too numerous to receive attention here and the reader is referred to recent reviews (Gargaglioni and Milsom, 2007; Baghdadwala et al., 2016). Instead, in the sense of the man looking for his keys, we will highlight just one substance to show the type of influence we are talking about: namely the indolamine serotonin, also known as 5-hydroxytryptamine, or 5-HT.

The role of 5-HT in the control of breathing is a hot topic that definitely needs more attention. To begin with, it is a ubiquitous molecule, being present in protists, plants, and in every animal clade, and even may have been pivotal in the origin of life itself. This indolamine is chemically related to growth hormones (auxins) in plants and serves as a neurotransmitter, neuromodulator, and hormone in animals. With regard to the control of breathing, it is interesting to note that it not only is involved both in the response to ambient and blood-borne PO_2 levels, but also can be released in the brain in response to high PCO_2/low pH levels, as we shall see in the sections that follow.

13.3.1 Peripheral organs that sense oxygen

In fish, special paracrine cells known as neuroendocrine cells (NECs) located in the gills release 5-HT in response to low PO_2 in the ventilated water. The number of NECs increases in fish exposed to chronic hypoxia and decreases under chronic hyperoxia (Jonz and Nurse, 2009). Whereas NECs on gills are well positioned to detect ambient oxygen levels in the water, for responding to oxygen levels in the blood one particularly strategic location is the spiracle, which, as mentioned previously, is used for inspiration in some fish groups and—unlike gills—is supplied with blood that has already passed through the gills on its way to the most oxygen-dependent organs of a fish: the eyes and brain. In most ray-finned fish, the spiracle no longer opens to the outside, and thus detects only oxygen levels in the blood. Nevertheless, it keeps its original blood supply and often a gill-like structure (Waser et al., 2005), and is appropriately renamed 'pseudobranch'. Through NECs in the pseudobranchs, a fish has immediate feedback regarding how well its gills are doing in oxygenating the blood passing through them. It is tempting to speculate that the pseudobranch gave rise to the carotid bodies in tetrapods, which also contain NECs that release 5-HT in response to hypoxaemia (Milsom and Burleson, 2007). But the homology remains uncertain because the carotid bodies are not located at the site of the pseudobranch, but further upstream at the bifurcation of the common carotid artery into the internal and external carotid.

The location of NECs in the adult animal notwithstanding, they are derivatives of ectodermal neural crest cells, which embryologically are migratory, so they can end up almost anywhere in the body. They are found, for example, in all tetrapod lineages, embedded in the clearance epithelium of the terminal airways, where they form paracrine organs known as neuroepithelial bodies (NEBs). NEBs (Figure 13.1) are found on the surface of the smooth muscular trabeculae of amphibian lungs and also in turtles, crocodiles, lizards, and in snakes, where they can form prominent dome-like structures (Luchtel and Kardong, 1981; Perry, 1998). In mammals, NEBs release 5-HT from basally located dense-core vesicles in response to hypoxia but not to hypoxaemia (Lauweryns et al., 1978; Pan et al., 2004). And in birds they are found both in the airways and in the air sacs, where they form sensory organs called racemes. These NEBs are innervated by the vagus nerve (X). It remains to be demonstrated to what extent the 5-HT released on exposure to hypoxia has local action on the pulmonary smooth muscle, stimulates vagal afferents, or has endocrine activity by entering the bloodstream. Although the final word is not yet out, suffice it to say here that NEBs also release neuropeptides, the exact function of which also cries out for further study.

But how do NECs actually detect changing oxygen levels? At least in the fish gill it appears to involve membrane-bound K^+ channels in which an ion leak, analogous to that in the sinus pacemaker cells of the heart, is present. Technically, this is referred to as a K^+ (K_B) channel leak, and is similar to the TASK (TWIK-related tandem pore domain acid-sensitive K^+ channel)-type ion channel in mammalian carotid body. Hypoxia inhibits the K^+ channels, depolarizes the NEC membrane, and is believed to initiate a transient rise in intracellular Ca^{2+}, inducing release of dense-core granules with their contained 5-HT and neuropeptides, and activating vagal postsynaptic receptors (Jonz et al., 2016).

Other proposed oxygen chemoreceptors outside of the gills, such as those of the skin in embryos and the NEC-like cells of the ABOs in adult air-breathing fish, are structurally and biochemically similar to gill NECs, but are not well defined as oxygen sensors. The NECs of the skin of embryonic zebra fish *Danio rerio*, which lack functional gills, appear to be similar.

Relatively little work has been done on the central control of breathing in teleost fish, but in the goldfish, as in lampreys and sharks, the control centres for breathing appear to be distributed along the entire brainstem (Duchcherer et al. 2010). One study indicates that the stimulus for initiation of swim bladder air breathing in the gar pike *Lepisosteus osseus*, an obligatory air breather, is PCO_2/pH, which has no influence on fictive gill breathing in reduced preparations (Wilson et al., 2000; Perry et al., 2001).

13.3.2 Organs that sense PCO_2/pH

The simple answer is: there are none. Or are there many? Aside from the fact that carotid bodies are somehow able to respond to hypercapnia, no identifiable peripheral or central PCO_2/pH-receptor organs equivalent to NEBs have been identified.

The peripherally elicited carbon dioxide/pH response depends on the location and method of application of the stimulus. Applying weak acid, carbon dioxide, or irritants to the nasal epithelium (see Coates, 2001) and the larynx in tetrapods, for example, results in cessation of breathing. But on

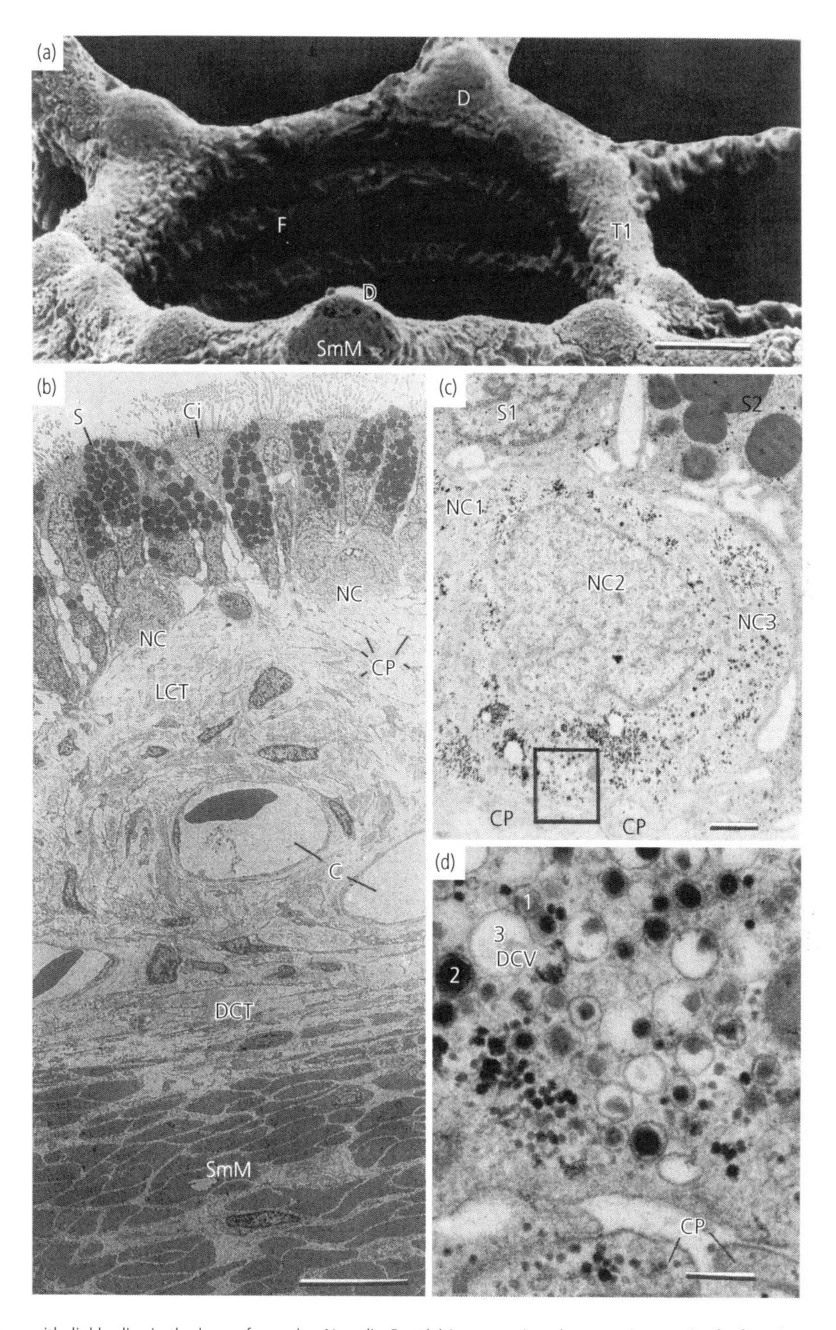

Figure 13.1 Neuroepithelial bodies in the lung of a snake, *Nerodia*. Part (a) is a scanning electron micrograph of a faveolar opening (F) showing the domes (D). First-order trabecula (T1) contain smooth muscle (SmM). Parts (b)-(d) show transmission electron micrographs of a neuroepithelial body and its neuroendocrine cells, NC, indicated in part (c) as NC1, 2 and 3. The square in part (c) is shown in detail in part (d), where basally situated, 5-HT-containing dense core vesicles (DCV) are shown in a putative developmental sequence, DCV 1,2 and 3. Other labled structures are capillaries, C; cell processes that contact the neuroendocrine cells, CP; dense connective tissue, DCT; loose connective tissue, LCT; serous secretory cells, S; and ciliated cells, Ci, of the trabecular clearance epithelium. Scale bars are 100 µm in part A, 10 µm in part B, 1 µm in part (c) and 0.2 µm in part (d). Perry, 1998.

the other hand, inhaling carbon dioxide also lowers the blood pH and dramatically increases breathing frequency and amplitude, due to central carbon dioxide/pH response. In the lungs of amniotes, carbon dioxide detection can be associated with stretch reception, the plesiomorphic condition being carbon dioxide-sensitive stretch receptors, with carbon dioxide sensitivity being maintained in birds and stretch reception, in mammals.

Morphological identification of central carbon dioxide receptors has eluded researchers, but multiple brainstem regions within the reticular formation have been identified that are extremely sensitive to carbon dioxide and can be excited by carbon dioxide application in isolated brainstem preparations in mammals (Nattie, 1999; Nattie and Li, 2009) and frogs. Even a partial list of locations showing carbon dioxide sensitivity sounds like a road map of Italy: retrotrapezoid nucleus, *nucleus isthmi*, *nucleus coeruleus*, *nucleus tractus solitarius*, and the raphe nuclei.

With the raphe nuclei we are back to 5-HT. The serotonergic raphe nuclei (raphe means 'seam') are a row of up to eight anatomically distinct regions, which make up the medial-most part of the reticular formation. It borders medially on the ciliated, ependyma-lined neural canal and laterally on the pons, and ventral respiratory group (VRG), which, includes parts of the *nucleus ambiguus*, the *nucleus retroambiguus*, and the *nucleus retrofacialis* as well as, in mammals, the putative pacemaker neurons of the pre-Bötzinger complex (see also section 13.4). At any rate, central carbon dioxide/pH reception is part and parcel of the control of air breathing.

13.4 Rhythm generation in the craniote brain

Cyclic occurrences, such as saccadic eye movement, shivering, or limb tremor involve neuronal networks, in which groups of neurons that innervate antagonistic muscles are interconnected by mutually inhibitory axons. If one of the groups is activated, it causes contraction of 'its' muscle while at the same time inhibiting those that would stimulate the antagonistic muscle or muscles. After the inhibition has passed, a second group of neurons then inhibits the first one and all others involved in the circuit, while stimulating 'its' muscle. This wave of outwardly directed activity and inwardly directed inhibition continues until it reaches the starting point and begins anew. It has long been assumed that suckling, hiccuping, and other cyclic phenomena that involve the respiratory system are coordinated by such neuronal networks, termed 'half centre models'. This may at least in part be the case and this phenomenon may also be involved in the cyclic nature of breathing.

Since the 1990s, much attention has focused on characterizing the control of breathing in anuran amphibians—in particular, the bullfrog *Lithobates catesbeianus* and the northern leopard frog *L. pipiens*)—and localization of centres in the brainstem that regulate ventilatory activity for gills in the tadpole and for lung breathing in the postmetamorphic frogs (Figure 13.2). It was very soon revealed that even in the adult frog there are two centres—one called the 'buccal' centre and the other, the 'lung' centre—which in intact as well as in isolated brainstem preparations responds to change in the composition of the bathing medium (McLean et al., 1995; Milsom et al., 1999).

Decreasing pH or increasing carbon dioxide in the bathing medium results in increasing the frequency of spontaneous 'lung' bursting activity in the branchiomeric cranial nerves V, VII, IX, and X, just as an intact frog responds by increasing the frequency of breathing episodes when the blood pH decreases. On the other hand, the 'gill' bursts are independent of the pH of the bath but change amplitude depending on the PO_2. The characterization of the neurons involved goes beyond the scope of this book and the reader is referred to recent reviews (Gargaglioni and Milsom, 2007; Baghdadwala et al., 2016) and to the experimental studies cited therein, but suffice it to say that microapplication of glutamate has a strong stimulatory effect and GABA is inhibitory for the 'lung' rhythm (Kogo and Remmers, 1994; Kogo et al., 1997; McLean and Remmers, 1997). The effect of 5-HT is more difficult to interpret: in the isolated brainstem it increases the amplitude of 'gill' bursts dramatically but completely inhibits 'lung' bursts, but see also Kinkead et al. (2002).

Based on what we know about the craniocaudal sequence of branchiomeric cranial nerves in the shark and the broad distribution of rhythm generation in

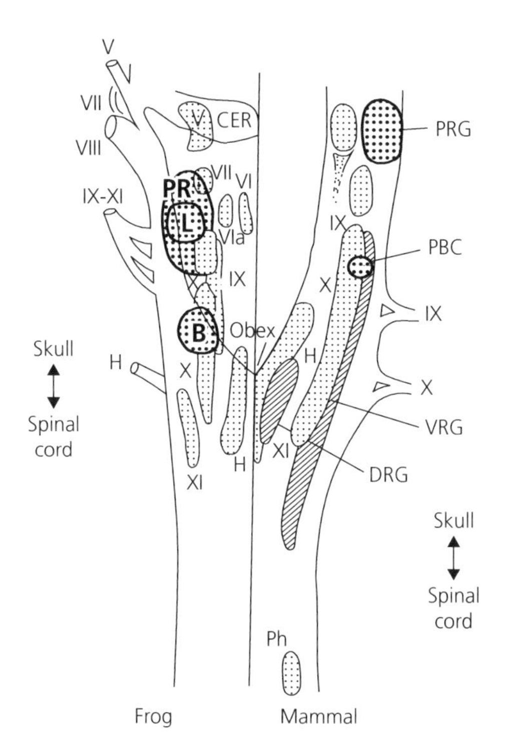

Figure 13.2 Semi-schematic frontal (coronal) section through the brainstem comparing a frog (left-hand side) with a mammal (right-hand side). Roman numerals indicate cranial nerves and their motor nuclei. Note that the hypoglossal nerve is indicated by the letter H. The 12th cranial nerve in mammals, the hypoglossal, is actually the second spinal nerve and the frog, due to its shorter skull. Lining up the profiles using the posterior end of the fourth ventricle (obex) as an orientation point, one sees a similar anterior to posterior sequence in the motor nuclei and also in the structures attributed to stimulation of respiratory muscles: The buccal centre (B), lung centre (L) and the priming centre (PR) in the frog and the Pre-Bötzinger Complex (PBC) in the mammal. Other structures indicated are the cerebellum (CER) in the frog and in the mammal, the dorsal respiratory group (DRG), Ph (phrenic motor nucleus), pontine respiratory group (PRG), and ventral respiratory group (VRG). Original, diverse sources.

the fish brain (Taylor et al., 2010), it was expected that the 'gill' centre would lie cranially, near the motor nucleus of the trigeminal nerve (V), but it actually lies in the most posterior division (rhombomere 8) of the brainstem, near the motor nucleus of the hypoglossal nerve (XII), which in the frog is actually a spinal nerve. In keeping with the insensitivity of the 'buccal' rhythm to pH, this brainstem region is carbon dioxide insensitive. The 'lung' burst generator, however, lies much farther anterior, in rhombomere 5 near the motor nuclei of nerves IX and X (Baghdadwala et al., 2016). Although 'lung' bursts in reduced preparations often begin coupled with a 'buccal' burst, transection of the brainstem between the two centres allows each to carry on independently.

We mentioned earlier that the breathing cycle in craniotes is not two-cycled, but actually three-cycled, with a separate post-inspiratory phase. Recently a 'priming centre' was discovered in the frog brainstem, surrounding the 'lung' centre. Its activity corresponds to the post-inspiratory phase, preparing the brain for the power stroke, which forces air into the lungs (Baghdadwala et al., 2016).

Parallel to these studies on frogs and tadpoles, the neonatal rat was being studied. In the mammalian brain a region of the anterior part of the VRG exists, which Jack Feldman named after a bottle of Bötzinger wine. Adjacent to this 'Bötzinger complex', in the 'pre-Bötzinger complex', a group of pacemaker neurons was found, which initiated the inspiratory act (Smith et al., 1991). More recently, adjacent to the pre-Bötzinger complex, a group of neurons called the post-inspiratory complex was discovered, the activity of which coincides with the post-inspiratory phase of ventilation (Anderson et al., 2016). Perhaps not coincidentally, the lung centre and priming centre in the frog and the pre-Bötzinger centre and the post-inspiratory complex lie in the same part of the brainstem: rhombomeres 7 and 8. In this area of active research, we are far from knowing all the answers. There is some doubt, for example, that the pre-Bötzinger centre is really a spontaneous pacemaker, but rather might consist of neurons that are associated with gasping in response to hypoxia of the preparation (St. John, 2009).

CHAPTER 14

The evolution of water-breathing respiratory faculties in craniotes

One of the characters that distinguished basal chordates from their nearest relatives, the Enteropneusta (acorn worms), was the ciliated branchial basket in the anterior part of their gut. As explained in Chapter 11, the ventral position of the heart(s) as well as its/their location upstream from the gills and the obligatory flow of perfusing fluid through the gill arches before being distributed directly to the body are plesiomorphic characters of craniotes that were already present in non-craniote chordates.

As in Tunicata and Cephalochordata, gas exchange can at least theoretically take place in the gills and some pelagic tunicates even use the water stream produced by the branchial cilia for locomotion by jet propulsion. But the primary function of the branchial basket is undoubtedly filter feeding. The ciliated epithelium provides a constant water flow, but this epithelium in and of itself is so thick as to present a disadvantageously long diffusion distance between the water and the perfusing fluid. On the other hand, at least in the amphioxus, the extremely thin epithelium of the atrium as well as the proximity of extensive coelomic cavities results in an effective site for gas exchange that is comparable with the respiratory mantle of the lamellibranch bivalve molluscs.

But we should not look askance at our humble relatives. Filter feeders do not feed all the time at full speed and the metabolic machinery has to be coordinated with presence of something to be metabolized. And if metabolism and the ciliary activity are aerobic processes, at least for a sessile animal, an automatic coupling of feeding and ventilation of respiratory surfaces is meaningful. The mucous sheet used for snaring suspended particles in amphioxus and tunicates is swallowed complete with mucus, which contains an organic iodine molecule that is homologous to thyroid hormone. Like the latter it stimulates metabolism. 5-HT, which in craniotes stimulates mucus production, is present in the CNS of amphioxus. Now, 5-HT has also long been known to stimulate ciliary activity in a wide variety of animals from protozoans to mammals (Maruyama et al., 1984; Castrodad et al., 1988; Yoshihiro et al., 1992; Wada et al., 1997; Nguyen et al., 2001; König et al., 2009; Walentek et al., 2014); and in amphioxus, ciliary beat is linked to water flow across the gas exchange surfaces. In addition, at least in tunicates, ciliary beat is under central nervous control (Goldschmid, 2007). Connecting the dots we see that from the very beginning of the craniote lineage, ventilatory control paralleled metabolic changes associated with feeding, and 5-HT and possibly even central nervous control have always been associated with control of breathing.

Okay, amphioxus and tunicates are not our direct ancestors, but are probably the closest living relatives of craniotes. A mismatch between the metabolic demand of muscular locomotion and the limited plasticity and maximal ventilatory capability of a ciliary respiratory pump could have been critical in limiting amphioxus and tunicates to a

Respiratory Biology of Animals: Evolutionary and Functional Morphology. Steven F. Perry, Markus Lambertz, Anke Schmitz, Oxford University Press (2019). © Steven F. Perry, Markus Lambertz, & Anke Schmitz.
DOI: 10.1093/oso/9780199238460.001.0001

relatively small size, and planktonic or sessile life histories. Eliminating this mismatch through the origin of muscle-powered gill ventilation could have been a pivotal factor in releasing an evolutionary cascade that resulted in conquest of the aquatic realm by craniotes. Ventilating the gills by muscle power also allowed the branchial epithelium to become thin, since cilia were no longer required, and the gills were relieved of their direct role in food gathering.

14.1 Origin and diversity of craniotes

The origin of craniotes still is somewhat shrouded in mystery and the closest possibly related thing are the conodonts. Conodonts are a radiation of jawless animals that first appeared in the Cambrian and became extinct toward the end of the Triassic (DeSantis and Brett, 2011). They may have evolved from a position on the craniote stem line between our common non-craniote ancestor and extant craniotes. Their paired anterior eyes and wear patterns on the strange teeth and as well as the sheer size (up to 40 cm) of these animals suggest that they may have been predators or at least scavengers rather than sedentary filter feeders (Purnell et al., 1995). So it is likely that the gills of these serpentine animals were dedicated respiratory organs and that a muscle-powered pump under central nervous control was already present before extant craniotes arose. Unfortunately, there are no fossil impressions of conodont gills and it is not known if branchial muscles were present. Until direct evidence is found, we can only speculate that conodonts were at least part of a radiation that used a muscle-powered ventilatory apparatus.

Let us now take a quick look at the extant craniote groups and their possible relationships to each other. Craniota consists of three extant monophyletic groups: Myxinoida (hagfish), Petromyzontida (lampreys), and Gnathostomata (jawed craniotes). We will not concern ourselves with the subgroups of hagfish or lampreys here or whether these jawless creatures together might also form a monophylum. The jawed craniotes fall into two groups: Chondrichthyes (cartilaginous fish) and Osteognathostomata (bony, jawed craniotes), previously called Osteichthyes. The cartilaginous fish have two extant lines: Holocephali (chimaeras) and the Neoselachii (extant sharks and rays), previously called Elasmobranchii. Bony craniotes also consist of two extant lineages: the Actinopterygii (ray-finned fish) and Sarcopterygii (lobe-finned craniotes). The latter have three extant lines: Actinistia (coelacanths), Dipnoi (lungfish), and Tetrapoda (tetrapods), which include the Lissamphibia (extant amphibians) and Amniota (terrestrial craniotes, including ourselves).

14.2 Myxinoida

Assuming that hagfish are indeed a relict branch of the most basal radiation of craniotes, character traits that its respiratory faculty shares with other craniotes are listed here:

A. Gill structure and function.
Pharyngeal gills that serve as gas exchange organs.
Gas-exchange structures with pillar cells.
Counter-current gas-exchange model.
B. Ventilatory mechanism
Ventilated by musculature rather than cilia.
C. Circulatory component
Red blood cells containing haemoglobin.
Gills receive deoxygenated blood from a central (gill) heart.
D. Control of breathing
Sequential, caudally-progressing oscillators in the brainstem.
Sequence starting at the level of the trigeminal cranial nerve (V) for velum activity.
Rhythm modulated by ambient oxygen levels.

As discussed in Chapter 11, hagfish differ so much in practically all of these traits from other craniotes that one could almost postulate their separate origin from a proto-craniote ancestor. But let us take a closer look at these traits one at a time, starting with 'Gill structure and function'. Now, perforation of the pharynx and a ventilatory stream through the perforations were already present in non-craniote chordates, so at least an exaptation by which functional gills could evolve as a plesiomorphy of craniotes is imaginable, even if they were are not used for gas exchange in non-craniote ancestors. But a closer look at the gills brings out some puzzling characteristics. The histological structure of the gas exchange folds with pillar cells in hagfish is very similar to what is seen in other craniotes. But the overall structure with parallel and branching folds is quite different: possibly representing a structural simplification reflecting the life history of these scavengers, which sometimes bury themselves completely in their decomposing prey. This might clog a more complicated gill structure.

Let us now move on to part B, the ventilatory mechanism. It does not involve cilia. The velum as a ventilatory motor in extant adult animals is seen only in hagfish, but see larval lampreys in section 14.3. Hagfish completely lack a gill skeleton in the septa between the gill pouches and an external cartilaginous support in the gill region is present only in some species. Then again they really don't need one since the velum is a positive-pressure pump, quite unlike the ventilatory mechanism in other craniotes, where the filling of the branchial spaces—at least in part—is a negative-pressure phenomenon caused by passive elastic or active muscular dilation of the branchial region. The velum respiratory pump and weak or lacking special superficial branchial constrictor muscles is something that hagfish appear to have in common with Palaeozoic pteraspid armoured fish (Jarvik, 1965), and could in fact be a very basal trait that was conserved in hagfish. The counter-current model might be counted as a physiological synapomorphy of all craniotes, but the way in which it is achieved is not. Of course, the counter-current model as such was not 'invented' by craniotes: we have already seen it independently turning up in molluscs, crustaceans, and maybe even in sand dollars. Its repeated appearance is only evidence for the efficiency of this system and the importance of energy-saving mechanisms in the evolution of a respiratory faculty in general. But as pointed out in Chapter 11, its persistence in spite of major differences in gill structure and ventilatory mechanisms among hagfish, lampreys, cartilaginous fish, and bony fish is striking.

Now let us look at part C, the circulatory component. The presence of red blood cells may be a synapomorphy of all craniotes, but the structure of hagfish haemoglobin differs considerably from that of other craniotes. The fact that a heart pumps deoxygenated blood into the gills is plesiomorphic for craniotes since this arrangement is also present in non-craniote chordates, but the heart itself in hagfish is also quite different from what is seen in other craniotes and must be considered an autapomorphy of that group: the atrium and ventricle are spatially separated and connected by a duct. Also, strictly speaking, the heart should probably be considered a gill heart, since hagfish also have other hearts: paired caudal hearts, paired anterior cardinal hearts, and an unpaired hepatic portal one.

Moving now to part D, control of breathing. Hagfish do not use cilia to ventilate their gills. The respiratory motor, the velum, is innervated by the trigeminal nerve, which in lampreys innervates the most anterior gills and in jawed fished, the jaws. It is not difficult to imagine that this pattern of innervation is ancestral and was exapted in fish lineages that later evolved. The end result—namely unidirectional ventilation of the gills—was maintained while the branchial skeleton, jaws, and associated musculature evolved and indeed the entire ventilatory mechanism changed and became refined in subsequent lineages. A respiratory role of 5-HT is already established in the hagfish. Also, the sequentially and caudally progressing oscillators of the brainstem are carried on all the way to teleosts. Stimulation of the serotonergic raphe nuclei results not only in increases in breathing frequency, but also the 5-HT released there stimulates ependymal ciliary beat and thereby increases circulation of cerebrospinal fluid in the parts of the brain associated with the control of breathing (Stokes, 1997).

14.3 Petromyzontida

Larval lampreys use a velum for gill ventilation (Rovainen, 1996). Although these filter feeders possess ciliated epithelia, they are used only for transport of the alimentary mucus sheet and not for water propulsion. For the time being we can assume that regulation of ciliary beat and its coordination with metabolism is via iodine-containing molecules and 5-HT, as in the amphioxus, but what about muscle-powered ventilation and locomotion? A recent study showed that contraction of muscles that surround the branchial region in response to central PCO_2/pH stimuli that would result when muscles exercise anaerobically, is present in addition to the velum, which powers normal larval breathing (Hoffman et al., 2016). In other words, a respiratory-related neuronal pathway for activating these muscles in response to changes in the internal milieu exists in these filter feeders that are part of a basally derived craniote lineage. The significance of this discovery will become evident when we come to air-breathing craniotes and we shall return to it then.

As described in Chapter 11, gill structure of the adult lamprey is comprised of filaments oriented parallel to the ventilatory water flow with a row of lamellae on each side (Marinelli and Strenger, 1959; Kempton, 1969; Mallatt, 1996). This is similar in larval and adult lampreys in spite of their different feeding behaviour and ventilation mechanisms (Youson and Freeman, 1976), and also very similar to that of sharks. These gills are fundamentally different from those of hagfish though, in which the folds on the filaments run parallel to the filamentar axis.

The final word has not yet been spoken on the phylogenetic relationship among hagfish, lampreys, and gnathostomes, but there is no reason to assume that the ancestor of the latter two groups ever possessed hagfish-like gills. The more parsimonious conclusion is that hagfish gill structure is an autapomorphy of that group, as discussed previously. Impressions of the gill region of Palaeozoic, jawless, cephalaspid fishes, revealing fern-like gills and superficial constrictor muscles, are consistent with this hypothesis and led Jarvik (1965) to even suggest that lampreys are indeed the last survivors of that group. Since a velum is lacking in gnathostomes and present in the other two groups, one may furthermore conclude that the plesiomorphic condition for craniotes was a velum.

The breathing cycle in the adult lamprey appears to begin in the para-trigeminal respiratory group in the vicinity of the motor nucleus (Bongianni et al., 2016) of cranial nerve V. This is responsible for closing the jaw in jawed fishes, but in the jawless lamprey it innervates the first gill pouches. This region of the brain is stimulated by glutamate and is inhibited by GABA and opioids. The lamprey has these characteristics in common with tetrapods.

Although the importance of the brainstem in generating breathing rhythm is well documented, pacemaker neurons have not yet been found there in the lamprey. The reverberating circuit that results in cyclic breathing appears to be distributed along the entire reticular formation of the brainstem, with concentrations near the branchiomeric nuclei V, VII, IX, and X.

14.4 Gnathostomata

Jawed fishes differ from hagfish and lampreys not only in that they have moveable jaws, as the name implies, but perhaps more importantly, in their five-part gill skeleton. Although the embryological origin of this structure as a derivative of the neural crest—those mysterious ectodermal mesenchyme-like cells that arise on the edge of the neural tube—is well known, the precursor of the gnathostome gill skeleton remains a mystery.

14.4.1 Chondrichthyes

Sharks/rays and chimaeras have a curious combination of lamprey-like gills and a completely gnathostome, five-part gill skeleton. Complete holobranch septa are present in both groups, but only in neoselachains does it reach the surface of the body, resulting in separate gill slits for the respiratory units. Embryologically, the gill region in sharks remains uncovered until shortly before hatching and the gills develop filamentous 'hairs' that serve in gas exchange before hatching in oviparous species. So it is not really clear whether the separate gill pockets or the unified gill region, covered by a single membrane, as seen in chimaeras, is the plesiomorphic condition.

The structure and function of the red blood cells and their haemoglobin—in spite of its myoglobin-like dissociation curve—is more similar to that of the Osteognathostomata than of lampreys and hagfish, and particularly the white blood cells and immunoglobulins bear a strong similarity to what we see in bony fishes and tetrapods.

Gill ventilation in sharks begins with closure of the mouth, constriction of the most anterior gill arches, and a wave of constriction moving in an anterior-to-posterior sequence along the gill arches, followed by opening the mouth, and dilation of the branchial region (see Taylor et al., 2010). In spontaneously breathing, resting sharks, gill beat is coupled with heartbeat. Here the motor neurons not of the trigeminal but of the dorsal vagal nucleus (DVN) appear to be positively driven by the activity of neighbouring respiratory neurons (Leite et al., 2014) and the possibility of an afferent cardiogenic stimulus cannot be excluded. We do not see a major change from the condition in the lamprey, since the entire neuronal complex is linked, it is just a question of which part sets the oscillator in motion. In exercising or excited animals the one-to-one coupling is broken, implying that breathing is stimulated by some other source.

14.5 Osteognathostomata

Although a functional analogue is present in chimaeras, a bony operculum must be considered an autapomorphy of bony, jawed fishes, for which it played (and still plays) a crucial role for the ventilatory mechanism. Another shared derived trait of osteognathostomatans is the central venous sinus of their gills.

14.5.1 Actinopterygii

Reduction of the holobranch septum is probably the most far-reaching evolutionary trend one can encounter within this lineage. In the bichirs and sturgeons, it is limited to exposure of the very tip of the gill filaments, since cartilaginous support of individual filaments (lepidotrichia) is lacking, and first appears in the Neopterygii: the sister group to the sturgeons. This group includes the Halecomorphi (the only extant species being the bowfin, *Amia calva*), the Ginglymodi (gars), and the Teleostei, which make up about 96 per cent of all extant fish species.

Teleosts arose, presumably as a marine group, during the Triassic and experienced a dramatic radiation during the Mesozoic, with some groups such as the Osteoglossomorpha and Ostariophysi completely and several others partially recolonizing fresh water. Although all teleost groups experienced some septal reduction, the holobranch septa are completely reduced in Protacanthopterygii, Paracanthopterygii, and Perciformes, the latter alone accounting for 40 per cent of all known fish species. The Ostariophysi, which include carps and catfish, show a septal reduction that is nearly as complete as in the Perciformes and have also separately developed filament adductor muscles. In addition, several groups have the ability to expose or bury their lamellae, depending on the oxygen content of the water. Whether this trait is plesiomorphic or has developed separately among various groups remains to be established. Ironically, some sharks and teleosts have evolved obligatory ram ventilation (see also Chapter 11), whereby the ventilatory muscles are reduced. In addition, convergently in tunas and billfishes, the filaments are immobilized by crossbeams, creating a sieve-like gill structure.

As discussed in Chapter 4, the Root effect also appears to have evolved convergently in physoclist fishes, whereby the physoclist state (loss of connection of the swim bladder to the pharynx) has developed separately in different fish groups.

14.5.2 Sarcopterygii

The lobe-finned fish appear to have started with the same basic endowment as the ray-finned fish did (Figure 14.1), and their gill morphology correlates well with the current phylogenetic consensus: the lungfish are the sister taxon of the tetrapods. The extant actinistians (*Latimeria*)—the sole survivors of the basal-most branching lineage of sarcopterygians—appear to have maintained a central venous sinus in their gills. Such a secondary circulatory system is absent in both lungfish and tetrapods, which instead appear to have evolved a true lymphatic system. The holobranch septa of *Latimeria* and the Dipnoi experienced some reduction similar to that of the actinopterygians, in particular in the coelacanth (Hughes, 1998). However, the gills of *Latimeria* exhibit a rather low surface area, matching the sluggish lifestyle of these deepwater fish (Hughes, 1995). The gills of the dipnoans are utilized differently among the three extant lineages, being almost obsolete in terms of oxygen uptake for the obligatory air-breathing South American lungfish (Moraes et al., 2005).

14.5.2.1 Lissamphibia

The first tetrapods retained their spiracles as adults (Clack, 2012), although it is not known entirely to what extent they were involved in gas exchange with water or air (Graham et al., 2014). The array of gill morphology among larval lissamphibians is great—from simple bumps on the gill arch in the secondary gills of ranid frogs to elaborate, bell-shaped structures that envelop the pre-hatching larva and serve as a pseudoplacenta in *Gastrotheca*.

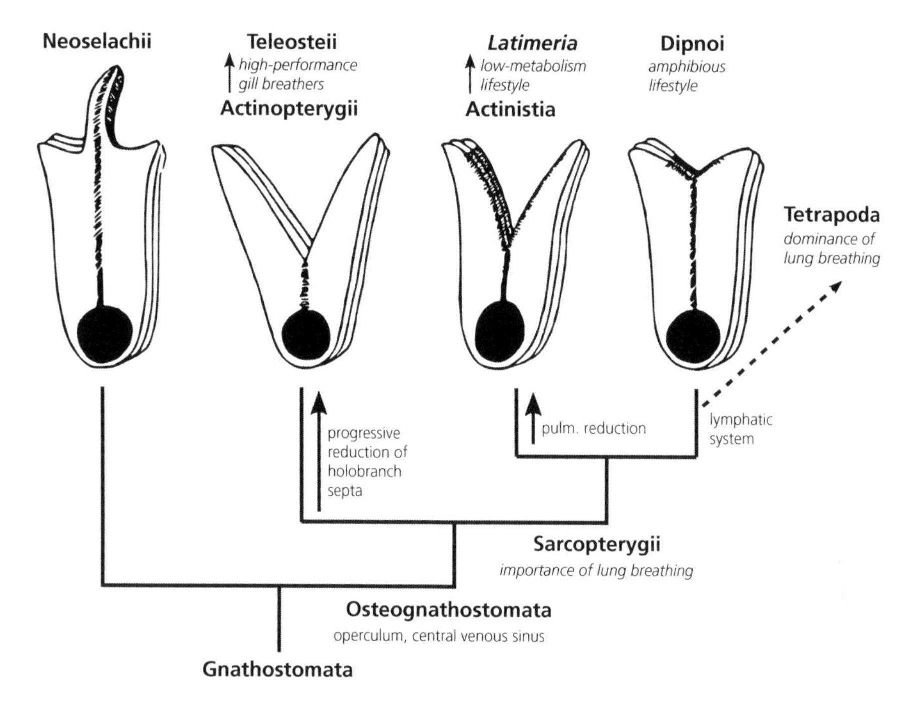

Fig. 14.1 Scenario for the principle evolution of gill structure among jawed fish. Neoselachians maintained a complete holobranchial gill architecture with projecting distal septal ends as integral structural components of their ventilatory mechanism. Their chondrichthyan sister clade Holocephali (not shown) evolved an operculum independently of the main radiation: Osteognathostomata. Within the latter group, a central venous sinus (CVS) is present and we see several reductions of the holobranchial septa, increasing the respiratory surface area potentially available. Within the ray-finned fish, the teleosts show the most advanced septal reductions and a dominance of high-performance gill breathers is found. For the lobe-finned fish, on the other hand, one may assume generally an increased importance of lung breathing. However, concurrent with a secondary reduction of functional lungs, actinistians, represented by the extant coelacanth, evolved a remarkable low-metabolism strategy. On the other side of the sarcopterygian phylogenetic tree, the CVS became reduced and a true lymphatic system appears to have replaced the secondary circulatory system. The bimodal breathing of lungfish allowed for an amphibious lifestyle, in which pulmonary air breathing even can be obligatory. Within tetrapods, gill-based aquatic respiration first became restricted to early developmental stages and lung breathing eventually fully dominant. Potential apomorphic traits are written in normal font; functional interpretations in italics. Schematic drawings of holobranch morphology after Hughes (1998).

Although the gill arches of lissamphibians are homologous with those of fishes, no extant lissamphibians have lamellae and countercurrent gas exchange. This suggests that functional adult gills were lost in lissamphibian ancestors and the larval and perennibranch gills evolved from larval rather than adult fish gills. In extant lungfish and in also in the bichir (Cladistia), the external larval gills resemble those of lissamphibians, in particular of gymnophionans (see Bartsch, 2015).

In addition to gills, cutaneous gas exchange could have been an important mechanism for carbon dioxide release among basal tetrapods. A potential extant analogue, for instance, is the South American lungfish *Lepidosiren paradoxa*. Even the presence of bony scales in stem line tetrapods (Romer, 1972) does not speak against cutaneous gas exchange, as demonstrated by the South American lungfish, in which the scales are covered by an epithelium that indeed is significant in carbon dioxide release (Moraes et al., 2005).

An extreme evolutionary change intimately associated with skin breathing is the repeated loss of lungs, as seen in all three extant clades of lissamphibians. Plethodontid salamanders, for instance, presumably originated in cold streams of the Appalachian Mountains during the Pleistocene (Goin et al., 1978). Under these frigid conditions, dissolved oxygen was abundant and air-filled lungs would have provided disadvantageous buoyancy in fast-flowing water. Since that time plethodontids have become the most species-rich salamander group, with more than 400 species, and have extended their range west to California and south to Honduras. Less is known about the evolutionary cascade that led to lungless frogs and caecilians.

14.5.2.2 Amniotes

Among Amniota, gas exchange with the water is more of a curiosity than an evolutionary trend. In Chapter 11 we saw several examples of sea snakes and aquatic turtles that convergently have adapted to carry out at least some gas exchange with water. The sea snake adjusts its circulatory system using a right–left shunt, preventing loss of pulmonary oxygen to the water during a dive. But several species of turtles from distantly related clades independently developed the ability to extract oxygen from the water using buccopharyngeal papillae, the skin, or the even cloaca as a water lung.

CHAPTER 15

The evolution of air-breathing respiratory faculties in craniotes

Before we set off on our journey through the evolution of air-breathing respiratory faculties in craniotes, we should take a brief look at the systematics within and surrounding the tetrapods, which will be the focus of this chapter. Most investigators agree that within the sarcopterygian (lobe-finned) line, two groups of fishes still exist: the Actinistia (coelacanths), with one genus, *Latimeria*, and the Dipnoi (lungfishes). The latter has two extant groups: Neoceratodidae with one extant species, *Neoceratodus forsteri* (Australian lungfish), and the Lepidosirenidae, with one species of South American lungfish (*Lepidosiren paradoxa*) and several African lungfish within the genus *Protopterus*. The general consensus is that Dipnoi are the sister taxon to the Tetrapoda. Especially during the Devonian, a large number of tetrapods evolved that were more or less capable of walking on land. We will not concern ourselves with the persisting debate about the details of tetrapod phylogeny and are satisfied with recognizing the extant Lissamphibia as the sister group of the true terrestrial craniotes, the Amniota. Three lissamphibian taxa exist (Anura, Caudata, and Gymnophiona), whereby it is generally accepted that the latter group is a sister taxon of the two former ones. Within the amniotes we find a principal subdivision into the Synapsida, the last survivors of which are us Mammalia, and the Sauropsida. Sauropsids include the traditional reptiles as well as the birds. Several competing hypotheses for this line of amniotes still persist, but we are inclined to follow the assumption that the Lepidosauria (*Sphenodon* plus Squamata, snakes, and lizards) are the sister taxon of a clade comprised of the Testudines (turtles) and the Archosauria (crocodilians and the last remaining dinosaurs: birds).

There is no doubt that the lungs are the primary site of aerial gas exchange for all tetrapods. So the first question when discussing the evolution of air breathing in this line of craniotes should be: where do the lungs come from? This question usually is intimately associated with the swim bladder, so let us begin with a critical look at swim bladders and their evolution and then move on to lungs.

As pointed out in Chapter 12, air breathing originated not only among sarcopterygians but also dozens of times among ray-finned fishes. There the ABOs range from gills and skin to highly derived structures such as the labyrinth organs, and virtually everything else that can be brought in contact with air. But in spite of impressive morphological specializations, no actinopterygian has evolved a separate venous return of oxygenated blood to the heart and some separation of oxygenated and deoxygenated blood masses, which characterize the plesiomorphic state for lungfishes and tetrapods. Generally speaking, surface skimming together with air-gulping behaviour, which has been observed in many unrelated species of fish, may have represented an initial stage in the evolution of craniote air breathing.

Respiratory Biology of Animals: Evolutionary and Functional Morphology. Steven F. Perry, Markus Lambertz, Anke Schmitz, Oxford University Press (2019).
DOI: 10.1093/oso/9780199238460.001.0001

15.1 Swim bladder evolution

The problems associated with the origin of swim bladders and their relationship to lungs will be discussed in the next section. But first it might be useful to know a bit more about the structure, function and evolution of swim bladders among the Actinopterygii. To begin with we can assume that this structure is an unpaired dorsal outgrowth of the posterior pharynx. It served as a flotation device from the beginning and was present at least at the base of the Actinopteri: all ray-finned fish except for the basally branching Cladistia. But within the Actinopteri, the structure and function of the swim bladder may differ substantially, which somewhat confounds understanding of its evolution. The second basally branching lineage, the sturgeons (Chondrostei), have developed a non-respiratory swim bladder independently of other fish, but other non-teleost Actinopteri, the bowfin *Amia calva* (Halecomorphi) and the gars (Ginglymodi), possess a so-called pulmonoid swim bladder, which is capable of aerial gas exchange. This organ, just like tetrapod lungs, receives a bilateral blood supply from the sixth branchial arch. Within teleosts, this vascular supply is replaced by an unpaired branch of the dorsal aorta (swim bladder artery), but several do have a respiratory swim bladder. Liem (1989) has proposed that the first teleosts evolved in an oxygen-rich marine habitat, where the branchial arch blood supply was eliminated and the pulmonoid swim bladder was replaced by a non-respiratory one. This organ became secondarily respiratory in fresh-water forms and remains so in many species of the basally branching teleosts, the Osteoglossiformes, whereas others evolved an oxygen-secretion mechanism. A particularly interesting variant is found within the Otophysi (carps, catfish, characids, and their close relatives). Among the characids, a two-chambered respiratory swim bladder originated: the anterior chamber being connected to the otic region of the skull by a series of bones (Weberian apparatus) enhances under-water hearing, while the posterior chamber is for air breathing. Finally the physoclist condition evolved with complete loss of a pneumatic duct that connects the swim bladder with the outside world. But it must be assumed that the physostome condition (pneumatic duct present) is ancestral for teleosts (Faenge, 1983).

Although admittedly a slight digression from respiration, perhaps a few words on the how and why of oxygen secretion are in order. To begin with, it is coupled with the presence of the Root effect in the blood (see Chapter 4), which results in a large release of haemoglobin-bound oxygen in response to a slight lowering of the blood pH. The latter occurs in the gas gland by secretion of lactic acid. The Root effect is also instrumental in supplying the retina with oxygen, and is still present in fishes such as sole and tuna, which have completely lost the swim bladder. Molecular studies indicate that the Root effect and oxygen secretion in the retina date back to at least the Triassic origin of the Halecostomi (Halecomorphi plus Teleostei). But since then, oxygen-secreting swim bladders appear to have arisen several times: within the Elopomorphi (in eels but not in tarpuns), in the Paracanthopterygii (in salmonids but not in smelts), among Otophysi (in the carp lineage but not in electric eels and knifefish, characins, and catfish), and again in the Acanthopterygii (Berenbrink et al., 2005). The latter group—which includes the Percomorpha and thus more than 40 per cent of all fish species—are as adults physoclist, so the swim bladder must secrete gas.

15.2 The origin of tetrapod lungs

Now back on course. With a few exceptions seen among amphibians in which the lungs have been reduced, they are present in all tetrapods and even in some fish. So lungs must be a plesiomorphy of tetrapods or maybe they are even an autapomorphy of Osteognathostomata or Gnathostomata. Right? But where did they come from? This is where the swim bladder usually comes into play. However, in spite of about two centuries of investigation it remains unclear if the lungs originated from the swim bladder, the swim bladder from the lungs, or neither. Recent reviews and textbooks (e.g. Liem, 1988; Kardong, 2002; Roux, 2002; Bartsch, 2004; Mickoleit, 2004) favour the scenario that lungs were present in the earliest osteognathostomes and migrated dorsally to form the swim bladder within the ray-finned fishes. But how sure are we, really? Let us begin with a time-machine trip back some 400 million years and look around.

During the Devonian, the PO_2 of the air was probably about 15 kPa, or 25 per cent lower than today, and in shallow, fresh-water habitats of equatorial regions of present-day Europe, North America, and Russia this would have led to low levels of dissolved oxygen. To make this situation a bit clearer: back then, the sea-level PO_2 would have been about the same as at an altitude of about 2500 metres today. If one takes into account subtropical temperatures and oxygen consumption by aquatic plants and bacteria, night-time oxygen levels could have been quite critical. So these fresh-water habitats could be reasonably compared with those in equatorial regions today: exactly where we still find a plethora of air-breathing fish, but at an elevation of 2500 m, making the conditions for hypoxia more extreme.

During the Silurian and into the Devonian, craniote life consisted primarily of bottom-dwelling, jawless ostracoderms and among gnathostomes, heavily armoured placoderms and the palaeoselachian progenitors of sharks and chimaeras. Sharks—initially fresh-water dwellers—appear to have dealt with fluctuating conditions by becoming marine, and now-extinct acanthodian fishes, which predate the modern Osteognathostomata, were already marine during the Silurian.

In the Devonian, placoderms became dominant predators. At least one placoderm (*Bothriolepis canadensis*) had paired gas bladders (Denison. 1941), which brings us back to another solution for dealing with hypoxic conditions: just to abide in this nutrient-rich environment and breathe air. We do not know for sure if these structures served in breathing (Myers, 1942), but they likely did contain gas. Indeed, it is hard to imagine how heavily armoured creatures such as placoderms could have inhabited the water column without some sort of flotation device. But even if the gas bladders did serve in air breathing, their origin in the anterior pharynx rather than in a posterior location speaks against their homology with the lungs of sarcopterygians.

Fast forward to the here and now and knowing what we do about swim bladders and Devonian fauna, let us ask again: did lungs evolve from the swim bladder, the swim bladder from the lungs, or neither of these? This topic has been dealt with in some detail elsewhere (Perry, 2007; Lambertz and Perry, 2015), so we will just summarize the most salient points here and hope to stimulate the reader to form his/her own conclusion. Or, better yet, to do the critical experiments and conclusively test the lung–swim bladder homology hypothesis using state-of-the-art methods, since many arguments are based on work done in the late nineteenth and early twentieth centuries.

Looking at the extant groups, chondrichthyans have neither lungs nor a swim bladder and embryological studies show little evidence that they ever possessed any, so we most likely can eliminate them. But in bony fishes, the most basally branching groups—among extant ray-finned fish the Cladistia, and among extant lobe-finned fishes, the Actinistia and Dipnoi—have or at least may have had paired organs that originate in the floor of the posterior pharynx (Lambertz, 2017). Swim bladders, on the other hand, with few exceptions originate ontogenetically in the roof of the pharynx. The ABO in cladistian fish, lungfish, and tetrapods have in common a ventral, posterior pharyngeal origin, but only tetrapods have retained the ancestral, complete ipsilateral nerve and blood supply. And even within that group there is no consensus regarding whether lungs are derived from gill pouches 6, 7, or 8 (Wiedersheim, 1909; Makuschok, 1913; Marcus, 1923), or from pouches at all (Greil, 1914).

Now to summarize the arguments about the origin of lungs in a historical context: Goette (1875) proposed, based on embryological studies of anuran embryos, that tetrapod lungs develop ontogenetically and evolved phylogenetically from modified gill pouches in the posterior pharynx. He then went on to speculate that the lungs actually may be derived from the piscine swim bladder. Sagemehl (1885) reversed this hypothesis and Dean (1895) later popularized the idea that ventral lungs could evolve into an unpaired, dorsal swim bladder. To do this, he placed *Polypterus*, *Neoceratodus*, *Erythrinus*, and *Lepisosteus*/*Amia* in a sequence that has no relation to the phylogenetic position of the groups (Liem, 1988), but presumably demonstrated that the pneumatic duct (and, therefore, the lungs) can migrate around the pharynx.

In the first quarter of the twentieth century, hot debates ensued, characterized by different researchers interpreting the same data in different ways. At the

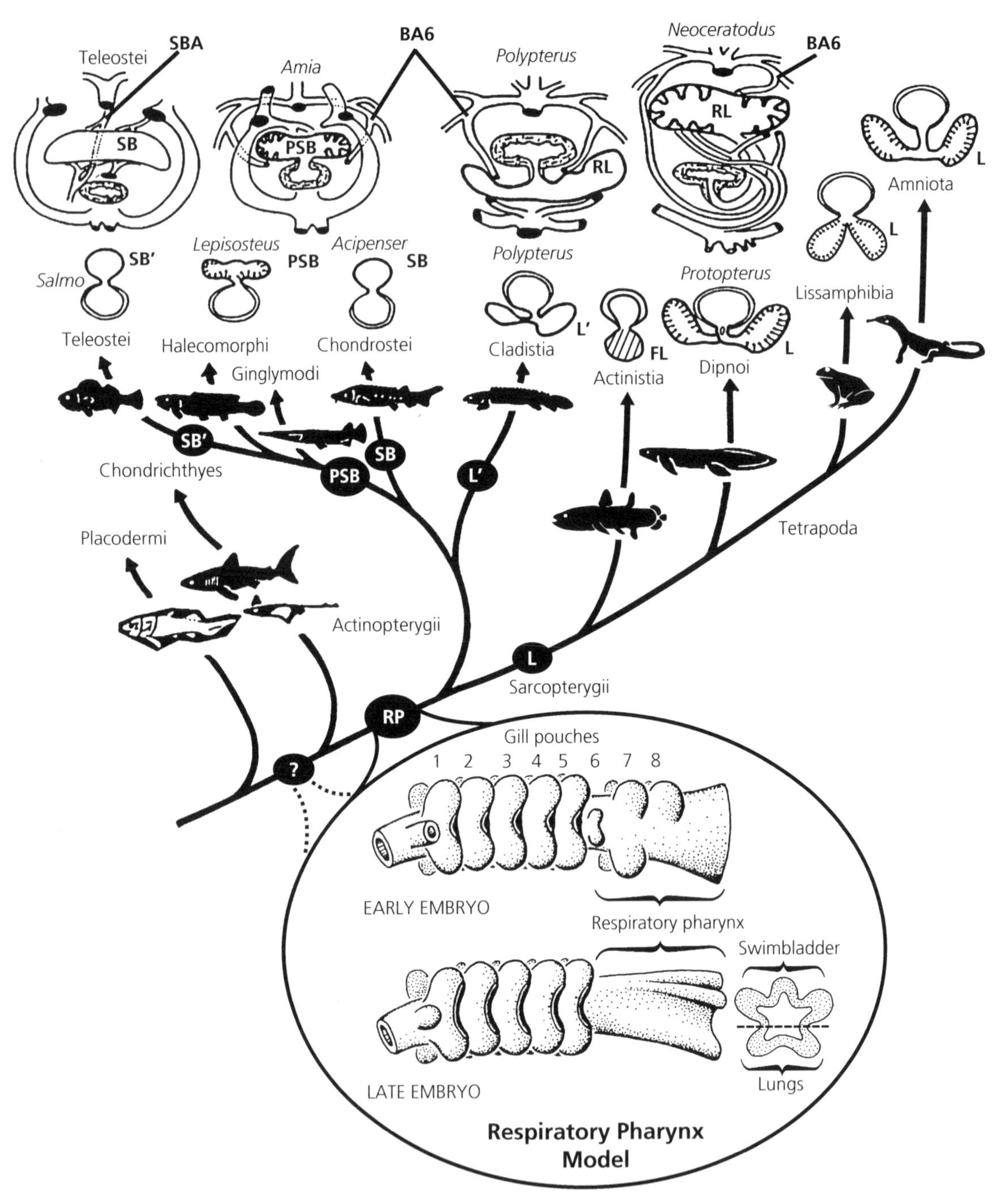

Figure 15.1 Evolutionary directions of craniote lungs and swim bladders. A respiratory pharynx (encircled inset), is seen as plesiomorphic either for all extant gnathostomes (dotted pathway) or for all osteognathostomes (solid pathway to RP, respiratory pharynx, on cladogram). It gives rise ventrally among actinopterygians only in cladistians to lung-like organs (L′), which lack pulmonary venous return to the heart. Among further actinopterygians the respiratory pharynx gives rise dorsally to the pulmonoid swim bladder (PSB), which retains the plesiomorphic blood supply from the 6th branchial arch (BA6), and independently to swim bladders (SB) with arterial supply from a new swim bladder artery (SBA). Among sarcopterygians the ventral part gives rise to lungs (L), in which pulmonary venous return and oxygenated/deoxygenated blood separation at the heart (see text) are present. Partial reduction of the left lung in the actinopterygian *Polypterus* is similar to Australian lungfish, *Neoceratodus*, in which only the right lung remains in adults. In other lungfish groups (here, *Protopterus*) both lungs are equally developed. The actinistian *Latimeria* has an unpaired fat-filled ventral organ (FL) and lacks functional lungs. From Hsia et al. (2013).

end of the day, one camp believed that lungs and the actinopterygian swim bladder were either one and the same organ (Greil, 1905, 1914), or that lungs gave rise to the swim bladder (Piper, 1902), although comparative embryological studies (Moser, 1904) had failed to demonstrate how the lung primordium moves from a ventrolateral to a dorsomedial position. The suggested migration remained speculative and could not be demonstrated to reflect any ontogenetic continuum, and to date, the relationship between lungs and swim bladders remains shrouded in mystery.

A reasonable escape route from this dilemma is the so-called respiratory pharynx hypothesis (Figure 15.1), first developed by Neumayer (1930) and Wassnetzov (1932). The paired, ventral ABO in the basally branching actinopterygian Cladistia and the sarcopterygians, as well as the dorsal unpaired pulmonoid swim bladders, show the same blood supply, innervation, and exhibit a comparable mechanism for the control of air breathing. According to the respiratory pharynx hypothesis, this is because all of these organs are derived from the posterior pharynx, which may have had a respiratory function in basal Osteognathostomata even before it evolved into the organs we see today (Perry et al., 2001). Lungs accordingly formed from a ventral and the swim bladder from a dorsal anlage, respectively. Lacking any hint of continuity at the embryological level, the only way to verify with any degree of certainty if sarcopterygian lungs, cladistian 'lungs' and actinopterygian swim bladders are homologous to each other is to trace the identity of the gene regulatory networks throughout development and we are a long way from that (Lambertz and Perry, 2015).

In conclusion, the question of lung/swim bladder homology boils down to the definition or concept of homology one uses. Lungs and the swim bladder show the same general location of origin and could be construed to be 'homologous' as derivatives of the posterior pharynx. This remains unchallenged and in principle is what the respiratory pharynx hypothesis suggests. However, conclusive evidence that one is actually derived from the other remains to be shown.

Tetrapods and their lungs probably arose during the relatively oxygen-poor Devonian. Probably most fish that did not escape to a cooler marine habitat and persisted inhabiting warm, shallow, vegetation- and microorganism-rich coastal regions had some kind of ABO. Since the pharynx possibly was already being used for breathing and the oxygen-secreting swim bladder had not yet evolved, it stands to reason that pharyngeal derivatives, including some that would later become a swim bladder or lungs, were used for air breathing. There is also some indication that breathing with internal gills and lung breathing were both possible in several early tetrapod lines (Clack, 2012). A neck first came into existence later with the Carboniferous advent of the amniotes. Before that, the head was contiguous with the trunk and the posterior pharyngeal region could easily have given rise more than once to dorsal or ventrolateral pouch-derived grooves, which, in turn, were exapted in the body cavity to form a swim bladder or lungs, respectively. In the water and the tangled, Devonian shorelines, the paired lateral solution would have had the advantage of allowing more manoeuvrability whereas the dorsal swim bladder provides more stability and automatic righting: an advantage in open water.

Remember the air-breathing catfish, *Heteropneustus*, mentioned in Chapter 12, in which unbranched, finger-like projections from the gill chambers penetrate deep into the musculature (Moussa, 1956; Munshi, 1961; Hughes and Munshi, 1973; Maina and Maloiy, 1986)? If you think about it, this could provide some insight into how lungs originated in tetrapods: possibly as a posterior extension of the gill chamber lining. This idea is also in keeping with the proposed embryological origin of amphibian lungs from gill pouches (Goette, 1875) and is also reminiscent of some remodelling sequences we have seen in completely unrelated groups such as molluscs.

15.3 General principles of the respiratory faculties in amniotes

You can read in most standard zoology or comparative anatomy textbook that amniote lungs began as simple sacks with a waffle-like internal structure and independently evolved to the more complex types seen in turtles, crocodilians, birds, and mammals. As appealing as this story is, just the reverse appears to be the case (Figure 15.2). The first amniote lungs almost certainly contained an intrapulmonary

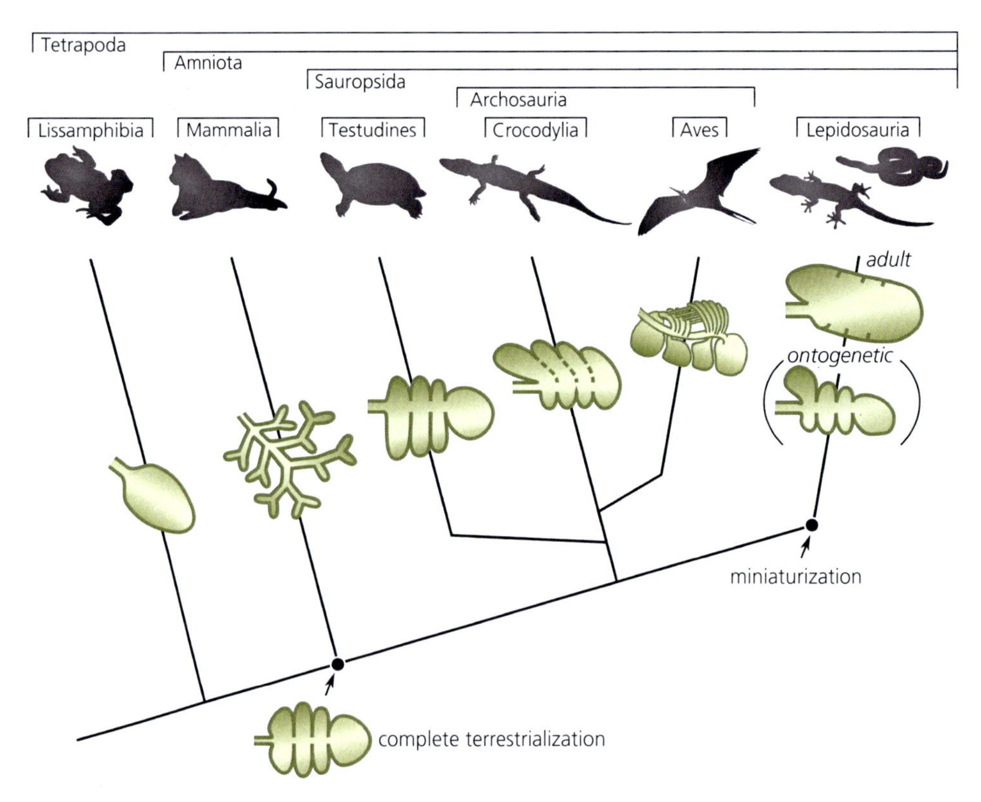

Figure 15.2 Hypothesized sequence of pulmonary evolution among amniotes. Starting with a multichambered, superficially turtle-like lung, we see several taxon-specific modifications and most remarkably, a secondary simplification among lepidosaurs, which is interpreted to have co-occured with a general miniaturization of body size. After Lambertz et al. (2015).

bronchus that supplied two rows of branched chambers, and therefore a rather complex, multi-chambered bauplan (Lambertz et al., 2015).

Amniotes arose during the end of the Carboniferous and looked much like modern lizards. They were of modest size: probably less than 20 cm in length. To put them in context, let us board the time machine again. Looking at the world back then, diadectamorph and seymouriamorph amphibians up to 2 m in length were the dominant terrestrial craniotes, and probably preyed on other amphibians or on fish that ventured onto or near the land. But the relatively small amniotes had many advantages: aided by claws and a flexible neck they could find and catch small, terrestrial invertebrates. Dry-land adaptations such as the lack of aquatic larvae and highly keratinized skin as well as vaulted, narrow jaws meant that from hatching onward they could inhabit upland regions (we are now talking about a time when the oxygen content of the air was more than twice as high as during the Devonian), and climb and dig in search of prey, which they could bite and hold fast with (relative to their small size) powerful jaws. Their hypothesized multichambered lungs could have been decisive in running down prey, in struggling with larger prey or predators, or fleeing from predators. Insects already had evolved a tracheal system, so if early amniotes hunted them, the amniotes would also have benefited from a respectable combination of locomotor ability, cardio-respiratory maximal capability and stamina, and predatory adaptations, also including keen sense organs.

In summary, we hypothesize a multichambered lung structure with a cartilage-reinforced intrapulmonary bronchus that opened to a series of chambers, disposed in medial and lateral rows. However, we should also not forget that—as mentioned in Chapter 12—amniote lungs differ from all other lungs and lung-like organs also at the microscopical level, namely by expressing two distinct types of pneumocytes. These took over different functions, which previously were assumed by a single type of cell only. The type I pneumocytes of amniotes constitute the gas exchange surface, whereas the type II cells are responsible for the production of surfactant. Whether this cell-type differentiation is intimately connected to the intrapulmonary branching remains entirely speculative, but may be worth further examination.

Breathing in amniotes was and still is primarily costal-aspirational, which is brought about by a very complex array of trunk muscles, particularly those of the hypaxial body wall. It may have been supported by some degree of gular pumping. Gastralia (belly ribs) stiffened the ventral body wall, preventing paradoxical inward movement during costal inspiration and enhancing its efficiency.

From this point, as indicated in Chapter 12, amniote respiratory evolution appears to have followed two fundamentally different directions. In general, one can observe (1) a low-compliance, high work of breathing strategy and (2) a high-compliance, low work of breathing one. The first was taken by the Synapsida, the group in which mammals later arose. This seems to have happened without booking a return ticket: once their complex respiratory apparatus evolved, it was and still is maintained. The second direction appears to have occurred repeatedly and in different manifestations. One example is embodied in lepidosaurs, which experienced a secondary simplification of their lung structure (see section 15.3.1.), but here we also see deviations from this strategy. Another example is the spatial and functional separation of ventilatory and gas exchange structures in the avian lung–air sac system.

15.3.1 Lepidosauria

The first lepidosaurs were only a few centimetres long (Evans and Borsuk-Bialynicka, 2009). They survived as the tuatara on a small number of New Zealand islands, and as the most species-rich radiation of reptiles: lizards and snakes (squamates). The tuatara and most squamates have a superficially simple lung structure. But among the lizards, the monitors (*Varanus*) and their close relatives the gila lizards (*Heloderma*) have a multichambered lung structure. So it is easy to construct—as mentioned earlier—a sequence of increasing complexity beginning with a presumably frog-like *Sphenodon* lung, passing through a gila/monitor stage, and ending up with the multichambered lungs of turtles and crocodiles, or even the more highly derived ones of birds and mammals. However, that is too easy to be true. If one takes the ontogenetic formation of even the simple lungs of, for instance, a gecko into account, one sees that they indeed undergo an early branching phase that is highly reminiscent of those of the other amniotes with more complex lungs (see Chapter 12).

We consequently hypothesize that the prevalence of a rather simple, almost sac-like structure of the adult lungs in the majority of lepidosaurs is related to the above-mentioned small body size of the clade's earliest representatives. In animals that size, surface tension problems (see Chapter 3) would have been extremely critical if complexly branching lungs were just miniaturized. A secondary simplification of pulmonary morphology, on the other hand, would have been an elegant way out of this dilemma and opened the door to the evolution of small lizards. Such animals, with a low metabolic demand, would have had an advantage in an environment with low oxygen levels as it persisted during parts of the Mesozoic, and would also have been able to exploit different ecological licences than the rather large-bodied synapsids or archosaurs (see also section 15.3.4.). Extremely small mammals such as the Etruscan shrew *Suncus etruscus* obviously exist as well. So why would body size be an argument here? Well, these shrews can inflate their lungs only because they have completely closed pleural cavities and a diaphragm that can generate the pressures needed to counteract surface tension phenomena. But lepidosaurs do not have these structures. Monitor and gila lizards therefore appear to have built upon the fetally complex structural template and secondarily 'reinvented' a complex lung. Such a re-evolution using an old bauplan is not so unusual: look at amphibian larval gills, for example, and also functional wings that have repeatedly reappeared in highly derived stick insects whose ancestors had entirely lost them (Whiting et al., 2003).

Lepidosaurs in general retained the ancestral multichambered lung during ontogeny and eventually, when the world became a more pleasant place to live, the only giant forms that could evolve with their secondary single-chambered lungs were snakes. The single-chambered lung is ideally suited for them, though, and snakes make up almost half of all reptilian species. Others, as already mentioned, seem to have re-constructed a multichambered lung based on their retained fetal anlage and consequently gave rise to giant Mesozoic sea-faring mosasaurs as well as to extant varanoid lizards. Komodo dragons (*Varanus komodoensis*) can reach a length of over 3 m and weigh 90 kg, and the extinct *V. priscus* was more than twice as large.

Although the tuatara has a very complex body wall structure with ribs that have uncinate processes and several layers of muscles (Figure 15.3), presumably representing something close to the plesiomorphic amniote condition, the thoracic structure of many lizards is simplified (Maurer, 1896; Gasc, 1981). The uncinate processes are lacking as is for instance the superficial layer of external intercostal muscles. Snakes and lizards use costal, aspiration breathing, which as described in Chapter 12, becomes a problem when the animals are walking. The phenomenon known as Carrier's constraint results, because the same hypaxial body wall muscles are used in breathing and in locomotion, and just when extra lung ventilation is most needed, it would be constrained. One main factor in the evolution of the respiratory faculty in amniotes in general and squamates in particular is the mechanisms they have evolved that constrain this constraint, as it were.

Among lizards, species-specific points of adhesion of the mesentery often restrict the movement of the viscera, thereby increasing the efficiency of breathing (Klein et al., 2005). Restriction of visceral movement means that the lungs can be exposed to greater negative pressure during inflation without obstruction, which in turn means that the lungs can develop more internal surfaces, finally resulting in decreased compliance but greater aerobic capabilities for the animal. This viscous circle culminates among teiioid lizards in the mesentery of the tegu lizard *Salvator merianae*, forming posterior to the lungs and liver an almost complete, diaphragm-like

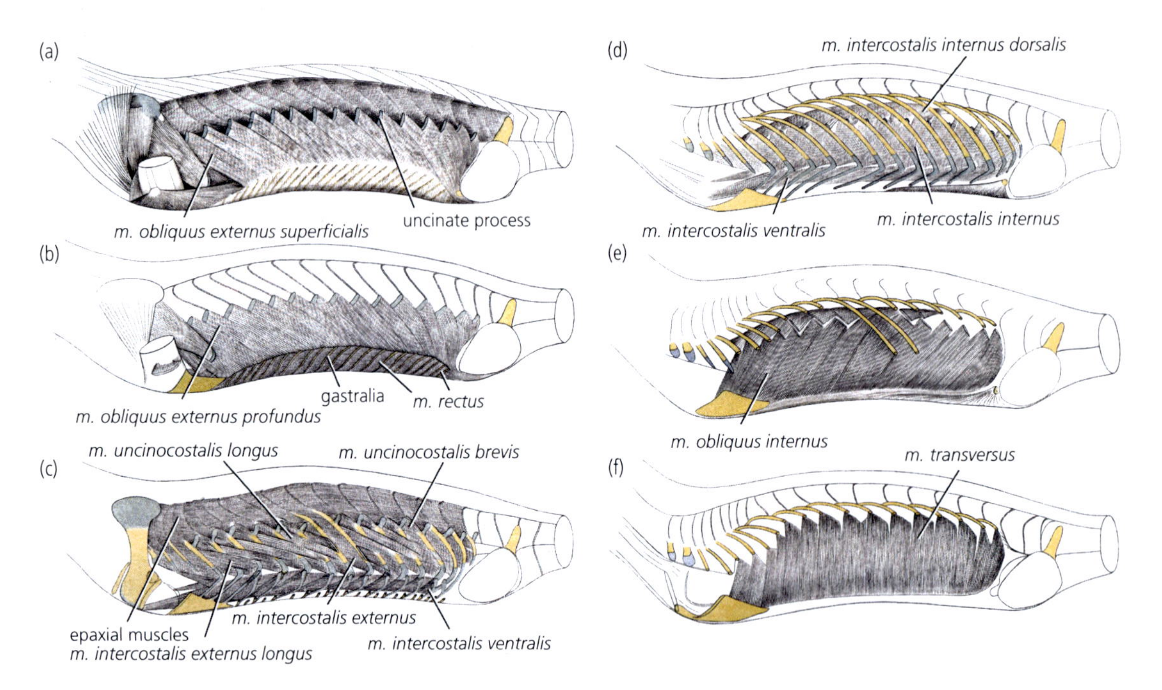

Figure 15.3 Hypaxial body wall muscles of the tuatara, *Sphenodon punctatus*, viewed from the left side. This representative of the most basal extant radiation of lepidosaurs shows the most complete body wall musculature of any amniote, being reduced or specialised in other groups. Sequential dissections of body wall layers, beginning (a) with the most superficial and ending (f) with the deepest. Note the presence of cartilaginous uncinate processes, indicated in blue, and the associated musculature (gray tones) on the ribs (yellow) as well as gastralia (see part (b)): structures associated with postural, locomotor and respiratory function. After Maurer (1896).

wall called the post-hepatic septum. While this septum cannot contract with each breath, it does contain significant smooth muscle, which presumably can adjust its tension.

Snakes went their own direction with regard to lung and body wall evolution. Only representatives of some radiations such as pythons and boas have retained two lungs. Most others have only the right lung along with its vascularization and innervation, whereas the left lung does not develop at all or remains vestigial. Snake lungs are also typically highly heterogeneous, with the posterior part and the most anterior part being thin-walled sacs with no pulmonary vasculature. They serve in air storage, as a buoyancy organ, a 'pneumo-skeleton', a place holder for accommodating ingested prey and in communication. The heart usually lies where the tracheal bifurcation used to be, and anterior to that many species have modified the trachea to contain deep faveolar parenchyma that is identical to that of the lung proper, viz. the tracheal lung mentioned in Chapter 12.

Snakes are of course particularly prone to Carrier's constraint, but their lungs usually are firmly attached to the body wall musculature and directly affected by every rib movement. Although serpentine locomotion can also move air in and out of the lungs, it will probably also move stored air to and fro (Donnelly and Woolcock, 1978).

15.3.2 Testudines

The lungs of turtles probably are the closest approximation of the ancestral amniote pulmonary bauplan (see Chapter 12), which makes them of great interest for understanding pulmonary morphology and evolution in amniotes in general. However, also among turtles the respiratory system tells an interesting evolutionary story, which begins with the extrapulmonary airways. The plesiomorphic arrangement of the trachea is in a mid-ventral cervical position and also the Pleurodira (side-necked turtles) maintain this condition. The Cryptodira (hide-necked turtles), on the other hand, exhibit a displacement of the trachea to one side of the neck. This must be assumed to have evolved in concert with the neck retraction mechanism as an adaptation that keeps the trachea patent. Tortoises truly are the experts in retracting their neck and the length of the trachea even can be greatly reduced (e.g. *Testudo*), whereby the elongated extrapulmonary bronchi run along both sides of the neck.

However, the lungs themselves are also of interest in an evolutionary context, especially with regard to inferring the phylogenetic relationships within the Testudines. The earlier-mentioned multichambered bauplan notwithstanding, a huge structural diversity of this organ exists among turtles. The principal branching pattern of the lungs seems to be taxon specific, highly conserved, and thus phylogenetically informative.

As an example (Figure 15.4), based on a combination of information on the developmental and adult anatomy of the respiratory system, one long-lasting controversial issue in chelonian systematics recently received the first morphological support for the robust molecular-based consensus topology. The sole extant representative of the Platysternidae, the Asian big-headed turtle *Platysternon megacephalum*, shows an intrapulmonary branching pattern that more strongly resembles that of the Testudinoidea (tortoises and pond turtles) than Chelydridae (snapping turtles) (Lambertz et al., 2010). This is in full agreement with the molecular data concerning the placement of *P. megacephalum*, but contradicts the majority of osteology-based morphological evidence.

Our knowledge about the comparative pulmonary anatomy of turtles still rests on the early studies of Milani (1897) and Gräper (1931). Since both authors lacked our present understanding of evolutionary character transformation, a modern systematic revision is long overdue. The high degree of anatomical diversity in chelonian lungs makes them a potential source of additional phylogenetically pertinent data, which could to help resolve conflicting molecular versus traditional osteology-based conclusions, but this has yet to be fully exploited. Particularly the latter approach is known to exhibit a high degree of homoplasy. Ongoing anatomical studies on chelonian lungs (M.L.) appear to corroborate this assumption and research continues.

In contrast to the exchanger, which is assumed to come close to the plesiomorphic amniote condition, the active pump has diverged dramatically from the ancestral state (see Chapter 12). As we shall see, this modification was not so much a result of the

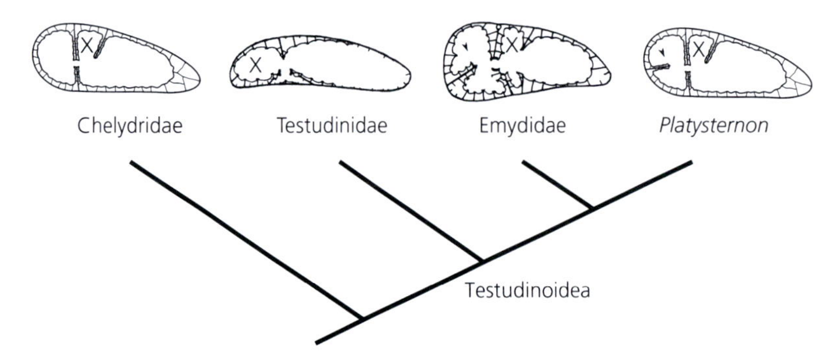

Figure 15.4 Phylogenetic relationships of selected cryptodiran turtles as inferred from pulmonary morphology. The X in the schematic cross sections indicates the medial lobe of the lateral chambers and is used for reference. Note the septum in the medial chambers (indicated by arrow heads) of the Emydidae and the Asian big-headed turtle (*Platysternon megacephalum*), suggesting a closer affinity due to a shared developmental branching pattern. After Lambertz et al. (2010).

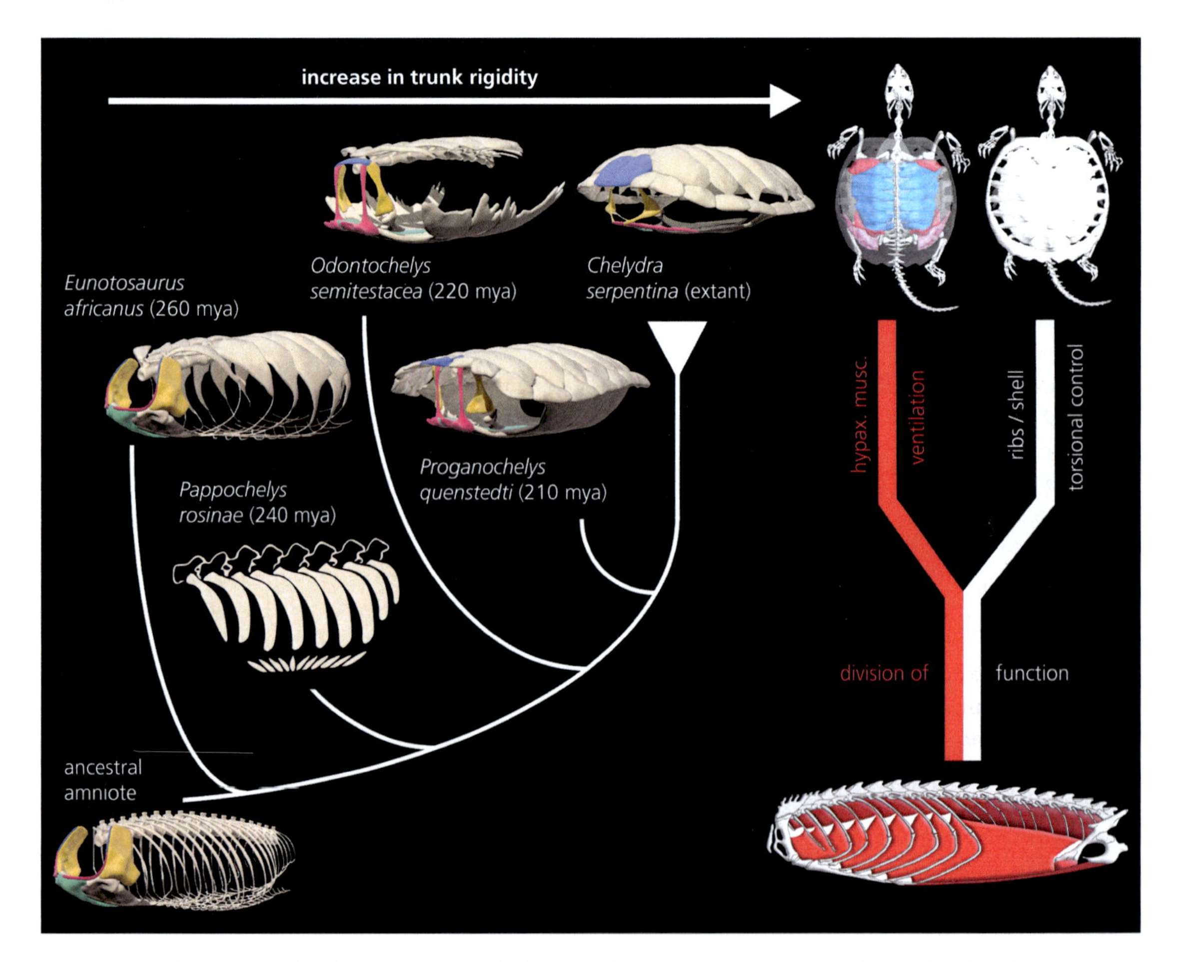

Figure 15.5 Pathway in the origin and early evolution of turtles (left-hand diagram) showing the sequence in the evolution of the shell. Starting with broadened but still individual ribs, the first element of the shell that is formed is the plastron (ventral part), which is assumed to have been beneficial for *m. obliquus abdominis*-based inspiration. On the right is a biomechanical interpretation of these events, illustrating the division of function of the body wall, in which the trunk muscles take over a ventilatory role and the ribs a mechanical in stabilizing the trunk with regard to torsional control. After Lyson et al. (2014) and Lambertz (2016).

shelled bauplan, but rather it is more likely that it allowed for the evolution of the shell in the first place.

Turtles must have started out as animals with individual ribs, and the shell evolved progressively. The scenario for further evolution begins in the Permian of South Africa with the potentially earliest stem-line turtle *Eunotosaurus africanus* (Figure 15.5). This species, whose ribs—albeit pronouncedly broadened—are still free, already had a two-bellied transverse muscle and lacked intercostal muscles (Lyson et al., 2014). So a structural and functional divergence from the plesiomorphic state of the trunk had already occurred during the dawn of turtles. According to this scenario, the broadened ribs stabilized the trunk, counteracting torsion during locomotion, whereas the hypaxial body wall muscles became primarily responsible for breathing, thereby evading 'Carrier's constraint'. As we see in the Triassic toothed turtle *Odontochelys semitestacea*, the first component of the shell to evolve was the ventral part: the plastron. From a purely functional point of view, this 'makes perfect sense'. The increased rigidity of the ventral trunk would have facilitated the inspiratory function of the oblique abdominal musculature, which in extant turtles inserts on the caudal edge of the plastron (Lambertz, 2016b). So the early modification of the ventilatory system of turtles with an adaptive advantage of uncoupling locomotion and respiration can be assumed actually to have paved the way to the evolution of the iconic shell, and not vice versa.

15.3.3 Mammalia

Let us now take a closer look at the mammalian respiratory faculty. It is unlikely that the long catalogue of characters that differentiates synapsids—in particular mammals—from all other amniotes arose together, but mammals were already present as relatively small creatures that lived a clandestine nocturnal and subterranean life throughout the Mesozoic. We have already spoken of miniaturization and its potential role for the evolution of lepidosaurs. Something similar appears to have happened to the synapsids. When non-avian dinosaurs became extinct around 65 mya, mammals surfaced and rapidly evolved, so that by the end of the Palaeocene some 10 million years later all of the major mammalian groups that exist today in addition to several now extinct ones were present. In general, mammals represent high-performance metabolic regulators, characterized by a constant, high metabolic rate and in some cases an impressive aerobic scope of over 30. The same bauplan encompasses shrews with a body mass of a few grams up to the blue whale, the largest animal ever to exist. Volant forms, marine divers, and arctic-to-tropical and desert-adapted species have colonized virtually every habitat on earth.

In mammals, key elements of the cardiorespiratory faculty, including the four-chambered heart, highly adapted red blood cells, the bronchoalveolar lung in closed pleural cavities, highly effective lung surfactant, and costal-diaphragmatic breathing mechanisms combine synergistically with an efficient locomotor apparatus and astute CNS. From its general advantageous beginning, each mammalian group has experienced its own evolution, including that of the respiratory faculty. The branching pattern in mammalian lungs described in Chapter 12, for example, is group specific, and has been used in rodent systematics (Wallau et al., 2000). And in artiodactyls the left lung has a separate bronchus anterior to the major bronchial bifurcation to the two lungs. Nobody really knows what this is good for, but it is interesting that whales also have this strange branching pattern, which is consistent with the hypothesis that they are highly derived (cet-) artiodactyls. But here we shall concern ourselves only with a scenario to explain how the basic mammalian respiratory faculty got there.

In general, the branching pattern of mammalian lungs is most reminiscent of that seen in turtles, particularly sea turtles (see later in this section). In both cases, the main intrapulmonary bronchus in its plesiomorphic condition is wider at the beginning than at the end and is supported over its entire length by cartilage. Monopodially branching off the main bronchus are two rows of second-order bronchi. In mammals, two rows are only identifiable at early stages in lung development, later the rows become obscured. The further branching of the airways is dichotomous, giving rise to sequentially smaller airways. Comparing late fetal lung anatomy in the sea turtle *Caretta caretta* and a mammal (Perry et al., 1989), it is not difficult to envision the conversion of the proximal chamber regions to second-order bronchi in a basal mammalian lung simply by retention of the fetal condition. A similar transitional stage is seen in new-born marsupials (Mortola et al., 1999; Frappell and Mortola, 2000).

The evolution of tiny mammalian alveoli resulted in a low compliance and high work of breathing. This situation only 'makes sense' if the aerobic activity of the individuals can be maintained and this is where the diaphragm comes in. Indeed, the evolution of the bronchoalveolar lung as part of a high-performance functional unit would not been have possible at all without the coupled evolution of the diaphragm. Experimental or therapeutic paralysis of the diaphragm results in greatly diminished respiratory function. Similarly, during the third trimester of human pregnancy, when the diaphragm is practically immobilized and only costal breathing is possible, the aerobic capacity is also severely reduced.

Recent studies have indicated that a progenitor of the mammalian diaphragm already may have been present in early synapsids that diverged from the amniote baseline very early on (Lambertz et al., 2016). During the Palaeozoic, large-bodied synapsids, such as *Cotylorhynchus* and *Dimetrodon*, became the dominant tetrapod fauna. A lung anatomy not unlike that of sea turtles already may have been present in these basally branching synapsid groups. The proposed diaphragm-like muscle would have allowed an uncoupling of locomotor and respiratory movements and release from Carrier's constraint, and the evolution of the low-compliance, high-performance mammalian respiratory faculty could blossom.

But let us not look at the bronchoalveolar lung as a faulty construction that, like the helpless girl tied to the log in the sawmill in silent film melodramas, can only be saved from certain destruction by a *deus ex machina*: the diaphragm. On the contrary, this combination of an effective gas exchanger, and cardiovascular system together with an efficient ventilatory apparatus, a highly developed central nervous and sensory system, and an adaptable locomotor apparatus together precipitated an evolutionary cascade that ended up in mammals (Perry and Carrier, 2006; Perry et al., 2010).

15.3.4 Archosauria

A third group within the sauropsids, the archosaurs, arose during the Triassic, quickly became the dominant terrestrial craniotes as crocodilians and related species, and dominated again during the Jurassic and Cretaceous as dinosaurs. Even now archosaurs—with more than 10,000 species of birds—almost outnumber all other terrestrial craniote groups.

As opposed to the low-compliance, high-work strategy of mammals, the avian lung–air sac system follows a different biophysical path, characterized by extremely high compliance and maintenance of costal breathing. But where did the avian respiratory system come from? It is possible to envision how a consensus lung, combining the characteristics that the crocodilian and avian lungs have in common, could not only have given rise to these two lung types, but also to a lung–air sac system that the largest terrestrial animals ever to live—the sauropod dinosaurs—also had.

Starting with sauropods and working backwards, these gigantic animals have a problem: avoiding the 'gravitational constraint' that would result if you have a lung that is a metre or more high. How would a lung cope with pulmonary blood pressure high enough to reach all exchange surfaces without causing oedema in the lower parts? The trick was that the lungs were probably relatively flat, adhered to the dorsolateral body wall much as in birds and were connected to large air sacs that lacked pulmonary vasculature. This anatomy killed several birds with one stone, as it were: it made the dinosaur lighter than it would appear to be, it provided a large tidal volume, allowing a huge dead space and a long neck, where the wet tracheal surface provided evaporational cooling during inspiration, and on expiration, water vapour could condense on the cooled surface and reduce water loss (Perry et al., 2011; Sverdlova et al., 2012; Henderson, 2013). Whether sauropods had a flow-through lung like we find in birds is not known. That would depend on whether parabronchi developed.

Coming back to the question of a consensus archosaur lung: what would such a lung look like that could give rise to those of crocodiles, sauropods, and birds? Crocodiles have multichambered lungs that at least in the fetal stage somewhat resemble those of birds. Also the lungs have pores connecting one chamber to another and unidirectional air flow through these pores has been demonstrated (Perry, 1991; Farmer and Sanders, 2010). So can we envision a crocodile lung giving rise to an avian or dinosaur lung–air sac system? It is a good idea, but there are some problems.

First of all, at least the abdominal and interclavicular air sacs are absolutely essential for birds to breathe. Air sacs are not present in crocodilian lungs, but they must have been present in saurischian dinosaurs, because of the gravitational constraint problem. Secondly, the parabronchi of the avian paleopulmo connect secondary bronchus groups from the front and the back of the lungs with each other, but not adjacent chambers, as do the pores in crocodilian lungs. And embryologically this does not happen by initially adjacent secondary bronchial anlagen becoming pushed apart by parabronchial development. On the contrary, widely separated secondary bronchial groups become connected by the parabronchi growing together and meeting in the plane of anastomosis. This does not happen in crocodiles and we do not know about dinosaurs. Finally, the respiratory mechanics (discussed in Chapter 12 in more detail) in crocodilians and birds are completely different. Here again, the sauropod may be closer to birds than to crocodiles, with their mobile pubic bone and hepatic piston.

The really big dinosaurs are so different from birds that it is hard to get a respiratory handle on them, so let us look at some dinosaurs that are a bit closer to birds and see if we can piece together a scenario on how to get from the consensus archosaur lung described in Chapter 12 to the avian lung–air sac system.

We postulate an evolutionary cascade that revolved around the cardiorespiratory faculty and locomotor apparatus in small, theropod dinosaurs. These highly mobile animals had a compressed, stiff trunk and relied on rib breathing. In fact, some, such as *Velociraptor*, had enormous uncinate processes, which are consistent with this hypothesis (Codd et al. 2005). This anatomy would have cramped the middle part of the stiff lung, bringing anterior and posterior regions closer together, and at the same time displaced the more flexible, distal parts, exacerbating the development of air sacs somewhere else. Even in its initial stages long before the division of saurischian dinosaurs into sauropods and theropods, cardiogenic undirectional air movement during non-ventilatory periods could have favoured the anatomical division into gas exchange and high-compliance mechanical regions in the lung, which were later exapted, giving rise to the lung–air sac system. Each step in this sequence would lead to more efficient lungs, a superior aerobic lifestyle, and birds. In addition, Carrier's constraint becomes sequentially eliminated as locomotion is moved from the trunk to the extremities.

So did the avian lung–air sac system develop from the crocodilian lung? Not likely, but they did most likely arise from a common source. Comparing crocodilian and avian lungs, though, even the specific homology of any structure except the main intrapulmonary bronchus and the second-order branches (secondary bronchi and chambers, respectively) remains speculative.

We have focused here on the structure and function of the lung–air sac system, but a respiratory faculty has circulatory, locomotor, and control components. However, as discussed previously regarding mammals, a synergistic combination of respiration, locomotion, and central nervous competence has again resulted in an evolutionary cascade. Birds now occupy virtually every biotope imaginable: swifts can spend almost their entire adult life in the air whereas others, such as penguins and 'ratites', are marine divers or terrestrial pedestrians.

15.4 Evolution of the control of breathing in craniotes: a final kick at the cat

Since the neuronal control of breathing in a way encompasses the entirety of the craniote respiratory faculty, we have saved a look at its evolution for this last part. Beginning with filter feeding, in non-craniote chordates we already see a central control of the ciliary beating, which is responsible for ventilation of the branchial basket. Here, indolamines and catecholamines play from the very beginning a regulatory role. We pick the indolamine serotonin (5-HT) and check its involvement throughout the evolution of the respiratory faculty as we progress in basally derived chordate lineages from ciliary to muscle-powered ventilation in craniotes, well knowing that a host of other substances could also be followed. Given the gaps in the phylogenetic record it is not possible to recreate a clear linear evolutionary baseline progression but comparing the control elements of respiratory faculties, we can piece together parts of a scenario.

In craniotes, the gills consisted from the very beginning of a linear sequence of openings in the pharynx, through which ventilatory water is unidirectionally propelled, beginning at the anterior opening and moving caudally. Regardless of whether the motor is a velum, superficial constrictor muscles, or highly differentiated branchiomeric muscles, a fore-to-aft sequence of cranial nerve activity is maintained from hagfish to teleosts.

Jumping now from water breathing to air breathing, we see one important change. Although the branchiomeric cranial nerves V, VII, IX, X, XI, and XII (S-2 in amphibians) continue to be crucially involved, in keeping with dramatic changes in the structure and function of the ventilatory apparatus, the sequence of firing is now controlled from nodes. In 'the frog' these are called the priming centre, the lung centre, and the gill centre (Baghdadwala et al., 2016). This is consistent with the hypothesis that the control of breathing comes to lie posterior rather than anterior in the brainstem in these amphibians. In the gill centre, the frequency of fictive air breathing events, but not of rhythm, is stimulated both by application of acidic carbon dioxide-free and hydrogencarbonate-acidified baths in reduced preparations. And in the obligatory air-breathing actinopterygian gar pike *Lepisosteus*, fictive air breathing episodes in reduced preparations are also initiated by lowering the pH of the bathing fluid (Perry et al., 2001). So it is possible that the neurological control of air breathing even predates the separation of actinopterygian and sarcopterygian lines. More research is urgently needed here and also in lungfish, to clarify if such control nodes are present and, if so, if they are homologous to those seen in tetrapods.

Frogs and perhaps tetrapods in general appear to use a neuronal network of mutually inhibitory neurons (gill centre in the frog, mentioned previously), located in the *nucleus ambiguus* and surrounding the motor nucleus of cranial nerve X, to initiate air breathing, although a more anterior region near the nucleus of cranial nerve V (priming and lung centres mentioned previously), also exists. But the lung oscillator did not evolve from the gill oscillator: on the contrary. Separate neuronal pathways for response to oxygen or carbon dioxide/pH may have existed even before muscle-powered breathing, and afferents may have always been important in coordinating the initiation of air breathing. The importance of 5-HT in simulating breathing by detecting oxygen levels peripherally in the carotid bodies as well as the presence of 5-HT oxygen-sensitive neuroepithelial bodies in the lungs are consistent with this hypothesis. And looking at the possible orchestrating function of the serotonergic raphe nuclei (including stimulation of ependymal ciliary activity) in the posterior region of the brainstem, it appears that this region together with exaptation of nuclei of cranial nerves X and XII may have been pivotal control elements during the evolution of air breathing in tetrapods.

Looking now to amniotes such as turtles and mammals, we also see a respiratory control area 'hard wired' in a region of the brainstem similar to that of the priming and lung centres in the frog (Douse and Mitchell, 1990). In amniotes it is called the ventral respiratory group (VRG). In mammals the VRG is near the motor nuclei of cranial nerves V and VII, not far from what would be the lung oscillator and priming oscillator in the frog. This is also the location of the mammalian respiratory (pre-Bötzinger) neurons (Smith et al., 1991) and the post-inspiratory center (Anderson et al., 2016). It remains unclear if the pre-Bötzinger and the respiratory parafacial neurons (Pisanski, 2018) next to the cranial nerve VII motor nucleus in mammals evolved from lung or priming oscillators or their precursors, or if the frog is just a 'red herring' (Gans, 1970) and its control of breathing is relevant only to the frog and closely related amphibians.

Having said this, it is perhaps significant to note that frogs and toads are buccal breathers and amniotes are aspiration breathers, so one could even assume that the control of breathing might be completely different in these groups. What appears to be a two-phase breathing mechanism in extant amphibians is easily derived from the sequence in activity of jaw and hypobranchial muscles seen in the lungfish, *Protopterus* (McMahon, 1969), again suggesting that the neurological sensing and control pathway may be older than the actual tetrapod breathing mechanism. In *Protopterus*, the buccal pump sequence of (1) opening the glottis and the mouth, allowing air to leave the lungs and mouth, then (2) closing the mouth and—whilst the operculum is held closed—forcing the trapped air through the open glottis into the lungs, easily can be derived from the sequence the lungfish used for ventilating its gills with water (McMahon, 1969). The main difference is that during water breathing the glottis is kept closed and the operculum is open when water is pressed across the gills. While it is easiest to make a comparison of lungfish with perennibranch caudate amphibians, recent studies have extended it to frogs and even mammals. In frogs (*Lithobates* and *Rana*), mouth closing takes place earlier, because the nares in these anurans are opened by a curious mechanism, whereby pressing the jaws closed rotates the premaxillary bone in the upper jaw. Also, neurobiological studies clearly demonstrate three rather than two phases because the complete or partial closure of the glottis following inspiration is recognized as a separate, post-inspiratory phase (Gargaglioni and Milsom, 2007).

In other words, the respiratory cycle in all tetrapods consists of three parts: inspiration, in which the lungs are filled; post-inspiration, in which the glottis is closed; and expiration, when the glottis is opened and air is expelled from the lungs. Amphibians and reptiles are usually episodic breathers, so the post-inspiratory phase consists of the relatively simple act of closing the glottis and keeping it closed until the next breathing episode. Mammals and birds, on the other hand, usually do not close the glottis completely between breaths. Air oscillation during post-inspiration is under fine control and often associated with sound production. Nevertheless, the persistence of the three-part breathing cycle in spite of major anatomical differences suggests that it was there before the divergence of lissamphibian and amniote lineages. Additional focused studies on lungfish could push the origin of the three-phase control of breathing even further back in the sarcopterygian lineage.

However, in looking for insight into evolutionary pathways we should not just look back. In birds, for example, the stimuli driving respiration are basically similar to what is seen in mammals, except that pulmonary stretch receptors appear to be reduced and the avian brain is much more tolerant to hypoxia and hypercapnia than in mammals. In Chapter 5, we saw how these adaptations may enable birds to fly at altitudes where mammals can barely survive for few hours, and learning how they do it may help us to better understand the multidimensional aspects of the evolution of the control of breathing in general.

To put the big pieces together in a speculative scenario, it is possible that for craniotes in general the main oxygen-regulated control entity lies in the posterior brainstem and projects forward to the region of cranial nerves V and VII, where the sequence of gill breathing begins. This mechanism may be older than the origin of tetrapods or even of sarcopterygians. In air breathers, the response of the anterior centres is usually inhibited, but becomes responsive when the pH is low. The ascending input would then set activity patterns in motion for air breathing using the same branchiomeric cranial nerves that were responsible for gill breathing in fish, but are now innervating highly modified anatomy. Using the lungfish as a model, one can envision the origin of aspiration breathing. And since aspiration has also been reported in a lungfish and is not mutually exclusive with buccal breathing, parallel existence of these two breathing modes in the ancestors of amniotes is also possible (McMahon, 1969; Lomholt et al., 1975). We could gain some insight into this question if we could show that a homologue of the piscine gill oscillator in amniotes—which never have functional gills—is still there ticking along but just never surfacing. There is some evidence that cyclic occurrences such as gular pumping in some reptiles, suckling in infant mammals, or hiccups may represent the de-inhibition of a gill oscillator (Straus et al., 2003).

CHAPTER 16

The bottom line

The endosymbiosis of mitochondria by eukaryotic cells together with photosynthetic oxygen production opened the door to evolutionary pathways in diverse lineages, often ending in high- performance animals that operate at high energy levels, but also resulted in several intermediate metabolic plateaus. Animals with elevated oxygen requirements also need specialized respiratory faculties, consisting of a gas exchanger, a ventilatory pump, a physical link to some mechanism—usually including at least one heart and oxygen-transporting proteins—for internal distribution of respiratory gases and a control element that coordinates supply and removal of respiratory gases with metabolic demand.

These multifaceted respiratory faculties that exist in time and space arrived at their present condition through evolution, introducing a fifth dimension to the concept of the respiratory faculty—namely the transmission of genetic information from generation to generation—in addition to the three spatial ones plus time. Let us now look back at some of the major events in the evolution of respiratory faculties in animals.

Life began in the water so it is reasonable to assume that the oldest form of gas exchange was directly between the surface of the animal and respiratory gases dissolved in water. Stepping back and looking at the evolution of respiratory faculties, it is impressive how just those parameters that exacerbate diffusing capacity have repeatedly emerged: large surface area, thin diffusion barrier, maintenance of large driving pressures for respiratory gases through ventilation, perfusion, and oxygen-binding proteins. And at the end of the chain, cells with enormous differences in oxidative demand even within the same organism are supplied with just the right amount of oxygen to keep ATP production flowing, while at the same time not stifling the initial glycolytic steps.

Aside from the surface of the animal as a gas exchanger, ventilation probably was one of the first elements of a respiratory faculty to appear. Of course, living in flowing water is the most obvious way to achieve ventilation without metabolic cost. But many animals can also move about. Cilia and flagella can be used to move through the water, exposing the creatures to fresh environments, but if the animals are stationary, these organelles can also be used to move water over them. So a cilia-driven water current, such as seen in sponges and many filter-feeding organisms like bivalve molluscs, tunicates, and amphioxus, can be considered a step in the direction of a respiratory faculty, even if its 'primary' function is usually seen as feeding. As animals evolved larger-bodied, more mobile, and even predatory forms, moving thin sheets of water by cilia was no longer appropriate and various forms of muscle-powered mass ventilation are seen: the siphon in cephalopod molluscs, a modified leg—the scaphognathite—in crustaceans, and intrinsic and extrinsic gill musculature in craniotes to name a few. Indeed, in craniotes, muscle-powered gill ventilation supplanting cilia provided a breakthrough that precipitated an evolutionary cascade ending in teleost

Respiratory Biology of Animals: Evolutionary and Functional Morphology. Steven F. Perry, Markus Lambertz, Anke Schmitz, Oxford University Press (2019). © Steven F. Perry, Markus Lambertz, & Anke Schmitz.
DOI: 10.1093/oso/9780199238460.001.0001

fish. Probably the most notable evolutionary trend is the establishment of unidirectional ventilation of the gills allowing for counter-current gas exchange right from the beginning of the craniote lineage, and its maintenance and refinement in spite of dramatic changes in the active pump mechanism.

Directed water flow is more efficient in ventilating respiratory surfaces than undirected turbulence, and counter-current is the most efficient gas exchange model. Can it be any wonder that directed flow of the respiratory medium has arisen in every major group of animals, and many also have cross-current or counter-current systems? But once lost, counter-current systems do not seem to re-emerge. A wide variety of aqueous gas exchange mechanisms exist in adult tetrapods, ranging from well-developed external gills in perennibranch salamanders to survival aids in the buccal cavity, skin, and even the urinary bladder of some aquatic turtles. But in spite of this repeated origin of water breathing, no tetrapods have evolved directed water flow or a vein for the separate return of oxygenated blood from the gills or skin to the heart. The oxygen-enriched blood is always returned to the posterior cardinal veins or the posterior vena cava, resulting in an admixture with oxygen-poor venous blood. Incidentally, this also applies conversely to air-breathing fishes: no air-breathing actinopterygian fish has ever evolved a mechanism for keeping oxygen-poor and oxygen-rich blood separated in the heart.

The limited solubility of oxygen in water lends an energetic advantage to any animal that can breathe air: at the same partial pressure, a litre of air contains roughly 30 times much oxygen as a litre of water. But life on land is not just a bed of roses; it also presents limiting factors such as desiccation, surface tension phenomena and collapse of respiratory surfaces, and potential physical damage. Also, because of the very different physical properties of water and air, ventilatory mechanisms that evolved in water rarely work well for moving air. Accordingly, we often see parallel systems evolving, whereby gills are maintained for water breathing and lungs, mantle epithelium, tracheae, or other surfaces are used in air. Craniotes, molluscs, and arthropods provide numerous examples. Going one conceptual level deeper: because oxygen is too easy to come by on land, we confront buffering problems associated with hypoventilation and accumulation of waste gases. These are problems every amphibious animal must face. The solutions may involve radically different structures in different animal groups, but physiologically the results converge.

One curious terrestrial evolutionary direction that could be instructive is seen in spiders. Animals that have perfectly good book lungs for breathing air evolved a tracheal system, also for breathing air. Closer analysis of this phenomenon reveals that the book lungs oxygenate the haemolymph in general, whereas the tracheae directly supply specific, high-metabolic-need organs. If we look around we see this principle occurring in other organisms or organs. Think of the mammalian liver, where the hepatic artery opens into sinusoids, providing the hepatocytes with mixed—rather than pure arterial—blood. But then there is the heart with its dedicated coronary arteries supplying the working muscle with blood with a PO_2 of around 11 kPa.

Anatomical constraints such as a cuticle that has a very low permeability for oxygen result not only in limitations, but also in possibilities. In many arthropods and their close relatives, the cuticle forms the most efficient gas exchange system of all, the tracheal system. It takes advantage of the extremely high diffusion rate of gas in gas and appeared probably independently in velvet worms, arachnids, isopods, myriapods, and hexapods. Insect tracheae conduct air to the terminal locus of gas exchange with little loss of oxygen to the haemolymph, whereas those of most other tracheate arthropods have tracheal lungs, which release oxygen to the respiratory protein-containing haemolymph.

Among craniotes, surfactant prevents delicate gas exchange surfaces from collapsing and sticking together, and directed airflow and auxiliary breathing mechanisms repeatedly appeared. Surfactant seems to have been there from the very beginning, but the auxiliary breathing mechanisms may well have arisen from exaptation of jaw and body wall muscles, resulting in preservation of a function (e.g. counter-current exchange of craniote gills or lung ventilation) while the mechanisms that bring it about may change.

The multiple origin and evolution of circulatory systems, which interface with the respiratory faculty, is certainly of great interest, but dealing with them in detail is not justified in the present book. Suffice it to say that, compared with invertebrates, craniotes are unique in that the heart lies upstream from the gills and pumps deoxygenated blood into them. For most fish this is not a problem and the residual blood pressure downstream from the gills is sufficient to perfuse the body. Now returning to terrestrialization: the relatively high vascular resistance in craniote lungs does pose a problem, which lungfish and tetrapods (but not ray-finned fish) have perhaps separately solved by developing a pulmonary vein or veins that return the oxygenated blood to the heart. The evolution of mechanisms that allow or prevent admixture of oxygenated and deoxygenated blood in the heart go beyond the scope of this book, but the return of oxygenated blood to the heart for circulation to the body certainly was a key factor in the evolutionary cascade concomitant with the terrestrial radiation of amniotes.

And while on the subject of craniote terrestrialization, the origin and evolution of ABOs and indeed of the entire respiratory faculty, is a veritable mine field. The appearance of ABOs dozens of times in different groups of teleost fishes—a group of highly specialized water breathers—speaks volumes and perhaps gives us some insight into the Palaeozoic origin of lungs. The advantages of air breathing in oxygen-poor waters have been mentioned earlier. We speculate that pharyngeal ABOs were widespread among Devonian fishes and ventilation of the pharynx with air probably preceded the origin of lungs and the respiratory swim bladder. The lung-like organs of the cladistian ray-finned fish *Polypterus* appear to be a relic of those times. But the lack of pulmonary venous return to the heart and also the lack of surfactant-producing cells in the parenchyma suggest that these ABOs are an autapomorphy of the Cladistia. The similar vascular supply and innervation of these ABOs and tetrapod lungs indicates only that these structures are all derivatives of the posterior pharynx. Among ray-finned fishes, swim bladders appear to have originated once as a pulmonoid swim bladder with a double blood supply from the sixth branchial arches, similar to that seen in *Amia* and *Lepisosteus*. Air breathing was later separately eliminated in ancestral teleosts and sturgeons. A swim bladder with a single arterial blood supply from the dorsal aorta arose among marine teleosts, and then experienced varied evolution in different groups, including respiratory swim bladders on the one hand, and oxygen-secreting physoclist ones on the other hand.

The use of PCO_2/pH in the blood rather than PO_2 in the control of air breathing appears already to have been in place before the separation of actinopterygian and sarcopterygian lines. So the two (actually three)-phase breathing cycle may have originated in basal sarcopterygians before the development of lungs and was preserved in lungfish and the ancestors of extant amphibians and amniotes, notwithstanding the possible separate origin of lungs in these groups. Expiration was aided by the transverse abdominal musculature, lacking in ray-finned fishes, and remains to this day.

The lungs of extant amphibians as derivatives of pharyngeal structures as well as the breathing mechanisms apparently were already present when the ancestors of the Lissamphibia arose. But amniotes? We now have evidence that the first amniote lungs probably had a complex structure, possibly not unlike that seen in basally branching lineages of turtles, and the three-phase breathing cycle is also plesiomorphic. Furthermore, aspiration breathing may be more ancient than once thought. Also, that the lungs are derivatives of the posterior pharynx is belied not only by their embryological origin, but also by their vascular supply and innervation. But exactly which pharyngeal structures gave rise to them and their ontogenetic relationship to the lungs of lungfish and lissamphibians remains unclear.

One of the most fascinating aspects of the evolution of the respiratory faculty in animals is the parallel emergence of respiratory proteins in diverse taxa. Oxygen-binding organic molecules such as myoglobin and haemoglobin may have been around even before life as we know it existed, and were more than likely key factors in the origin of life and certainly in the evolution of aerobic metabolism. Their role in sequestering oxygen and releasing it again under metabolically meaningful conditions is deeply embedded in life and certainly will remain the focus of research for decades to come. We need only to think of the multiple crucial functions of cytochromes to impress this upon us, but the more obvious function of oxygen transport is not to be overlooked as being crucial to the survival of active, complex organisms.

Unlike globins, haemocyanins originated as phenoloxidase enzymes and later exapted into the role of oxygen transport molecules, convergent with haemoglobin. In spite of a completely different molecular structure, the physiological properties of craniote haemoglobin and the haemocyanins of arthropods and molluscs show astounding similarities, including oxygen uptake, transport and release, cooperativity, Bohr effect, Root effect, and the modulation of these properties by external effectors. Some milestones in respiratory biology that involve respiratory proteins include (1) the development of intracellular haemoglobins and haemerythrins; (2) the independent evolution of molluscan and arthropod haemocyanins, which offered advantages over the haemoglobins in the more advanced molluscan and arthropod cardiovascular systems; and (3) the evolution of porphyrin-containing extracellular globins, which evolved independently in several groups and serve a plethora of functions, many of which have yet to be discovered. In addition, our analysis of respiratory proteins points up some interesting principles in the evolutionary biology of respiration in animals in general. Certainly a major step in the evolution of the respiratory faculty in craniotes was the packaging of haemoglobin in erythrocytes, which gave the flexibility of a relatively small oxygen transporter with none of the disadvantages, and at the same time freed up the plasma for other functions. Somehow craniotes appear to have just dropped from the sky complete with red blood cells, immunoglobulins, and a blood coagulation cascade.

In general, the role of respiratory proteins is different for oxygen and carbon dioxide and also differs between animals with single and multiple respiratory systems. There is a complex trade-off in animals that breathe with a tracheal system (using terminal diffusion) but also possess respiratory proteins, and in bimodal breathers with lungs and gills or lungs and tracheae.

Although the world is populated by recognizable species, these species are not static but are subject to evolution. One major factor related to respiratory faculties is the ability of animals to acclimatize or adapt without evolving new species. Particularly with respect to the response to the availability of oxygen we find astounding convergence not only in the molecular mechanisms but also in the behavioural results of exposure to hypoxia and anoxia in insects and other invertebrate groups, and also in craniotes. Although HIF and HSP are key factors in orchestrating these responses in all species, there is no reason to presume *a priori* that the results would be convergent, since the developmental biology, anatomy, and respiratory physiology of the groups involved are totally different.

Little evolutionary consistency is recognizable in metabolic control mechanisms. On the contrary, the control mechanisms appear to display on the one hand a surprising degree of variability within lineages and on the other hand, convergence among only distantly related groups. In the end effect, animals do whatever they can that is consistent with the internal constraints of their anatomical bauplan, their physiological capabilities, behavioural repertoire, and survival. When they enter the realm of environmental extremes they are forced to converge. We are dealing with a complex, multidimensional network and this is probably the key to its success.

Finally let us reconsider the control mechanisms. To begin with they are more or less self-regulating. Cilia beat faster in warm water, and the metabolic rate and oxygen demand also increase. More particulates are caught in the filtering net, which—at least in amphioxus—contains cilia-stimulating 5-HT and a homologue of metabolism-stimulating thyroid hormone. In craniotes, the brainstem with its direct access to the branchiomeric cranial nerves drives muscle-powered breathing, using dissolved oxygen as a signal. Water-breathing crustaceans and molluscs also adjust ventilation according to the oxygen content of the water. To some extent this is also self-regulating: since warm water contains less oxygen than cold water, more water has to be pumped at the higher temperature in order to extract the same amount of oxygen and the metabolic rate may also rise. So comparing fishes with invertebrate taxa, we see amazing parallel respiratory faculties for water breathing. But we should not miss the forest for the trees. As metabolic demand increases, the anatomical and physiological solutions for satisfying it must become more and more similar, simply because there are fewer alternatives. It is no coincidence that the counter-current model has evolved independently several times, reaching its peak in teleost fish and cephalopod molluscs, but it is often overlooked that this is just the most effective biophysical solution to a common problem: effective aquatic gas exchange.

In terrestrial animals, the PCO_2/pH of the circulating internal medium rather than ambient oxygen is used as a signal, both in invertebrates such as crabs and terrestrial snails, as well as in craniotes. What sense does this make? Why is it not enough to know the PO_2 of the haemolymph/blood leaving the respiratory organ for air breathing? The PO_2 in water can vary wildly and very quickly with time and location, so it is of major survival value for an aquatic animal to be able to detect ambient PO_2 and quickly adjust its behaviour. Using the PCO_2/pH stimulus is not a good option, since due to its high solubility, carbon dioxide is quickly lost and is not a reliable signal. But on land, oxygen is usually abundant, so quick reactions are not necessary. Only under severe hypoxia is a gasping reflex elicited in terrestrial craniotes.

On the other hand, sticking to amniote examples, the capillary endothelium in the lungs, unlike the pillar cells of gills, does not contain carbonic anhydrase. Carbon dioxide is difficult to release and is an important part of a buffer system. By their very nature, buffer systems show a certain latency and smooth the reaction to changes in metabolic rate. The anatomy of the respiratory system takes this latency into account. If the reaction is too fast or the anatomical latency is too great, an oscillating breathing rhythm results. Of course, a signal that is also part of a buffer system complicates matters, since hydrogencarbonate is not the only buffer in blood and the form of nutrition, involvement of anaerobic metabolism, and so on, all influence the pH. But, as evidenced by the repeated occurrence of PCO_2/pH-induced air breathing in animals, this signal seems to do the job. It enables bar-headed geese to fly over the Himalayan Mountains and cheetahs to run at speeds in excess of 100 km h^{-1} and recover.

We are a long way from knowing all the answers, but although the goal is desirable, the fun is really in the journey. So let us end on a positive note from that famous comparative anatomist, Johann Wolfgang von Goethe (1749–1832):

> O! glücklich! wer noch hoffen kann
> Aus diesem Meer des Irrthums aufzutauchen.
> Was man nicht weiß das eben brauchte man,
> Und was man weiß kann man nicht brauchen.
> Doch laß uns dieser Stunde schönes Gut,
> Durch solchen Trübsinn, nicht verkümmern!

> *How fortunate is he who can still hope*
> *To emerge from these uncharted waters.*
> *What we don't know is really what we need,*
> *And what we know is of no use at all.*
> *But let the radiance of this hour,*
> *Not be marred by such dark thoughts!*

Goethe JW (1808). *Faust—Eine Tragödie*. Tübingen: J.G. Cotta'sche Buchhandlung, p. 71.

References

Adamczewska, A. M. and Morris, S. (1994). Exercise in the terrestrial Christmas Island red crab *Gecarcoidea natalis* I. Blood gas transport. Journal of Experimental Biology 188, 235–56.

Adamczewska, A. M. and Morris, S. (1996). The respiratory gas transport, acid-base state, ion and metabolite status of the Chrismas Island blue crab, *Cardisoma hirtipes* (Dana) assessed in situ with respect to immersion. Physiological Zoology 69, 67–92.

Adamczewska, A. M. and Morris, S. (1998). The functioning of the haemocyanin of the terrestrial Christmas Island red crab *Gecarcoidea natalis* and roles for organic modulators. Journal of Experimental Biology 201, 3233–44.

Adamczewska, A. M. and Morris, S. (2000). Locomotion, respiratory physiology, and energetics of amphibious and terrestrial crabs. Physiological and Biochemical Zoology 73, 706–25.

Affolter, M. and Shilo, B. Z. (2000). Genetic control of branching morphogenesis during *Drosophila* tracheal development. Current Opinion in Cell Biology 12, 731–5.

Al-Ghamdi, M., Jones, J. F. X. and Taylor, E. W. (2001). Evidence of a functional role in lung inflation for the buccal pump in the agamid lizard *Uromastyx aegyptius microlepis*. Journal of Experimental Biology 204, 521–31.

Altman, P. L. and Dittmer, D. S. (1971). Respiration and Circulation. Federation of American Societies for Experimental Biology, Bethesda.

Amaral-Silva, L. D., Lambertz, M., Zara, F. J., Klein, W., Gargaglioni, L. H. and Bícego, K. C. (2019). Parabronchial remodeling in chicks in response to embryonic hypoxia. Journal of Experimental Biology. 222, jeb197979.

Anderson, J. F. and Prestwich, K. N. (1975). The fluid pressure pumps of spiders (Chelicerata, Araneae). Zeitschrift für Morphologie der Tiere 81, 257–77.

Anderson, J. F. and Prestwich, K. N. (1985). The physiology of exercise at and above maximal aerobic capacity in a theraphosid (tarantula) spider, *Brachypelma smithi*. Journal of Comparative Physiology B: Biochemical, Systemic, and Environmental Physiology 155, 529–39.

Anderson, T. M., Garcia, A. J., 3rd, Baertsch, N. A., Pollak, J., Bloom, J. C., Wei, A. D., Rai, K. G., Ramirez, J. M. (2016). A novel excitatory network for the control of breathing. Nature 536, 76–80.

Angersbach, D. (1978). Oxygen transport in the blood of the tarantula *Eurypelma californicum*: pO_2 and pH during rest, activity and recovery. Journal of Comparative Physiology 123, 113–25.

Anzenbacher, P. and Anzenbacherova, E. (2001). Cytochromes P450 and metabolism of xenobiotics. Cellular and Molecular Life Sciences 58, 737–47.

Arieli, R. and Lehrer, C. (1988). Recording of locust breathing frequency by barometric method exemplified by hypoxic exposure. Journal of Insect Physiology 34, 325–8.

Baghdadwala, M. I., Duchcherer, M., Trask, W. M., Gray, P. A. and Wilson, R. J. A. (2016). Diving into the mammalian swamp of respiratory rhythm generation with the bullfrog. Respiratory Physiology & Neurobiology 224, 37–51.

Bailly, X., Chabasse, C., Hourdez, S., Dewilde, S., Martial, S., Moens, L. and Zal, F. (2007). Globin gene family evolution and functional diversification in annelids. FEBS Journal 274, 2641–52.

Barbariol, V. and Razouls, S. (2000). Experimental studies on the respiratory metabolism of *Mytilus galloprovincialis* (Mollusca bivalvia) from the Mediterranean Sea (Gulf of Lion). Vie et Milieu Life and Environment 50, 87–92.

Barej, M., Böhme, W., Perry, S. F., Wagner, P. and Schmitz, A. (2010). The hairy frog, a curly fighter? – A novel hypothesis on the function of hairs and claw-like terminal phalanges, including their biological and systematic significance (Anura: Arthroleptidae: *Trichobatrachus*). Revue Suisse de Zoologie 117, 243–63.

Barnes, R. D. (1980). Invertebrate Zoology. Saunders College, Philadelphia.

Bartholomew, G. A. and Casey, T. M. (1978). Oxygen consumption of moths during rest, pre-flight warm-up, and flight in relation to body size and wing morphology. Journal of Experimental Biology 76, 11–25.

Bartsch, P. (2004). Actinopterygii. In: Spezielle Zoologie. Teil 2: Wirbel- oder Schädeltiere (ed. R. W. Westheide and Rieger), pp. 226–87. Spektrum Akademischer Verlag, Heidelberg.

Bartsch, P. (2015). Craniota III. Actinopterygii, Strahelnflosser. In: Spezielle Zoologie, Teil 2: Wirbel- und Schädeltiere (eds. W. Westheide and G. Rieger), pp. 113–27. Springer Spectrum, Heidelberg.

Bäumer, C., Pirow, R. and Paul, R. J. (2002). Circulatory oxygen transport in the water flea *Daphnia magna*. Journal of Comparative Physiology B: Biochemical, Systemic, and Environmental Physiology 172, 275–85.

Beall, C. M. (2006). Andean, Tibetan, and Ethiopian patterns of adaptation to high-altitude hypoxia. Integrative and Comparative Biology 46, 18–24.

Beckel, W. E. and Schneiderman, H. A. (1956). The spiracle of the cecropia moth as an independent effector. Anatomical Record 125, 559–60.

Beckel, W. E. and Schneiderman, H. A. (1957). Insect spiracle as an independent effector. Science 126, 352–3.

Becker, H. O., Böhme, W. and Perry, S. F. (1989). Die Lungenmorphologie der Warane (Reptilia: Varanidae) und ihre systematisch-stammesgeschichtliche Bedeutung. Bonner zoologische Beiträge 40, 27–56.

Behr, M. (2010). Molecular aspects of respiratory and vascular tube development. Respiratory Physiology & Neurobiology 173, S33–6.

Belkin, D. A. (1968). Aquatic respiration and underwater survival of two freshwater turtle species. Respiration Physiology 4, 1–14.

Bell, H. J., Inoue, T. and Syed, N. I. (2008). A peripheral oxygen sensor provides direct activation of an identified respiratory CPG neuron in *Lymnaea*. In: Integration in Respiratory Control: From Genes to Systems, vol. 605, pp. 25–9. Springer-Verlag, Berlin.

Bentley, P. J. and Shield, J. W. (1973). Ventilation of toad lungs in absence of buccopharyngeal pump. Nature 243, 538–9.

Berenbrink, M., Koldkjaer, P., Kepp, O. and Cossins, A. R. (2005). Evolution of oxygen secretion in fishes and the emergence of a complex physiological system. Science 307, 1752–7.

Bergmann, S., Markl, J. and Lieb, B. (2007). The first complete cDNA sequence of the hemocyanin from a bivalve, the protobranch *Nucula nucleus*. Journal of Molecular Evolution 64, 500–10.

Berner, R. A., VandenBrooks, J. M. and Ward, P. D. (2007). Evolution – oxygen and evolution. Science 316, 557–8.

Bickford, D., Iskandar, D. and Barlian, A. (2008). A lungless frog discovered on Borneo. Current Biology 18, R374–5.

Bicudo, J. E. P. W. and Campiglia, S. (1985). A morphometric study of the tracheal system of *Peripatus acacioi* Marcus and Marcus (Onychophora). Respiratory Physiology 60, 75–82.

Bishop, C. M. (1999). The maximum oxygen consumption and aerobic scope of birds and mammals: getting to the heart of the matter. Proceedings of the Royal Society of London B: Biological Sciences 266, 2275–81.

Blank, M. and Burmester, T. (2012). Widespread occurrence of N-terminal acylation in animal globins and possible origin of respiratory globins from a membrane-bound ancestor. Molecular Biology and Evolution 29, 3553–61.

Bliss, D. (1968). The transition from water to land in decapod Crustaceans. American Zoologist 8, 355–92.

Bock, W. J. and von Wahlert, G. (1965). Adaptation and the form-function complex. Evolution 19, 269–99.

Bongianni, F., Mutolo, D., Cinelli, E. and Pantaleo, T. (2016). Neural mechanisms underlying respiratory rhythm generation in the lamprey. Respiratory Physiology & Neurobiology 224, 17–26.

Bouchet, P., Rocroi, J. P., Fryda, J., Hausdorf, B., Ponder, W., Valdes, A. and Waren, A. (2005). Classification and nomenclator of gastropod families. Malacologia 47, 1–368.

Bradley, T. J. (2006). Discontinous ventilation in insects: protecting tissues from O_2. Respiratory Physiology & Neurobiology 154, 30–6.

Bradley, T. J. (2007). Control of the respiratory pattern in insects. In: Hypoxia and the Circulation, vol. 618, pp. 211–20. Springer-Verlag, Berlin.

Brainerd, E. L. (1994a). The evolution of lung-gill bimodal breathing and the homology of vertebrate respiratory pumps. American Zoologist 34, 289–99.

Brainerd, E. L. (1994b). Mechanical design of polypterid fish integument for energy-storage during recoil aspiration. Journal of Zoology 232, 7–19.

Braun, F. (1931). Beiträge zur Biologie und Atmungsphysiologie der *Argyroneta aquatica* Cl. Zoologische Jahrbücher. Abteilung für Systematik 62, 175–262.

Breepoel, P. M., Kreuzer, F. and Hazevoet, M. (1980). Studies of the hemoglobins of the eel (*Anguilla anguilla* L). 1. Proton binding of stripped hemolysate: separation and properties of 2 major components. Comparative Biochemistry and Physiology Part A: Molecular & Integrative Physiology 65, 69–75.

Brett, S. S. and Shelton, G. (1979). Ventilatory mechanisms of the amphibian, *Xenopus laevis*: role of the buccal force pump. Journal of Experimental Biology 80, 251–69.

Bridges, C. R. (2001). Modulation of haemocyanin oxygen affinity: properties and physiological implications in a changing world. Journal of Experimental Biology 204, 1021–32.

Bridges, C. R., Berenbrink, M., Muller, R. and Waser, W. (1998). Physiology and biochemistry of the pseudobranch: an unanswered question? Comparative Biochemistry and Physiology Part A: Molecular & Integrative Physiology 119, 67–77.

Bridges, C. R. and Scheid, P. (1982). Buffering and CO_2 dissociation of body fluids in the pupa of the silkworm moth, *Hyalophora cecropia*. Respiration Physiology 48, 183–97.

Bromhall, C. (1987). Spider tracheal systems. Tissue & Cell 19, 793–807.

Buck, J. (1962). Some physical aspects of insect respiration. Annual Review of Entomology 7, 27–56.

Buck, J. and Friedman, S. (1958). Cyclic CO_2 release in diapausing pupae—III: CO_2 capacity of the blood: carbonic anhydrase. Journal of Insect Physiology 2, 52–60.

Buck, J. and Keister, M. (1955). Cyclic CO_2 relaese in diapausing *Agapema* pupae. Biological Bulletin 109, 144–63.

Buck, J. and Keister, M. (1958). Cyclic CO_2 release in diapausing pupae—II: Tracheal anatomy, volume and pCO_2; blood volume; interburst CO_2 release rate. Journal of Insect Physiology 1, 327–40.

Buck, J., Keister, M. and Specht, H. (1953). Discontinous respiration in diapausing *Agapema* pupae. Anatomical Record 117, 541.

Buck, J. B. (1958). Cyclic CO_2 release in insects IV. A theory of mechanism. Biological Bulletin 114, 118–40.

Burggren, W. and Mwalukoma, A. (1983). Respiration during chronic hypoxia and hyperoxia in larval and adult bullfrogs (*Rana catesbeiana*). 1. Morphological responses of lungs, skin and gills. Journal of Experimental Biology 105, 191–203.

Burggren, W. W. (1992). Respiration and circulation in land crabs: novel variations on the marine design. American Zoologist 32, 417–27.

Burkett, B. N. and Schneiderman, H. A. (1974). Roles of oxygen and carbon dioxide in the control of spiracular function in cecropia pupae. Biological Bulletin 147, 274–93.

Burmester, T. (1999). Identification, molecular cloning, and phylogenetic analysis of a non-respiratory pseudo-hemocyanin of *Homarus americanus*. Journal of Biological Chemistry 274, 13217–22.

Burmester, T. (2001). Molecular evolution of the arthropod hemocyanin superfamily. Molecular Biology and Evolution 18, 184–95.

Burmester, T. (2002). Origin and evolution of arthropod hemocyanins and related proteins. Journal of Comparative Physiology B – Biochemical Systemic and Environmental Physiology 172, 95–107.

Burmester, T. (2015). Expression and evolution of hexamerins from the tobacco hornworm, *Manduca sexta*, and other Lepidoptera. Insect Biochemistry and Molecular Biology 62, 226–34.

Burmester, T. and Hankeln, T. (1999). A globin gene of *Drosophila melanogaster*. Molecular Biology and Evolution 16, 1809–11.

Burmester, T. and Hankeln, T. (2007). The respiratory proteins of insects. Journal of Insect Physiology 53, 285–94.

Burmester, T. and Hankeln, T. (2014). Function and evolution of vertebrate globins. Acta Physiologica 211, 501–14.

Burri, P. H. (1974). Postnatal growth of rat lung. 3. Morphology. Anatomical Record 180, 77–98.

Burri, P. H. and Weibel, E. R. (1971). Morphometric estimation of pulmonary diffusion capacity II. Effect of pO_2 on the growing lung. Respiration Physiology 11, 247–64.

Bushnell, P. G. and Brill, R. W. (1991). Responses of swimming skipjack (*Katsuwonus pelamis*) and yellowfin (*Thunnus albacares*) tunas to acute hypoxia, and a model of their cardiorespiratory function. Physiological Zoology 64, 787–811.

Bushnell, P. G. and Brill, R. W. (1992). Oxygen-transport and cardiovascular-responses in skipjack tuna (*Katsuwonus pelamis*) and yellowfin tuna (*Thunnus albacares*) exposed to acute hypoxia. Journal of Comparative Physiology B – Biochemical Systemic and Environmental Physiology 162, 131–43.

Busse, K., Beek, P., Bott, U., Grube, W., Möser, T., Perry, S. F., B., U. and Valdebenito, I. (2006). Haemoglobin and size dependent constraints on swim bladder inflation in fish larvae. Verhandlungen der Gesellschaft für Ichthyologie 5, 35–53.

Bustamante, J., Bredeston, L., Malanga, G. and Mordoh, J. (1993). Role of melanin as a scavenger of active oxygen species. Pigment Cell Research 6, 348–53.

Bustami, H. P. and Hustert, R. (2000). Typical ventilatory pattern of the intact locust is produced by the isolated CNS. Journal of Insect Physiology 46, 1285–93.

Bustami, H. P., Harrison, J. F. and Hustert, R. (2002). Evidence for oxygen and carbon dioxide receptors in insect CNS influencing ventilation. Comparative Biochemistry and Physiology A – Molecular and Integrative Physiology 133, 595–604.

Butler, P. J. (1991). Exercise in birds. Journal of Experimental Biology 160, 233–62.

Butler, P. J. and Jones, D. R. (1997). Physiology of diving of birds and mammals. Physiological Reviews 77, 837–99.

Cameron, J. N. (1981). Brief introduction to the land crabs of the Palau Islands: stages in the transition to air breathing. Journal of Experimental Zoology 218, 1–5.

Cameron, J. N. and Mecklenburg, T. A. (1973). Aerial gas exchange in the coconut crab, Birgus latro. Respiration Physiology 19, 245–61.

Carrel, J. E. (1987). Heart rate and physiological ecology. In: Ecophysiology of Spiders (ed. W. Nentwig), pp. 95–110. Springer Verlag, Berlin.

Carrel, J. E. and Heathcote, R. D. (1976). Heart rate in spiders: influence of body size and foraging strategies. Science 193, 148–50.

Carrier, D. R. (1987). Lung ventilation during walking and running in 4 species of lizards. Experimental Biology 47, 33–42.

Case, J. F. (1956a). Carbon dioxide and oxygen effects on the spiracles of flies. Physiological Zoology 29, 163–71.

Case, J. F. (1956b). Neural basis of insect spiracular function. Anatomical Record 124, 270.

Case, J. F. (1957a). The median nerves and cockroach spiracular function. Journal of Insect Physiology 1, 85–94.

Case, J. F. (1957b). Differentiation of the effect of pH and CO_2 on the spiracular function of insects. Journal of Cellular and Comparative Physiology 49, 103–13.

Casey, T. M. (1992). Allometric scaling of muscle performance and metabolism: insects. In: Hypoxia and Mountain Medicine (eds. J. R. Sutton G. Coates and C. S. Houston), pp. 152–62. Pergamon Press, Oxford.

Casey, T. M., May M. L., and Morgan, K. R. (1985). Flight energetics of euglossine bees in relation to morphology and wing stroke frequency. Journal of Experimental Biology 116, 271–89.

Castrodad, F. A., Renaud, F. L., Ortiz, J. and Phillips, D. M. (1988). Biogenic amines stimulate regeneration of cilia in *Tetrahymena thermophila*. Journal of Protozoology 35, 260–4.

Chapelle, G. and Peck, L. S. (1999). Polar gigantism dictated by oxygen availability. Nature 399, 114–5.

Chen, Q. F., Ma, E., Behar, K. L., Xu, T. and Haddad, G. G. (2002). Role of trehalose phosphate synthase in anoxia tolerance and development in *Drosophila melanogaster*. Journal of Biological Chemistry 277, 3274–9.

Cheng, W. T., Liu, C. H. and Kuo, C. M. (2003). Effects of dissolved oxygen on hemolymph parameters of freshwater giant prawn, *Macrobrachium rosenbergii* (de Man). Aquaculture 220, 843–56.

Chown, S. L. (2002). Respiratory water loss in insects. Comparative Biochemistry and Physiology A – Molecular and Integrative Physiology 133, 791–804.

Chown, S. L. (2011). Discontinuous gas exchange: new perspectives on evolutionary origins and ecological implications. Functional Ecology 25, 1163–8.

Chown, S. L., Gibbs, A. G., Hetz, S. K., Klok, C. J., Lighton, J. R. B. and Marais, E. (2006). Discontinuous gas exchange in insects: a clarification of hypotheses and approaches. Physiological and Biochemical Zoology 79, 333–43.

Clack, J. A. (2012). Gaining Ground: The Origin and Evolution of Tetrapods, 2nd ed. Indiana University Press, Bloomington.

Coates, E. L. (2001). Olfactory CO_2 chemoreceptors. Respiration Physiology 129, 219–29.

Codd, J. R., Boggs, D. F., Perry, S. F. and Carrier, D. R. (2005). Activity of three muscles associated with the uncinate processes of the giant Canada goose (*Branta canadensis maximus*). Journal of Experimental Biology 208, 849–57.

Codd, J. R., Rose, K. A. R., Tickle, P. G., Sellers, W. I., Brocklehurst, R. J., Elsey, R. M., Crossley, D. A. and (2019). A novel accessory respiratory muscle in the American alligator (*Alligator mississippiensis*). Biology Letters, 15, 20190354.

Contreras, H. L. and Bradley, T. J. (2009). Metabolic rate controls respiratory pattern in insects. Journal of Experimental Biology 212, 424–8.

Corbari, L., Carbonel, P. and Massabuau, J. C. (2004). How a low tissue O-2 strategy could be conserved in early crustaceans: the example of the podocopid ostracods. Journal of Experimental Biology 207, 4415–25.

Corbari, L., Carbonel, P. and Massabuau, J.-C. (2005). The early life history of tissue oxygenation in crustaceans: the strategy of the myodocopid ostracod *Cylindroleberis mariae*. Journal of Experimental Biology 208, 661–70.

Crosfill, M. L. and Widdicombe, J. G. (1961). Physical characteristics of chest and lungs and work of breathing in different mammalian species. Journal of Physiology London 158, 1–14.

Culik, B. M. and McQueen, D. J. (1985). Monitoring respiration and activity in the spider *Geolycosa domifex* (Hancock) using time-lapse televison and CO_2-analysis. Canadian Journal of Zoology 63, 843–6.

Cuvier, G. (1812). Recherches sur les ossemens fossiles de quadrupèdes. Deterville, Paris.

Damen, W. G. M., Saridaki, T. and Averof, M. (2002). Diverse adaptations of an ancestral gill: a common evolutionary origin for wings, breathing organs, and spinnerets. Current Biology 12, 1711–6.

Daniels, C. B., Orgeig, S., Sullivan, L. C., Ling, N., Bennett, M. B., Schurch, S., Val, A. L. and Brauner, C. J. (2004). The origin and evolution of the surfactant system in fish: insights into the evolution of lungs and swim bladders. Physiological and Biochemical Zoology 77, 732–49.

Datry, T., Hervant, F., Malard, F., Vitry, L. and Gibert, J. (2003). Dynamics and adaptive responses of invertebrates to suboxia in contaminated sediments of a stormwater infiltration basin. Archiv Fur Hydrobiologie 156, 339–59.

De Maio, A. (1999). Heat shock proteins. Facts, thoughts, and dreams. Reply. Shock 12, 324–5.

de Miguel, C., Kinsler, F., Casanova, J. and Franch-Marro, X. (2016). Genetic basis for the evolution of organ morphogenesis. The case of spalt and cut in development of insect trachea. Development 143, 3615–22.

Dean, B. (1895). Fishes, Living and Fossil: An Outline of Their Forms and Possible Relationships. Macmillan, New York.

Decker, H. and Terwilliger, N. (2000). Cops and robbers: putative evolution of copper oxygen-binding proteins. Journal of Experimental Biology 203, 1777–82.

Decker, H. and Tuczek, F. (2000). Tyrosinase/catecholoxidase activity of hemocyanins: structural basis and molecular mechanism. Trends in Biochemical Sciences 25, 392–7.

Dela-Cruz, J. and Morris, S. (1997). Respiratory, acid-base, and metabolic responses of the christmas island blue crab, *Cardisoma hirtipes* (Dana), during simulated environmental conditions. Physiological Zoology 70, 100–15.

Delaney, R. G. and Fishman, A. P. (1977). Analysis of lung ventilation in aestivating lungfish *Protopterus aethiopicus*. American Journal of Physiology 233, R181–7.

Denison, R. H. (1941). The soft anatomy of *Bothriolepis*. Journal of Paleontology 15, 553–61.

DeSantis, M. K. and Brett, C. E. (2011). Late Eifelian (Middle Devonian) biocrises: timing and signature of the pre-Kacak Bakoven and Stony Hollow Events in eastern North America. Palaeogeography Palaeoclimatology Palaeoecology 304, 113–35.

Destoumieux-Garzon, D., Saulnier, D., Garnier, J., Jouffrey, C., Bulet, P. and Bachere, E. (2001). Crustacean immunity: antifungal peptides are generated from the C terminus of shrimp hemocyanin in response to microbial challenge. Journal of Biological Chemistry 276, 47070–7.

Diaz, H. and Rodriguez, G. (1977). The branchial chamber in terrestrial crabs: a comparative study. Biological Bulletin 153, 485–504.

Dickinson, M. H. and Lighton, J. R. B. (1995). Muscle efficiency and elastic storage in the flight motor of Drosophila. Science 268, 87–90.

Domenici, P., Lefrancois, C. and Shingles, A. (2007). Hypoxia and the antipredator behaviours of fishes. Philosophical Transactions of the Royal Society B – Biological Sciences 362, 2105–21.

Donnelly, P. M. and Woolcock, A. J. (1978). Stratification of inspired air in the elongated lungs of carpet python, *Morelia spilotes variegata*. Respiration Physiology 35, 301–15.

Douse, M. A. and Mitchell, G. S. (1990). Episodic respiratory related discharge in turtle cranial motoneurons: in vivo and in vitro studies. Brain Research 536, 297–300.

Duchcherer, M., Kottick, A. and Wilson, R. J. A. (2010). Evidence for a distributed respiratory rhythm generating network in the goldfish (*Carassius auratus*). In: New Frontiers in Respiratory Control – XIth Annual Oxford Conference on Modeling and Control of Breathing (eds. I. Homma, H. Onimaru and Y. Fukuchi), pp. 3–7. Springer, Berlin.

Dudley, R. (1998). Atmospheric oxygen, giant paleozoic insects and the evolution of aerial locomotor performance. Journal of Experimental Biology 201, 1043–50.

Duncker, H. -R. (1968). Die extracutanen Melanocyten der Echsen (Sauria). Ergebnisse der Anatomie und Entwicklungsgeschichte 40, 1–55.

Duncker, H. R. (1971). The lung air sac system of birds: a contribution to the functional anatomy of the respiratory apparatus. Ergebnisse der Anatomie und Entwicklungsgeschichte 45, 7–171.

Dunlop, J. A., Anderson, L. I., Kerp, H. and Hass, H. (2004). A harvestman (Arachnida: Opiliones) from the Early Devonian Rhynie cherts, Aberdeenshire, Scotland. Transactions of the Royal Society of Edinburgh Earth Sciences 94, 341–54.

Dunlop, J. A., Kamenz, C. and Scholtz, G. (2007). Reinterpreting the morphology of the Jurassic scorpion Liassoscorpionides. Arthropod Structure & Development 36, 245–52.

Dunlop, J. A. and Webster, M. (1999). Fossil evidence, terrestrialization and arachnid phylogeny. Journal of Arachnology 27, 86–93.

Dwarakanath, S. K., Gabriel, P. J. and Joseph, K. T. (1977). On the mode of anaerobiosis in millipede *Spirostreptus asthenes* (Pocock). Monitore Zoologico Italiano – Italian Journal of Zoology 11, 65–70.

Ebner, B., Panopoulou, G., Vinogradov, S. N., Kiger, L., Marden, M. C., Burmester, T. and Hankeln, T. (2010). The globin gene family of the cephalochordate amphioxus: implications for chordate globin evolution. BMC Evolutionary Biology 10, 370.

Eliakim, R., Goetz, G. S., Rubio, S., ChailleyHeu, B., Shao, J. S., Ducroc, R. and Alpers, D. H. (1997). Isolation and characterization of surfactant-like particles in rat and human colon. American Journal of Physiology – Gastrointestinal and Liver Physiology 272, G425–34.

Ellington, C. P., Machin, K. E. and Casey, T. M. (1990). Oxygen consumption of bumblebees in forward flight. Nature 347, 472–3.

Ellis, C. H. (1944). The mechanism of extension in the legs of spiders. Biological Bulletin 86, 41–50.

Erlichman, J. S. and Leiter, J. C. (1993). CO_2 chemoreception in the pulmonate snail, *Helix aspersa*. Respiration Physiology 93, 347–63.

Erlichman, J. S. and Leiter, J. C. (1997). Identification of CO_2 chemoreceptors in *Helix pomatia*. American Zoologist 37, 54–64.

Ertas, B., von Reumont, B. M., Wagele, J. W., Misof, B. and Burmester, T. (2009). Hemocyanin suggests a close relationship of Remipedia and Hexapoda. Molecular Biology and Evolution 26, 2711–8.

Evans, S. E. and Borsuk-Bialynicka, M. (2009). A small lepidosauromorph reptile from the early triassic of Poland. Acta Palaeonthology Polen 65, 179–202.

Faenge, R. (1983). Gas-exchange in fish swim bladder. Reviews of Physiology Biochemistry and Pharmacology 97, 111–58.

Fagernes, C. E., Stensløkken, K. -O., Røhr, A. K., Berenbrink, M., Ellefsen, S. and Nilsson, G. E. (2017). Extreme anoxia tolerance in crucian carp and goldfish through neofunctionalization of duplicated genes creating a new ethanol-producing pyruvate decarboxylase pathway. Scientific Reports 7, 7884.

Farahani, R. and Haddad, G. G. (2003). Understanding the molecular responses to hypoxia using *Drosophila* as a genetic model. Respiratory Physiology & Neurobiology 135, 221–9.

Farley, R. D. (1990). Regulation of air and blood flow through the booklungs of the desert scorpion, *Paruroctonus mesaensis*. Tissue and Cell 22, 547–69.

Farley, R. D. (2008). Development of respiratory structures in embryos and first and second instars of the bark scorpion, *Centruroides gracilis* (Scorpiones: Buthidae). Journal of Morphology 269, 1134–56.

Farley, R. D. and Case, E. F. (1968). Sensory modulation of ventilative pacemaker output in the cockroach *Periplaneta americana*. Journal of Insect Physiology 14, 591–601.

Farley, R. D., Case, J. F. and Roeder, K. D. (1967). Pacemaker for tracheal ventilation in the cockroach *Periplaneta americana*. Journal of Insect Physiology 13, 1713–28.

Farmer, C. G. (2010). The provenance of alveolar and parabronchial lungs: insights from paleoecology and the discovery of cardiogenic, unidirectional airflow in the American alligator (*Alligator mississippiensis*). Physiological and Biochemical Zoology 83, 561–75.

Farmer, C. G. (2015). Similarity of crocodilian and avian lungs indicates unidirectional flow is ancestral for archosaurs. Integrative and Comparative Biology 55, 962–71.

Farmer, C. G. and Carrier, D. R. (2000). Pelvic aspiration in the American alligator (*Alligator mississippiensis*). Journal of Experimental Biology 203, 1679–87.

Farmer, C. G. and Sanders, K. (2010). Unidirectional airflow in the lungs of alligators. Science 327, 338–40.

Farrelly, C. A. and Greenaway, P. (2005). The morphology and vasculature of the respiratory organs of terrestrial hermit crabs (*Coenobita* and *Birgus*): gills, branchiostegal lungs and abdominal lungs. Arthropod Structure & Development 34, 63–87.

Faucheux, M. J. (1974). Researches on tracheal system of adult Diptera. 3. Evolution of tracheal system. Annales De La Societe Entomologique De France 10, 99–121.

Feder, M. E. and Burggren, W. W. (1992). Environmental Physiology of the Amphibians. University of Chicago Press, Chicago.

Fernandes, M. N., Moron, S. E. and Sakuragui, M. M. (2007). Gill morphological adjustments to environment and the gas exchange function. In: Fish Respiration and Environment (eds. M. N. Fernandes, F. T. Rantin, M. L. Glass, and G. Kapoor), pp. 93–120. Science Publishers, Enfield.

Ferrera, F., Paoli, P. and Taiti, S. (1996/1997). An original respiratory structure in the xeric genus *Periscyphis* Gerstacker, 1873 (Crustacea: Oniscoidea: Eubelidae). Zoologischer Anzeiger 235, 147–56.

Fielden, L. J., Duncan, F. D., Rechav, Y. and Crewe, R. M. (1994). Respiratory gas exchange in the tick *Amblyomma hebraeum* (Acari: Ixodidae). Journal of Medical Entomology 31, 30–5.

Fielden, L. J., Jones, R. M., Goldberg, M. and Rechav, Y. (1999). Feeding and respiratory gas exchange in the American dog tick, *Dermacantor variabilis*. Journal of Insect Physiology 45, 297–304.

Finlay, B. J. and Esteban, G. F. (2009). Oxygen sensing drives predictable migrations in a microbial community. Environmental Microbiology 11, 81–5.

Finlayson, L. H. (1966). Sensory innervation of the spiracular muscle in the tsetse fly (*Glossina morsitans*) and the larva of the waxmoth (*Galleria mellonella*). Journal of Insect Physiology 12, 1451–4.

Fishman, A. P. and Richards D. W. (eds.) (1982). Circulation of the Blood: Men and Ideas. Williams and Wilkins/American Physiological Society, Baltimore.

Fitzgerald, R. S. and Cherniack, N. S. (2012). Historical perspectives on the control of breathing. Comprehensive Physiology 2, 915–32.

FitzGibbon, S. I. and Franklin, C. E. (2010). The importance of the cloacal bursae as the primary site of aquatic respiration in the freshwater turtle, *Elseya albagula*. Australian Zoologist 35, 276–82.

Forster, R. R. (1980). Evolution of the tarsal organ, the respiratory system and the female genitalia in spiders. International Congress of Arachnology 8, 269–84.

Franch-Marro, X., Martin, N., Averof, M. and Casanova, J. (2006). Association of tracheal placodes with leg primordia in *Drosophila* and implications for the origin of insect tracheal systems. Development 133, 785–90.

Frappell, P. B. and Mortola, J. P. (2000). Respiratory function in a newborn marsupial with skin gas exchange. Respiration Physiology 120, 35–45.

Frazier, M. R., Woods, H. A. and Harrison, J. F. (2001). Interactive effects of rearing temperature and oxygen on the development of *Drosophila melanogaster*. Physiological and Biochemical Zoology 74, 641–50.

Fridovich, I. (1998). Oxygen toxicity: a radical explanation. Journal of Experimental Biology 201, 1203–1209.

Furuta, H. and Kajita, A. (1983). Dimeric hemoglobin of the bivalve mollusc *Anadara broughtonii*: complete amino acid sequence of the globin chain. Biochemistry 22, 917–22.

Gäde, G. (1984). Anaerobic energy metabolism. In: Environmental Physiology and Biochemistry of Insects (ed. K. H. Hoffmann), pp. 119–36. Springer Verlag, Berlin.

Gale, J., Rachmilevitch, S., Reuveni, J. and Volokita, M. (2001). The high oxygen atmosphere toward the end-Cretaceous; a possible contributing factor to the K/T boundary extinctions and to the emergence of C-4 species. Journal of Experimental Botany 52, 801–9.

Gamperl, A. K., Milsom, W. K., Farrell, A. P. and Wang, T. (1999). Cardiorespiratory responses of the toad (*Bufo marinus*) to hypoxia at two different temperatures. Journal of Experimental Biology 202, 3647–58.

Gans, C. (1970). Respiration in early tetrapods – the frog is a red herring. Evolution 24, 723–34.

Gans, C. and Clark, B. (1976). Studies on ventilation of *Caiman crocodilus* (Crocodilia: Reptilia). Respiration Physiology 26, 285–301.

Gans, C., Dejongh, H. J. and Farber, J. (1969). Bullfrog (*Rana catesbeiana*) ventilation: how does the frog breathe? Science 163, 1223–5.

Gans, C. and Hughes, G. M. (1967). Mechanism of lung ventilation in tortoise *Testudo graeca* Linne. Journal of Experimental Biology 47, 1–20.

García-Párraga, D., Lorenzo, T., Wang, T., Ortiz, J. L., Ortega, J., Crespo-Picazo, J. L., Cortijo, J. and Fahlman, A. (2018). Deciphering function of the pulmonary arterial sphincters in loggerhead sea turtles (*Caretta caretta*). Journal of Experimental Biology 221, jeb179820.

Gargaglioni, L. H. and Milsom, W. K. (2007). Control of breathing in anuran amphibians. Comparative Biochemistry and Physiology Part A: Molecular & Integrative Physiology 147, 665–84.

Garland, T. and Huey, R. B. (1987). Testing symmorphosis: does structure match functional requirements? Evolution 41, 1404–9.

Gasc, J.-P. (1981). Axial musculature. In: Biology of the Reptilia, vol. 111 (eds. F. C. Gans and T. S. Parsons), pp. 355–435. Academic Press, New York.

Gaunt, A. S. and Gans, C. (1969a). Mechanics of respiration in snapping turtle, *Chelydra serpentina* (Linne). Journal of Morphology 128, 195–227.

Gaunt, A. S. and Gans, C. (1969b). Diving bradycardia and withdrawal bradycardia in *Caiman crocodilus*. Nature 223, 207–8.

George, J. C. and Shah, R. V. (1954). The occurence of a striated outer muscular sheath in the lungs of *Lissemys punctata granosa* Schoepf. Journal of Animal Morphology and Physiology 1, 13–16.

Ghabrial, A., Luschnig, S., Metzstein, M. M. and Krasnow, M. A. (2003). Branching morphogenesis of the *Drosophila* tracheal system. Annual Review of Cell and Developmental Biology 19, 623–47.

Gifford J. R., Garten, R. S., Nelson, A. S., Trinity, J. D., Layec, G., Witman, M. A. H., Weavil, J. C., Mangum, T., Hart, C., Etheredge, C., et al. (2016). Symmorphosis and skeletal muscle $\dot{V}O_2$max: in vivo and in vitro measures reveal differing constraints in the exercise-trained and untrained human. Journal of Physiology 594, 1741–51.

Gillis, J. A. and Tidswell, O. R. A. (2017). The origin of vertebrate gills. Current Biology 27, 729–32.

Girgis, S. (1961). Aquatic respiration in common Nile turtle, *Trionyx triunguis* (Forskal). Comparative Biochemistry and Physiology 3, 206–17.

Glenner, H., Thomsen, P. F., Hebsgaard, M. B., Sorensen, M. V. and Willerslev, E. (2006). The origin of insects. Science 314, 1883–4.

Goette, A. (1875). Die Entwickelungsgeschichte der Unke (*Bombinator igneus*) als Grundlage einer vergleichenden Morphologie der Wirbelthiere. Leopold Voss, Leipzig.

Goin, C. J., Goin, O. B. and Zug, G. R. (1978). Introduction to Herpetology. W. H. Freeman, San Francisco.

Goldschmid, A. (2007). Chordata. In: Spezielle Zoologie. Teil 2 (eds. W. Westheide and G. Rieger), pp. 1–13. Heidelberg: Spektrum.

Goodrich, E. S. (1930). Studies on the Structure & Development of Vertebrates. MacMillan and Co, London.

Graham, J. (1997). Air-breathing Fishes: Evolution, Diversity, and Adaptation. Academic Press, San Diego.

Graham, J. B., Dudley, R., Aguilar, N. M. and Gans, C. (1995). Implications of the late paleozoic oxygen pulse for physiology and evolution. Nature 375, 117–20.

Graham, J. B., Wegner, N. S., Miller, L. A., Jew, C. J., Lai, N. C., Berquist, R. M., Frank, L. R. and Long, J. A. (2014). Spiracular air breathing in polypterid fishes and its implications for aerial respiration in stem tetrapods. Nature Communications 5, 3022.

Gräper, L. (1931). Zur vergleichenden Anatomie der Schildkrötenlunge. Gegenbaurs Morpholgisches Jahrbuch 68, 325–374.

Gray, J. M., Karow, D. S., Lu, H., Chang, A. J., Chang, J. S., Ellis, R. E., Marletta, M. A. and Bargmann, C. I. (2004). Oxygen sensation and social feeding mediated by a *C. elegans* guanylate cyclase homologue. Nature 430, 317–22.

Green, B. J., Li, W. Y., Manhart, J. R., Fox, T. C., Summer, E. J., Kennedy, R. A., Pierce, S. K. and Rumpho, M. E. (2000). Mollusc-algal chloroplast endosymbiosis. Photosynthesis, thylakoid protein maintenance, and chloroplast gene expression continue for many months in the absence of the algal nucleus. Plant Physiology 124, 331–42.

Greenaway, P., Morris, S. and McMahon, B. (1988). Adaptations to a terrestrial existence by the robber crab *Birgus latro* II. In vivo respiratory gas exchange and transport. Journal of Experimental Biology 140, 493–509.

Greenaway, P., Morris, S., McMahon, B. R., Farrelly, C. A. and Gallagher, K. L. (1996). Air breathing by the purple shore crab, *Hemigrapsus nudus* I. Morphology, behaviour, and respiratory gas exchange. Physiological Zoology 69, 785–805.

Greenberg, S. and Ar, A. (1996). Effects of chronic hypoxia, normoxia and hyperoxia on larval development in the beetle *Tenebrio molitor*. Journal of Insect Physiology 42, 991–6.

Greenlee, K. J. and Harrison, J. F. (1998). Acid-base and respiratory responses to hypoxia in the grasshopper *Schistocerca americana*. Journal of Experimental Biology 201, 2843–55.

Greenlee, K. J. and Harrison, J. F. (2004). Development of respiratory function in the American locust *Schistocerca americana* I. Across-instar effects. Journal of Experimental Biology 207, 497–508.

Greenstone, M. H. and Bennett, A. F. (1980). Foraging strategy and metabolic rates in spiders. Ecology 61, 1255–9.

Greil, A. (1905). Über die Anlage der Lungen, sowie der ultimobranchialen (postbranchialen, supraperikardialen) Körper bei anuren Amphibien. Anatomische Hefte 29, 445–506.

Greil, A. (1914). Zur Frage der Phylogenese der Lungen bei den Wirbeltieren. Erwiderung an Herrn M. Makuschok. Anatomischer Anzeiger 39, 202–6.

Greven, H. and Rudolph, R. (1973). Histologie und Feinstruktur der larvalen Kiemenkammer von Aeshna cyanea. Zeitschrift für Morphologie der Tiere 76, 209–26.

Grocott, M. P. W., Martin, D. S., Levett, D. Z. H., McMorrow, R., Windsor, J., Montgomery, H. E. and Caudwell Xtreme Everest Research Group. (2009). Arterial blood gases and oxygen content in climbers on Mount Everest. New England Journal of Medicine 360, 140–9.

Gruner, H. -E., Moritz, M., and Dunger, W. (1993). Lehrbuch der speziellen Zoologie. Band I: Wirbellose Tiere. 4. Teil: Arthropoda (ohne Insecta), 4th ed. Gustav Fischer Verlag, Jena.

Guadagnoli, J. A., Braun, A. A., Roberts, S. P. and Reiber, C. L. (2005). Environmental hypoxia influences hemoglobin subunit composition in the branchiopod crustacean Triops longicaudatus. Journal of Experimental Biology 208, 3543–51.

Gunga, H. -C. (2009). Nathan Zuntz: His Life and Work in the Fields of High Altitude Physiology and Aviation Medicine. American Physiological Society/Academic Press, Burlington.

Hagner-Holler, S., Schoen, A., Erker, W., Marden, J. H., Rupprecht, R., Decker, H. and Burmester, T. (2004). A respiratory hemocyanin from an insect. Proceedings of the National Academy of Sciences of the United States of America 101, 871–4.

Halperin, J., Ansaldo, M., Pellerano, G. N. and Luquet, C. M. (2000). Bimodal breathing in the estuarine crab Chasmagnathus granulatus Dana 1851 – physiological and morphological studies. Comparative Biochemistry and Physiology – Part A: Molecular & Integrative Physiology 126, 341–9.

Hancock, J. T. and Whiteman, M. (2016). Hydrogen sulfide signaling: interactions with nitric oxide and reactive oxygen species. Respiratory Science, 1365, 5–14.

Hansemann, D. V. (1915). Die Lungenatmung der Schildkröten. Sitzungsberichte der königlich preussischen Akademie der Wissenschaften Berlin 1915, 661–72.

Hardison, R. C. (1996). A brief history of hemoglobins: Plant, animal, protist, and bacteria. Proceedings of the National Academy of Sciences of the United States of America 93, 5675–9.

Hargrove, J. L. (2005). Adipose energy stores, physical work, and the metabolic syndrome: lessons from hummingbirds. Nutrition Journal, 4, 36.

Harrison, J. F. (1989). Ventilatory frequency and hemolymph acid-base status during short-term hypercapnia in the locust, *Schistocerca nitens*. Journal of Insect Physiology 35, 809–14.

Harrison, J. F. (1997). Ventilatory mechanism and control in grasshoppers. American Zoologist 37, 73–81.

Harrison, J. F. (2001). Insect acid-base physiology. Annual Review of Entomology 46, 221–50.

Harrison, J. F. and Lighton, J. R. B. (1998). Oxygen-sensitive flight metabolism in the dragonfly *Erythemis simplicicollis*. Journal of Experimental Biology 201, 1739–44.

Harrison, J. F., Phillips, J. E. and Gleeson, T. T. (1991). Activity physiology of the two-striped grasshopper, *Melanoplus bivittatus*: gas exchange, hemolymph acid-base status, lactate production, and the effect of temperature. Physiological Zoology 64, 451–72.

Harrison, J. M. (1986). Caste-specific changes in honeybee flight capacity. Physiological Zoology 59, 175–87.

Hawkey, C. M., Bennett, P. M., Gascoyne, S. C., Hart, M. G. and Kirkwood, J. K. (1991). Erythrocyte size, number and hemoglobin content in vertebrates. British Journal of Haematology 77, 392–7.

Hazelhoff, E. H. (1927). Die Regulierung der Atmung bei Insekten und Spinnen. Zeitschrift für wissenschaftliche Biologie 5, 179–90.

Hebets, E. A. and Chapman, R. F. (2000). Surviving the flood: plastron respiration in the non-tracheate arthropod *Phrynus marginemaculatus* (Amblypygi: Arachnida). Journal of Insect Physiology 46, 13–9.

Heiss, E., Natchev, N., Beisser, C., Lemell, P. and Weisgram, J. (2010). The fish in the turtle: on the functionality of the oropharynx in the common musk turtle *Sternotherus odoratus* (Chelonia, Kinosternidae) concerning feeding and underwater respiration. Anatomical Record 293, 1416–24.

Heller, J. (1930). Sauerstoffverbrauch der Schmetterlingspuppen in Abhängigkeit von der Temperatur. Zeitschrift für vergleichende Physiologie 11, 448–60.

Henderson, D. M. (2013). Sauropod necks: are they really for heat loss? PLoS One 8, e77108.

Henry, J. R. and Harrison, J. F. (2004). Plastic and evolved responses of larval tracheae and mass to varying atmospheric oxygen content in *Drosophila melanogaster*. Journal of Experimental Biology 207, 3559–67.

Henry, R. P. (1994). Morphological, behavioral, and physiological characterization of bimodal breathing crustaceans. American Zoologist 34, 205–15.

Henry, R. P., McBride, C. J. and Williams, A. H. (1993). Responses of the marsh periwinkle, *Littoraria* (*Littorina*) irrorata to temperature, salinity and desiccation, and the potential physiological relationship to climbing behavior. Marine Behaviour and Physiology 24, 45–54.

Henry, R. P. and Wheatly, M. G. (1992). Interaction of respiration, ion regulation, amd acid-base balance in the everyday life of aquatic Crustaceans. American Zoologist 32, 407–16.

Hervant, F., Mathieu, J. and Messana, G. (1997). Locomotory, ventilatory and metabolic responses of the subterranean *Stenasellus virei* (Crustacea, Isopoda) to severe hypoxia and subsequent recovery. Comptes Rendus De L Academie Des Sciences Serie Iii-Sciences De La Vie-Life Sciences 320, 139–48.

Hetz, S. K. and Bradley, T. J. (2005). Insects breathe discontinuously to avoid oxygen toxicity. Nature 433, 516–9.

Hilken, G. (1997). Tracheal systems in Chilopoda: a comparison under phylogenetic aspects. Scandinavian Entomology Supplement 51, 49–60.

Hilken, G. (1998). Vergleich von Tracheensystemen unter phylogenetischen Aspekten. Naturwissenschaftlicher Verein in Hamburg 37, 5–94.

Himstedt, W. (1996). Die Blindwühlen. (Neue Brehm Bücherei Vol. 630). Westarp Wissenschaften, Magdeburg.

Hoback, W. W. and Stanley, D. W. (2001). Insects in hypoxia. Journal of Insect Physiology 47, 533–42.

Hoback, W. W., Podrabsky, J. E., Higley, L. G., Stanley, D. W. and Hand, S. C. (2000). Anoxia tolerance of confamilial tiger beetle larvae is associated with differences in energy flow and anaerobiosis. Journal of Comparative Physiology B – Biochemical Systemic and Environmental Physiology 170, 307–14.

Hoback, W. W., Stanley, D. W., Higley, L. G. and Barnhart, M. C. (1998). Survival of immersion and anoxia by larval tiger beetles, *Cicindela togeta*. American Midland Naturalist 140, 27–33.

Hochachka, P. W. (1998). Mechanism and evolution of hypoxia-tolerance in humans. Journal of Experimental Biology 201, 1243–54.

Hochachka, P. W., Fields, J. and Mustafa, T. (1973). Animal life without oxygen: basic biochemical mechanisms. American Zoologist 13, 543–55.

Hochachka, P. W., Nener, J. C., Hoar, J., Suarez, R. K. and Hand, S. C. (1993). Disconnecting metabolism from adenylate control during extreme oxygen limitation. Canadian Journal of Zoology 71, 1267–70.

Hochachka, P. W. and Somero, G. (2002). Biochemical Adaptation. Oxford University Press, Oxford.

Hodgson, A. N. (1999). The biology of siphonariid limpets (Gastropoda: Pulmonata). In: Oceanography and Marine Biology, Vol 37, pp. 245–314. Taylor & Francis Ltd, London.

Hoefer, A. M., Perry, S. F. and Schmitz, A. (2000). Respiratory system of arachnids II: morphology of the tracheal system of Leiobunum rotundum and *Nemastoma lugubre* (Arachnida, Opiliones). Arthropod Structure & Development 29, 13–21.

Hoese, B. (1982). Morphologie und Evolution der Lungen bei den terrestrischen Isopoden (Crustacea, Isopoda, Oniscoidea). Zoologische Jahrbücher, Abteilung für Anatomie und Ontogenie der Tiere 107, 396–422.

Hoffman, M., Taylor, B. E. and Harris, M. B. (2016). Evolution of lung breathing from a lungless primitive vertebrate. Respiratory Physiology & Neurobiology 224, 11–6.

Holter, P. (1994). Tolerance of dung insects to low oxygen and high carbon dioxide concentrations. European Journal of Soil Biology 30, 187–93.

Holter, P. and Spangenberg, A. (1997). Oxygen uptake in coprophilous beetles (*Aphodius*, *Geotrupes*, *Sphaeridium*) at low oxygen and high carbon dioxide concentrations. Physiological Entomology 22, 339–43.

Hoogewijs, D., Terwilliger, N. B., Webster, K. A., Powell-Coffman, J. A., Tokishita, S., Yamagata, H., Hankeln, T., Burmester, T., Rytkonen, K. T., Nikinmaa, M. et al. (2007). From critters to cancers: bridging comparative and clinical research on oxygen sensing, HIF signaling, and adaptations towards hypoxia. Integrative and Comparative Biology 47, 552–77.

Hoppeler, H. and Vogt, M. (2001). Muscle tissue adaptations to hypoxia. Journal of Experimental Biology 204, 3133–9.

Hourdez, S., Frederick, L. A., Schernecke, A. and Fisher, C. R. (2001). Functional respiratory anatomy of a deep-sea orbiniid polychaete from the Brine Pool NR-1 in the Gulf of Mexico. Invertebrate Biology 120, 29–40.

Hourdez, S., Weber, R. E., Green, B. N., Kenney, J. M. and Fisher, C. R. (2002). Respiratory adaptations in a deep-sea orbiniid polychaete from Gulf of Mexico brine pool NR-1: metabolic rates and hemoglobin structure/function relationships. Journal of Experimental Biology 205, 1669–81.

Hoyle, G. (1959). The neuromuscular mechanism of an insect spiracular muscle. Journal of Insect Physiology 3, 378–94.

Hsia, C. C., Schmitz, A., Lambertz, M., Perry, S. F. and Maina, J. N. (2013). Evolution of air breathing: oxygen homeostasis and the transitions from water to land and sky. Comprehensive Physiology 3, 849–915.

Huang, S. C. and Newell, R. I. E. (2002). Seasonal variations in the rates of aquatic and aerial respiration and ammonium excretion of the ribbed mussel, *Geukensia demissa* (Dillwyn). Journal of Experimental Marine Biology and Ecology 270, 241–55.

Hughes, G. M. (1972). Morphometrics of fish gills. Respiration Physiology 14, 1–25.

Hughes, G. M. (1984). General anatomy of fish gills. In: Fish Physiology, vol. 10A (eds. J. Hoar and D. J. Randall), pp. 1–72. Academic Press, Orlando.

Hughes, G. M. (1995). The gills of the coelacanth, *Latimeria chalumnae*, a study in relation to body size. Philosophical Transactions of the Royal Society of London B: Biological Sciences 347, 427–38.

Hughes, G. M. (1998). The gills of the coelacanth, *Latimeria chalumnae* Latimeriidae. What can they teach us? Italian Journal of Zoology 65 (Supplement), 425–9.

Hughes, G. M. and Morgan, M. (1973). Structure of fish gills in relation to their respiratory function. Biological Reviews of the Cambridge Philosophical Society 48, 419–75.

Hughes, G. M. and Munshi, J. S. D. (1973). Fine-structure of respiratory organs of the climbing perch, *Anabas testudineus* (Pisces: Anabantidae). Journal of Zoology 170, 201–25.

Hustert, R. (1975). Neuromuscular coordination and proprioceptive control of rhythmical abdominal ventilation in intact *Locusta migratoria migratorioides*. Journal of Comparative Physiology 97, 159–79.

Immesberger, A. and Burmester, T. (2004). Putative phenoloxidases in the tunicate *Ciona intestinalis* and the origin of the arthropod hemocyanin superfamily. Journal of Comparative Physiology B – Biochemical Systemic and Environmental Physiology 174, 169–80.

Innes, A. J. and Taylor, E. W. (1986). The evolution of air-breathing in crustaceans: a functional analysis of branchial, cutanous and pulmonary gas exchange. Comparative Biochemistry and Physiology 85A, 621–37.

Inoue, T., Haque, Z., Lukowiak, K. and Syed, N. I. (2001). Hypoxia-induced respiratory patterned activity in *Lymnaea* originates at the periphery. Journal of Neurophysiology 86, 156–63.

Ireland, L. C. and Gans, C. (1972). Adaptive significance of flexible shell of tortoise *Malacochersus tornieri*. Animal Behaviour 20, 778–81.

Jackson, D. C. (1969). Buoyancy control in freshwater turtle, *Pseudemys scripta elegans*. Science 166, 1649–51.

Jackson, D. C. (2013). Life in a Shell: A Physiologist'S View of a Turtle. Harvard University Press, Cambridge.

Jackson, D. C. and Prange, H. D. (1979). Ventilation and gas-exchange during rest and exercise in adult green sea turtles. Journal of Comparative Physiology 134, 315–9.

Jacobs, W. (1939). Die Lunge der Seeschildkröte *Caretta caretta* (L.) als Schwebeorgan. Zeitschrift vergleichende Physiologie 27, 1–18.

Jaenicke, E. and Decker, H. (2003). Tyrosinases from crustaceans form hexamers. Biochemical Journal 371, 515–23.

Jaenicke, E., Decker, H., Gebauer, W. A., Markl, J. and Burmester, T. (1999). Identification, structure, and properties of hemocyanins from diplopod myriapoda. Journal of Biological Chemistry 274, 29071–4.

Jarecki, J., Johnson, E. and Krasnow, M. A. (1999). Oxygen regulation of airway branching in *Drosophila* is mediated by branchless FGF. Cell 99, 211–20.

Jarvik, E. (1965). Specializations in early vertebrates. Annales de la Société royale zoologique de Belgique 94, 11–95.

Jeram, A. J. (1990). Book-lungs in a lower Carboniferous scorpion. Nature 343, 360–1.

Jeram, A. J., Selden, P. A. and Edwards, D. (1990). Land animals in the Silurian: arachnids and myriapods from Shropshire, England. Science 250, 658–61.

Jonz, M. G. and Nurse, C. A. (2009). Oxygen-sensitive neuroepithelial cells in the gills of aquatic vertebrates. In: Airway Chemoreceptors in the Vertebrates: Structure, Evolution and Function (eds. G. Zaccone, E. Cutz, D. Adriaensen, C. A. Nurse and A. Mauceri), pp. 1–30. Science Publishers, Enfield.

Jonz, M. G., Buck, L. T., Perry, S. F., Schwerte, T. and Zaccone, G. (2016). Sensing and surviving hypoxia in vertebrates. Respiratory Science, 1365, 43–58.

Joos, B., Lighton, J. R. B., Harrison, J. F., Suarez, R. K. and S.P., R. (1997). Effects of ambient oxygen tension on flight performance, metabolism, and water loos in the honeybee. Physiological Zoology 70, 167–74.

Jørgenson, C. B. (1998). Role of urinary and cloacal bladders inchelonian water economy : historical and comparative perspectives. Biological Reviews 73, 347–66.

Jürgens, K. D., Bartels, H. and Bartels, R. (1981). Blood-oxygen transport and organ weights of small bats and small non-flying mammals. Respiration Physiology 45, 243–60.

Kaestner, A. (1929). Bau und Funktion der Fächertracheen einiger Spinnen. Zeitschrift für Morphologie und Ökologie der Tiere 13, 463–558.

Kaiser, A., Klok, C. J., Socha, J. J., Lee, W. K., Quinlan, M. C. and Harrison, J. F. (2007). Increase in tracheal investment with beetle size supports hypothesis of oxygen limitation on insect gigantism. Proceedings of the National Academy of Sciences of the United States of America 104, 13198–203.

Kamenz, C., Dunlop, J. A. and Scholtz, G. (2005). Characters in the book lungs of Scorpiones (Chelicerata, Arachnida) revealed by scanning electron microscopy. Zoomorphology 124, 101–9.

Kamenz, C., Dunlop, J. A., Scholtz, G., Kerp, H. and Hass, H. (2008). Microanatomy of early devonian book lungs. Biology Letters 4, 212–5.

Kanatous, S. B., Hawke, T. J., Trumble, S. J., Pearson, L. E., Watson, R. R., Garry, D. J., Williams, T. M. and Davis, R. W. (2008). The ontogeny of aerobic and diving capacity in the skeletal muscles of Weddell seals. Journal of Experimental Biology 211, 2559–65.

Kardong, K. V. (2002). Vertebrates: Comparative Anatomy, Function, Evolution. McGraw-Hill, New York.

Keister, M. and Buck, J. (1961). Respiration of *Phormia regina* in relaton to temperature and oxygen. Journal of Insect Physiology 7, 51–72.

Kempter, B., Markl, J., Brenowitz, M., Bonaventura, C. and Bonaventura, J. (1985). Immunological correspondence between arthropod hemocyanin subunits. 2. Xiphosuran (*Limulus*) and spider (*Eurypelma*,

Cupiennius) hemocyanin. Biological Chemistry Hoppe-Seyler 366, 77–86.

Kempton, R. T. (1969). Morphological features of functional significance in gills of spiny dogfish *Squalus acanthias*. Biological Bulletin 136, 226–40.

Kestler, P. (1985). Respiration and respiratory water loss. In: Environmental Physiology and Biochemistry of Insects (ed. Klaus Hoffmann) 137–84. Springer Verlag, Berlin.

King, P. and Heatwole, H. (1994). Non-pulmonary respiratory surfaces of the Chelid turtle *Elseya latisternum*. Herpetologica 50, 262–5.

Kinkead, R., Belzile, O. and Gulemetova, R. (2002). Serotonergic modulation of respiratory motor output during tadpole development. Journal of Applied Physiology 93, 936–46.

Kirkton, S. D., Niska, J. A. and Harrison, J. E. (2005). Ontogenetic effects on aerobic and anaerobic metabolism during jumping in the American locust, *Schistocerca americana*. Journal of Experimental Biology 208, 3003–12.

Kjellesvig-Waering, E. N. (1986). A restudy of the fossil Scorpionida of the world. Palaeontographica Americana 55, 1–287.

Klein, W., Abe, A. and Perry, S. F. (2003a). Static lung compliance and body pressures in *Tupinambis merianae* with and without post-hepatic septum. Respiratory Physiology & Neurobiology 135, 73–86.

Klein, W., Abe, A. S., Andrade, D. V. and Perry, S. F. (2003b). Structure of the posthepatic septum and its influence on visceral topology in the tegu lizard, *Tupinambis merianae* (Teiidae: Reptilia). Journal of Morphology 258, 151–7.

Klein, W., Andrade, D. V., Abe, A. S. and Perry, S. F. (2003c). Role of the post-hepatic septum on breathing during locomotion in *Tupinambis merianae* (Reptilia: Teiidae). Journal of Experimental Biology 206, 2135–43.

Klein, W., Böhme, W. and Perry, S. F. (2000). The mesopneumonia and the post-hepatic septum of the Teiioidea (Reptilia: Squamata). Acta Zoologica 81, 109–19.

Klein, W., Reuter, C., Böhme, W. and Perry, S. F. (2005). Lungs and mesopneumonia of scincomorph lizards (Reptilia: Squamata). Organisms Diversity & Evolution 5, 47–57.

Klok, C. J., Mercer, R. D. and Chown, S. L. (2002). Discontonous gas-exchange in centipedes and its convergent evolution in tracheated arthropods. Journal of Experimental Biology 205, 1019–29.

Kobayashi, H., Fujiki, M. and Suzuki, T. (1988). Variation in oxygen-binding properties of *Daphnia magna* hemoglobin. Physiological Zoology 61, 415–9.

Koeppen, B. M. and Stanton, B. A. (2017). Berne & Levy Physiology, 7th ed. Elsevier, Amsterdam.

Kogo, N., Perry, S. F. and Remmers, J. E. (1997). Laryngeal motor control in frogs: Role of vagal and laryngeal feedback. Journal of Neurobiology 33, 213–22.

Kogo, N. and Remmers, J. E. (1994). Neural organization of the ventilatory activity in the frog, *Rana catesbeiana*. 2. Journal of Neurobiology 25, 1080–94.

Kohnert, S., Perry, S. F. and Schmitz, A. (2004). Morphometric analysis of the larval branchial chamber in the dragonfly *Aeshna cyanea* Muller (Insecta, Odonata, Anisoptera). Journal of Morphology 261, 81–91.

König, P., Krain, B., Krasteva, G. and Kummer, W. (2009). Serotonin increases cilia-driven particle transport via an acetylcholine-independent pathway in the mouse trachea. PloS One 4, e4938.

Kooyman, G. L. and Ponganis, P. J. (1997). The challanges of diving to depth. American Scientist 85, 530–9.

Kovac, D. (1982). Zur Überwinterung der Wasserwanze *Plea minutissima* (Heteroptera, Pleidae): Diapause mit Hilfe der Plastronatmung. Nachrichten des Entomologischen Vereins Apollo 3, 59–76.

Krishnan, S. N., Sun, Y. A., Mohsenin, A., Wyman, R. J. and Haddad, G. G. (1997). Behavioral and electrophysiologic responses of *Drosophila melanogaster* to prolonged periods of anoxia. Journal of Insect Physiology 43, 203–10.

Krogh, A. (1920a). Studien über Tracheenrespiration III. Die Kombination von mechanischer Ventilation mit Gasdiffusion nach Versuchen an Dytiscuslarven. Pflügers Archiv 179, 113–20.

Krogh, A. (1920b). Studien über Tracheenerespiration II. Über Gasdiffusion in den Tracheen. Pflügers Archiv 179, 95–112.

Krogh, A. (1941). The Comparative Physiology of Respiratory Mechanisms. Dover Publications. Inc, New York.

Krolikowski, K. and Harrison, J. F. (1996). Haemolymph acis-base status, tracheal gas levels and the control of post-exercise ventilation rate in grasshoppers. Journal of Experimental Biology 199, 391–9.

Kusche, K. and Burmester, T. (2001). Diplopod hemocyanin sequence and the phylogenetic position of the Myriapoda. Molecular Biology and Evolution 18, 1566–73.

Kusche, K., Hembach, A., Hagner-Holler, S., Gebauer, W. and Burmester, T. (2003). Complete subunit sequences, structure and evolution of the 6x6-mer hemocyanin from the common house centipede, *Scutigera coleoptrata*. European Journal of Biochemistry 270, 2860–8.

Kusche, K., Ruhberg, H. and Burmester, T. (2002). A hemocyanin from the Onychophora and the emergence of respiratory proteins. Proceedings of the National Academy of Sciences of the United States of America 99, 10545–8.

Lambertz, M. (2016a). Craniota vs. Craniata: arguments towards nomenclatural consistency. Journal of Zoological Systematics and Evolutionary Research 54, 174–6.

Lambertz, M. (2016b). Recent advances on the functional and evolutionary morphology of the amniote respiratory apparatus. Annals of the New York Academy of Sciences, 1365, 100–13.

Lambertz, M. (2017). The vestigial lung of the coelacanth and its implications for understanding pulmonary diversity among vertebrates: new perspectives and open questions. Royal Society Open Science 4, 171518.

Lambertz, M., Arenz, N. and Grommes, K. (2018). Variability in pulmonary reduction and asymmetry in a serpentiform lizard: the sheltopusik, *Pseudopus apodus* (Pallas, 1775). Vertebrate Zoology 68, 21–6.

Lambertz, M., Böhme, W. and Perry, S. F. (2010). The anatomy of the respiratory system in *Platysternon megacephalum* Gray, 1831 (Testudines: Cryptodira) and related species, and its phylogenetic implications. Comparative Biochemistry and Physiology Part A: Molecular & Integrative Physiology 156, 330–6.

Lambertz, M., Shelton, C. S., Spindler, F. and Perry, S. F. (2016). A caseian point for the evolution of a diaphragm homologue among the earliest synapsids. Annals of the New York Academy of Sciences, 1385, 3–20.

Lambertz, M., Grommes, K., Kohlsdorf, T. and Perry, S. F. (2015). Lungs of the first amniotes: why simple if they can be complex? Biology Letters 11, 20140848.

Lambertz, M. and Perry, S. F. (2015, "2016"). The lung-swimbladder issue: a simple case of homology – or not?. In: Zaccone, G., Dabrowski, K., Hedrick, M.S., Fernandes, J. M. O. and Icardo, J. M. (Eds.): Phylogeny, Anatomy and Physiology of Ancient Fishes. pp. 201–211, CRC Press, Boca Raton, FL.

Lamy, E. (1902). Les trachées des araignées. Annales des sciences naturelles, Zoologie 15, 149–280.

Landberg, T., Mailhot, J. D. and Brainerd, E. L. (2003). Lung ventilation during treadmill locomotion in a terrestrial turtle, *Terrapene carolina*. Journal of Experimental Biology 206, 3391–404.

Landberg, T., Mailhot, J. D. and Brainerd, E. L. (2009). Lung ventilation during treadmill locomotion in a semi-aquatic turtle, *Trachemys scripta*. Journal of Experimental Zoology Part a-Ecological Genetics and Physiology 311A, 551–62.

Lane, N. and Martin, W. F. (2012). The origin of membrane bioenergetics. Cell 151, 1406–16.

Lauweryns, J. M., Cokelaere, M., Lerut, T. and Theunynck, P. (1978). Cross-circulation studies on influence of hypoxia and hypoxemia on neuro-epithelial bodies in young rabbits. Cell and Tissue Research 193, 373–86.

Lavista-Llanos, S., Centanin, L., Irisarri, M., Russo, D. M., Gleadle, J. M., Bocca, S. N., Muzzopappa, M., Ratcliffe, P. J. and Wappner, P. (2002). Control of the hypoxic response in *Drosophila melanogaster* by the basic helix-loop-helix PAS protein similar. Molecular and Cellular Biology 22, 6842–53.

Law, S. H. W., Wu, R. S. S., Ng, P. K. S., Yu, R. M. K. and Kong, R. Y. C. (2006). Cloning and expression analysis of two distinct HIF-alpha isoforms – gcHIF-1alpha and gcHIF-4alpha – from the hypoxia-tolerant grass carp, *Ctenopharyngodon idellus*. BMC Molecular Biology 7, 15.

Lease, H. M., Wolf, B. O. and Harrison, J. F. (2006). Intraspecific variation in tracheal volume in the American locust, *Schistocerca americana*, measured by a new inert gas method. Journal of Experimental Biology 209, 3476–83.

Lechner, A. J. and Banchero, N. (1980). Lung morphometry in guinea-pigs acclimated to hypoxia during growth. Respiration Physiology 42, 155–69.

Lee, S. Y., Lee, B. L. and Soderhall, K. (2003). Processing of an antibacterial peptide from hemocyanin of the freshwater crayfish *Pacifastacus leniusculus*. Journal of Biological Chemistry 278, 7927–33.

Leite, C. A. C., Wang, T., Abe A. S. and Taylor, E. W. (2014). Autonomic control of the cardiorespiratory interactions in vertebrates. In: Abstracts of the Third International Congress of Respiratory Science (eds. M. Lambertz, K. Grommes, and S. F. Perry), p. 76. Franzbecker, Hildesheim.

Levi, H. W. (1967). Adaptations of respiratory systems of spiders. Evolution 21, 571–83.

Levi, H. W. (1976). On the evolution of tracheae in Arachnids. Bulletin of the British Arachnological Society 3, 187–8.

Lewis, G. W., Miller, P. L. and Mills, P. S. (1973). Neuromuscular mechanisms of abdominal pumping in the locust. Journal of Experimental Biology 59, 149–68.

Lieb, B., Dimitrova, K., Kang, H. S., Braun, S., Gebauer, W., Martin, A., Hanelt, B., Saenz, S. A., Adema, C. M. and Markl, J. (2006). Red blood with blue-blood ancestry: intriguing structure of a snail hemoglobin. Proceedings of the National Academy of Sciences of the United States of America 103, 12011–6.

Liebensteiner, M. G., Oosterkamp, M. J. and Stams, A. J. M. (2016). Microbial respiration with chlorine oxyanions: diversity and physiological and biochemical properties of chlorate- and perchlorate-reducing microorganisms. Respiratory Science, 1365, 59–72.

Liem, K. F. (1988). Form and function of lungs: the evolution of air breathing mechanisms. American Zoologist 28, 739–59.

Liem, K. F. (1989). Respiratory gas bladders in teleosts: functional conservatism and morphological diversity. American Zoologist 29, 333–52.

Lighton, J. R. B. (1996). Discontinous gas exchange in insects. Annual Review of Entomology 41, 309–24.

Lighton, J. R. B. (1998). Notes from underground: towards ultimate hypotheses of cyclic, discontinous gas-exchange in tracheate arthropods. American Zoologist 38, 483–91.

Lighton, J. R. B. (2007). Respiratory biology: why insects evolved discontinuous gas exchange. Current Biology 17, R645–7.

Lighton, J. R. B. and Berrigan, D. (1995). Questioning paradigms: casre-specific ventilation in harvester ants, *Messor pergandei* and *M. julianus* (Hymenoptera: Formicidae). Journal of Experimental Biology 198, 521–30.

Lighton, J. R. B. and Duncan, F. D. (1995). Standard and exercise metabolism and the dynamics of gas exchange in the giant red velvet mite, *Dinothrombium magnificum*. Journal of Insect Physiology 41, 877–84.

Lighton, J. R. B. and Fielden, L. J. (1996). Gas exchange in wind spiders (Arachnida, Solphugidae): independent evolution of convergent control strategies in solphugids and insects. Journal of Insect Physiology 42, 347–57.

Lighton, J. R. B., Fielden, L. J. and Rechav, Y. (1993). Discontinous ventilation in a non-insect, the tick *Amblyomma marmoreum* (Acari, Ixodidae): characterization and metabolic modulation. Journal of Experimental Biology 180, 229–45.

Lighton, J. R. B. and Joos, B. (2002). Discontinuous gas exchange in the pseudoscorpion *Garypus californicus* is regulated by hypoxia, not hypercapnia. Physiological and Biochemical Zoology 75, 345–9.

Lighton, J. R. B. and Ottesen, E. A. (2005). To DGC or not to DGC: oxygen guarding in the termite *Zootermopsis nevadensis* (Isoptera: Termopsidae). Journal of Experimental Biology 208, 4671–8.

Lighton, J. R. B. and Turner, R. J. (2008). The hygric hypothesis does not hold water: abolition of discontinuous gas exchange cycles does not affect water loss in the ant *Camponotus vicinus*. Journal of Experimental Biology 211, 563–7.

Lighton, J. R. B. and Wehner, R. (1993). Ventilation and respiratory metabolism in the thermophilic desert ant, *Cataglyphis bicolor* (Hymenoptera, Formicidae). Journal of Comparative Physiology B: Biochemical, Systemic, and Environmental Physiology 163, 11–7.

Locke, M. (1958a). The co-ordination of growth in the tracheal system of insects. Quarterly Journal of Microscopical Science 99, 373–91.

Locke, M. (1958b). The formation of tracheae and tracheoles in *Rhodnius prolixus*. Quarterly Journal of Microscopical Science 99, 29–46.

Locy, W. W. and Larsell, O. (1916a). The embryology of the bird's lung based of observations of the domestic fowl. Part 1. American Journal of Anatomy 19, 447–504.

Locy, W. W. and Larsell, O. (1916b). The embryology of the bird's lung based of observations of the domestic fowl. Part 2. American Journal of Anatomy 20, 1–44.

Lomholt, J. P., Johansen, K. and Maloiy, G. M. O. (1975). Is the aestivating lungfish the first vertebrate with suctional breathing. Nature 257, 787–8.

Loudon, C. (1988). Development of *Tenebrio molitor* in low oxygen levels. Journal of Insect Physiology 34, 97–103.

Loudon, C. (1989). Tracheal hypertrophy in mealworms: design and plasticity in oxygen supply systems. Journal of Experimental Biology 147, 217–35.

Luchtel, D. L. and Kardong, K. V. (1981). Ultrastructure of the lung of the rattlesnake, *Crotalus viridis oreganus*. Journal of Morphology 169, 29–47.

Lyson, T. R., Schachner, E. R., Botha-Brink, J., Scheyer, T. M., Lambertz, M., Bever, G. S., Rubidge, B. S. and de Queiroz, K. (2014). Origin of the unique ventilatory apparatus of turtles. Nature Communications 5, 5211.

Ma, E. and Haddad, G. G. (1999). Isolation and characterization of the hypoxia-inducible factor 1 beta in *Drosophila melanogaster*. Molecular Brain Research 73, 11–6.

Macintyre, D. H. and Toews, D. P. (1976). The mechanics of lung ventilation and the effects of hypercapnia on respiration in Bufo marinus. Canadian Journal of Zoology 54, 1364–74.

Madan, J. J. and Wells, M. J. (1996). Cutaneous respiration in Octopus vulgaris. Journal of Experimental Biology 199, 2477–83.

Maina, J. N. (2005). The Lung-Air Sac System of Birds. Springer, Berlin.

Maina, J. N. (ed.) (2017). The biology of the avian respiratory system - Evolution, development, structure and function. Springer, Cham.

Maina, J. N. and Maloiy, G. M. O. (1986). The morphology of the respiratory organs of the African air-breathing catfish (*Clarias mossambicus*): a light, electron and scanning microscopic study, with morphometric observations. Journal of Zoology 209, 421–45.

Makuschok, M. (1913). Über genetische Beziehung zwischen Schwimmblase und Lungen. Anatomischer Anzeiger 44, 33–55.

Mallatt, J. (1996). Ventilation and the origin of jawed vertebrates: a new mouth. Zoological Journal of the Linnean Society 117, 329–404.

Mallatt, J. and Paulsen, C. (1986). Gill ultrastructure of the Pacific hagfish *Eptatretus stouti*. American Journal of Anatomy 177, 243–69.

Mangum, C. P. (1986). Invertebrate oxygen carriers. In: Invertebrate Oxygen Carriers, (ed. B. Linzen), pp. 277–80. Springer Verlag, Berlin.

Mangum, C. P. (1998). Major events in the evolution of the oxygen carriers. American Zoologist 38, 1–13.

Manning, P. L. and Dunlop, J. A. (1995). The respiratory organs of Eurypterids. Palaeontology 38, 287–97.

Mans, C., Drees, R., Sladky, K. K., Hatt, J. -M. and Kircher, P. R. (2013). Effects of body position and extension of the neck and extremities on lung volume measured via computed tomography in red-eared slider turtles (*Trachemys scripta elegans*). Journal of the American Veterinary Medical Association 243, 1190–6.

Marais, E., Klok, C. J., Terblanche, J. S. and Chown, S. L. (2005). Insect gas exchange patterns: a phylogenetic perspective. Journal of Experimental Biology 208, 4495–507.

Marcus, H. (1923). Beitrag zur Kenntnis der Gymnophionen. VI. Über den Übergang von der Wasser- zur Luftatmung mit besonderer Berücksichtigung des Atemmechanismus von *Hypogeophis*. Zeitschschrift für Anatomie und Entwicklungsgeschichte 69, 328–43.

Marcus, H. (1937). Lungen. In: Handbuch der Vergleichenden Anatomie der Wirbertiere (eds. L. Bolk,

E. Göppert, E. Kallius, and W. Lubosch), pp. 909–88. Urban and Schwarzenberg, Berlin.

Marder, E. and Bucher, D. (2001). Central pattern generators and the control of rhythmic movements. Current Biology 11, R986–96.

Marinelli, W. and Strenger, A. (1959). Vergleichende Anatomie und Morphologie der Wirbeltiere I. *Squalus acanthias*. Deuticke, Wien.

Markl, J. (1986). Evolution and function of structurally diverse subunits in the respiratory protein hemocyanin from Arthropods. Biological Bulletin 171, 90–115.

Markl, J. (2013). Evolution of molluscan hemocyanin structures. Biochimica Et Biophysica Acta-Proteins and Proteomics 1834, 1840–52.

Markl, J. and Decker, H. (1992). Molecular structure of the arthropod hemocyanins. Advances in Comparative and Environmental Physiology, 13, 325–76.

Markl, J., Stöcker, W., Runzler, R. and Precht, E. (1986). Immunological correspondence between the hemocyanin subunits of 86 arthropods: evolution of a multigene protein family. In: Invertebrate Oxygen Carriers (ed. B. Linzen), pp. 281–99. Springer Verlag, Berlin.

Martinez, C. B. R., Alvares, E. P., Harris, R. R. and Santos, M. C. F. (1999). A morphological study on posterior gills of the mangrove crab Ucides cordatus. Tissue & Cell 31, 380–9.

Maruyama, I., Inagaki, M. and Momose, K. (1984). The role of serotonin in mucociliary transport system in the ciliated epithelium of frog palatine mucosa. European Journal of Pharmacology 106, 499–506.

Marxen, J. C., Pick, C., Oakley, T. H. and Burmester, T. (2014). Occurrence of hemocyanin in ostracod crustaceans. Journal of Molecular Evolution 79, 3–11.

Matthews, P. G. D. and Seymour, R. S. (2008). Haemoglobin as a buoyancy regulator and oxygen supply in the backswimmer (Notonectidae, *Anisops*). Journal of Experimental Biology 211, 3790–9.

Maurer, F. (1896). Die ventrale Rumpfmuskulatur einiger Reptilien - Eine vergleichend-anatomische Untersuchung. In: Festschrift zum siebzigsten Geburtstag von Carl Gegenbauer, pp. 181–256. Engelmann, Leipzig.

McClain, C. R. and Rex, M. A. (2001). The relationship between dissolved oxygen concentration and maximum size in deep-sea turrid gastropods: an application of quantile regression. Marine Biology 139, 681–5.

McComb, C., Meems, R., Syed, N. and Lukowiak, K. (2003). Electrophysiological differences in the CPG aerial respiratory behavior between juvenile and adult Lymnaea. Journal of Neurophysiology 90, 983–92.

McLean, H. A., Kimura, N., Kogo, N., Perry, S. F. and Remmers, J. E. (1995). Fictive respiratory rhythm in the isolated brain-stem of frogs. Journal of Comparative Physiology A: Neuroethology, Sensory, Neural, and Behavioral Physiology 176, 703–13.

McLean, H. A. and Remmers, J. E. (1997). Characterization of respiratory-related neurons in the isolated brainstem of the frog. Journal of Comparative Physiology A: Neuroethology, Sensory, Neural, and Behavioral Physiology 181, 153–9.

McMahon, B. R. (1969). A functional analysis of aquatic and aerial respiratory movements of an African lungfish, *Protopterus aethiopicus*, with reference to evolution of lungventilation mechanism in vertebrates. Journal of Experimental Biology 51, 407–30.

McMahon, B. R. and Burggren, W. W. (1979). Respiration and adaptation to the terrestrial habitat in the land hermit crab *Coenobita clypeatus*. Journal of Experimental Biology 79, 265–81.

McMahon, B. R. and Burggren, W. W. (1988). Respiration. In: Biology of the Land Crabs (eds. W. W. Burggren and B. R. McMahon), pp. 249–97. Cambridge University Press, New York.

McQueen, D. J. (1980). Active respiration rates for the burrowing wolf spider *Geolycosa domifex* (Hancock). Canadian Journal of Zoology 58, 1066–74.

Melzner, F., Bock, C. and Portner, H. O. (2007). Coordination between ventilatory pressure oscillations and venous return in the cephalopod Sepia officinalis under control conditions, spontaneous exercise and recovery. Journal of Comparative Physiology B – Biochemical Systemic and Environmental Physiology 177, 1–17.

Mendes, E. G. and Sawaya, P. (1958). The oxygen consumption of "Onychophora" and its relation to size, temperature and oxygen tension. Revista Brasileira de Biologia 18, 129–42.

Mendes, E. G. and Ulian, G. B. (1987). The influence of size, temperature and oxygen tension upon the respiratory metabolism of the terrestrial amphipod *Talitrus* (*Talitroides*) *pacificus* Hurley, 1955. Comparative Biochemistry and Physiology 86A, 155–62.

Mendez-Sanchez, J. F. and Burggren, W. W. (2017). Cardiorespiratory physiological phenotypic plasticity in developing air-breathing anabantid fishes (*Betta splendens* and *Trichopodus trichopterus*). Physiological Reports 5, e13359.

Messner, B. and Adis, J. (1995). Es gibt nur fakultative Plastronatmer unter den tauchenden Webspinnen. Deutsche Entomologische Zeitschrift N.F. 42, 453–9.

Metzger, R. J. and Krasnow, M. A. (1999). Development: genetic control of branching morphogenesis. Science 284, 1635–9.

Mickoleit, G. (2004). Phylogenetische Systematik der Wirbeltiere. Verlag Dr. Friedrich Pfeil, Munich.

Milani, A. (1897). Beiträge zur Kenntnis der Reptilienlunge II. Zoologische Jahrbücher, Abteilung für Anatomie und Ontogenie der Tiere 10, 93–153.

Mill, P. J. (1974). Respiration: aquatic insects. In: The Physiology of Insecta VI (ed. M. Rockstein), pp. 403–69. Academic Press, London.

Mill, P. J. (1985). Structure and physiology of the hemoespiratory system. In: Comprehensive Insect Physiology, Biochemistry And Pharmacology, Vol. 3, Integument, Respiration and Circulation, pp. 517–93. Pergamon Press: Oxford.

Mill, P. J. and Pickard, R. S. (1972). Anal valve movement and normal ventilation in aeshnid dragonfly larvae. Journal of Experimental Biology 56, 537–43.

Miller, J. E. and Pawson, D. L. (1989). Hansenothuria benti, new genus, new species (Echinodermata, Holothuroidea) from the Tropical Western Atlantic: a bathyal, epibenthic holothurian with swimming abilities. Proceedings of the Biological Society of Washington 102, 977–86.

Miller, P. L. (1962). Spiracle control in adult dragonflies (Odonata). Journal of Experimental Biology 39, 513–35.

Miller, P. L. (1964). Factors altering spiracle control in adult dragonflies: Hypoxia and temperature. Journal of Experimental Biology 41, 345–57.

Miller, P. L. (1966). The regulation of breathing in insects. Advances in Insect Physiology 3, 279–344.

Miller, P. L. (1969). Inhibitory nerves to the insect spiracles. Nature 221, 171–3.

Millidge, A. F. (1986). A revision of the tracheal structures of the Linyphiidae (Araneae). Bulletin of the British Arachnological Society 7, 57–61.

Milsom, W. K. and Johansen, K. (1975). Effect of buoyancy induced lung-volume changes on respiratory frequency in a chelonian (*Caretta caretta*). Journal of Comparative Physiology 98, 157–60.

Milsom, W. K. and Vitalis, T. Z. (1984). Pulmonary mechanics and the work of breathing in the lizard, *Gekko gecko*. Journal of Experimental Biology 113, 187–202.

Milsom, W. K. and Burleson, M. L. (2007). Peripheral arterial chemoreceptors and the evolution of the carotid body. Respiratory Physiology & Neurobiology 157, 4–11.

Milsom, W. K., Reid, S. G., Meier, J. T. and Kinkead, R. (1999). Central respiratory pattern generation in the bullfrog, *Rana catesbeiana*. Comparative Biochemistry and Physiology a-Molecular and Integrative Physiology 124, 253–64.

Mirceta, S., Signore, A. V., Burns, J. M., Cossins, A. R., Campbell, K. L. and Berenbrink, M. (2013). Evolution of mammalian diving capacity traced by myoglobin net surface charge. Science 340, 1234192.

Monge-Najera, J. and Hou, X. U. (2002). Experimental taphonomy of velvet worms (Onychophora) and implications for the Cambrian "explosion, disparity and decimation" model. Revista de Biologia Tropical 50, 1133–8.

Moore, S. J. (1976). Some spider organs as seen by the scanning electron microscope, with special reference to the book-lung. Bulletin of the British Arachnological Society 3, 177–87.

Moraes, M. F., Holler, S., Costa, O. T., Glass, M. L., Fernandes, M. N. and Perry, S. F. (2005). Morphometric comparison of the respiratory organs in the South American lungfish *Lepidosiren paradoxa* (Dipnoi). Physiological and Biochemical Zoology 78, 546–59.

Morris, S. (1991). Respiratory gas exchange and transport in crustaceans: ecological determinants. Memoirs of the Queensland Museum 31, 241–61.

Morris, S. (2002). The ecophysiology of air-breathing in crabs with special reference to *Gecarcoidea natalis*. Comparative Biochemistry and Physiology Part B: Biochemistry & Molecular Biology 131, 559–70.

Morris, S. and Callaghan, J. (1998). The emersion response of the Australian Yabby *Cherax destructor* to environmental hypoxia and the respiratory and metabolic responses to consequent air-breathing. Journal of Comparative Physiology B: Biochemical, Systemic, and Environmental Physiology 168, 389–98.

Morris, S., Greenaway, P. and McMahon, B. R. (1988). Adaptations to a terrestrial existence by the robber crab *Birgus latro* I. An in vitro investigation of blood gas transport. Journal of Experimental Biology 140, 477–91.

Morris, S., Greenaway, P. and McMahon, B. R. (1996). Air breathing by the purple shoe crab, *Hemigrapsus nudus* (Dana). IV. Aquatic hypoxia as an impetus for emersion? Oxygen uptake, respiratory gas transport, and acid-base state. Physiological Zoology 69, 864–86.

Mortola, J. P., Frappell, P. B. and Woolley, P. A. (1999). Breathing through skin in a newborn mammal. Nature 397, 660.

Morton, D. B. (2004). Atypical soluble guanylyl cyclases in *Drosophila* can function as molecular oxygen sensors. Journal of Biological Chemistry 279, 50651–3.

Moser, F. (1904). Beiträge zur vergleichenden Entwicklungsgeschichte der Schwimmblase. Archive der Mikroskopie, Anatomie und Entwicklungsgeschichte 63, 532–74.

Moussa, T. A. (1956). Morphology of the accessory air-breathing organs of the teleost, *Clarias lazera* (C. and V.). Journal of Morphology 98, 125–60.

Müller, G., Fago, A. and Weber, R. E. (2003). Water regulates oxygen binding in hagfish (*Myxine glutinosa*) hemoglobin. Journal of Experimental Biology 206, 1389–95.

Müller, J. (1831). Kiemenlöcher an einer jungen *Coecilia hypocyanea* im Museum der Naturgeschichte zu Leyden beobachtet. Isis von Oken 1831, 709–11.

Munns, S. L., Owerkowicz, T., Andrewartha, S. J. and Frappell, P. B. (2012). The accessory role of the diaphragmaticus muscle in lung ventilation in the estuarine crocodile *Crocodylus porosus*. Journal of Experimental Biology 215, 845–52.

Munshi, J. S. D. (1961). Accessory respiratory organs of *Clarias batrachus* (Linn). Journal of Morphology 109, 115–39.

Munshi, J. S. D. (1985). Structure, function and evolution of accessory breathing organs of air-breathing fishes of India. Fortschritte der Zoologie 30, 353–66.

Myers, G. S. (1942). The lungs of *Bothriolepis*. Ichthyological Bulletin 2, 134–6.

Myers, T. B. and Retzlaff, E. (1963). Localization and action of the centre of the cuban burrowing cockroach. Journal of Insect Physiology 9, 607–14.

Nattie, E. (1999). $CO(_2)$, brainstem chemoreceptors and breathing. Progress in Neurobiology 59, 299–331.

Nattie, E. and Li, A. H. (2009). Central chemoreception is a complex system function that involves multiple brain stem sites. Journal of Applied Physiology 106, 1464–6.

Nentwig, W. (2012). The species referred to as *Eurypelma californicum* (Theraphosidae) in more than 100 publications is likely to be *Aphonopelma hentzi*. Journal of Arachnology 40, 128–30.

Neumayer, L. (1930). Die Entwicklung des Darms von *Acipenser*. Acta Zoologica (Stockholm) 39, 1–151.

Nguyen, T., Chin, W. -C., O'Brien, J. A., Verdugo, P. and Berger, A. J. (2001). Intracellular pathways regulating ciliary beating of rat brain ependymal cells. Journal of Physiology 531, 131–40.

Nielsen, M. G. and Christian, K. A. (2007). The mangrove ant, *Camponotus anderseni*, switches to anaerobic respiration in response to elevated CO_2 levels. Journal of Insect Physiology 53, 505–508.

Nikam, T. B. and Khole, V. V. (1989). Insect Spiracular Systems. Ellis Horwood, Chichester.

Nilsson, G. E. and Lutz, P. L. (2004). Anoxia tolerant brains. Journal of Cerebral Blood Flow and Metabolism 24, 475–86.

Noble, G. K. (1931). The Biology of the Amphibia. McGraw-Hill, New York.

Nogge, G. (1976). Ventilationsbewegungen bei Solifugen. Zoologischer Anzeiger 196, 145–9.

Ong, K. J., Stevens, E. D. and Wright, P. A. (2007). Gill morphology of the mangrove killifish (*Kryptolebias marmoratus*) is plastic and changes in response to terrestrial air exposure. Journal of Experimental Biology 210, 1109–15.

Opell, B. D. (1987). The influence of web monitoring tactics on the tracheal systems of spiders in the family Uloboridae (Arachnida, Araneida). Zoomorphology 107, 255–9.

Opell, B. D. (1990). The relationships of book lung and tracheal systems in the spider family Uloboridae. Journal of Morphology 206, 211–6.

Opell, B. D. and Konur, D. C. (1992). Influence of web-monitoring tactics on the density of mitochondria in leg muscles of the spider family Uloboridae. Journal of Morphology 213, 341–7.

Orgeig, S., Morrison, J. L. and Daniels, C. B. (2016). Evolution, development, and function of the pulmonary surfactant system in normal and perturbed environments. Comprehensive Physiology 6, 363–422.

Otis, A. B. (1986). History of respiratory mechanics. In: Handbook of Physiology, Section 3: The respiratory System (eds. A. P. Fishman, P. T. Macklem, and J. Mead), pp. 1–12. American Physiological Society, Bethesda.

Owerkowicz, T., Farmer, C. G., Hicks, J. W. and Brainerd, E. L. (1999). Contribution of gular pumping to lung ventilation in monitor lizards. Science 284, 1661–3.

Pan, J., Yeger, H. and Cutz, E. (2004). Innervation of pulmonary neuroendocrine cells and neuroepithelial bodies in developing rabbit lung. Journal of Histochemistry & Cytochemistry 52, 379–89.

Paoli, P., Ferrara, F. and Taiti, S. (2002). Morphology and evolution of the respiratory apparatus in the family Eubelidae (Crustacea, Isopoda, Oniscidea). Journal of Morphology 253, 272–89.

Pape, H. -K., Kurtz, A. and Silbernagl, S. (eds.) (2018). Physiologie, 8th ed. Thieme, Stuttgart.

Parrino, V., Kraus, D. and Doeller, J. (2000). ATP production from the oxidation of sulfide in gill mitochondria of the ribbed mussel *Geukensia demissa*. Journal of Experimental Biology 203, 2209–18.

Paul, R. (1986). Gas exchange and gas transport in the tarantula *Eurypelma californicum*: an overview. In: Invertebrate Oxygen Carriers (ed. B. Linzen), pp. 321–6. Springer, Berlin.

Paul, R. and Fincke, T. (1989). Book lung function in arachnids II. Carbon dioxide and its relations to respiratory surface, water loss and heart frequency. Journal of Comparative Physiology 159, 419–32.

Paul, R., Fincke, T. and Linzen, B. (1987). Respiration in the tarantula *Eurypelma californicum*: evidence for diffusion lungs. Journal of Comparative Physiology B: Biochemical, Systemic, and Environmental Physiology 157, 209–17.

Paul, R., Fincke, T. and Linzen, B. (1989). Book lung function in arachnids. I. Oxygen uptake and respiratory quotient during rest, activity and recovery: relations to gas transport in the haemolymph. Journal of Comparative Physiology B: Biochemical, Systemic, and Environmental Physiology 159, 409–18.

Paul, R. J. (1991). Oxygen transport from book lungs to tissues: environmental physiology and metabolism in arachnids. Verhandlungen der Deutschen Zoologischen Gesellschaft 84, 9–14.

Paul, R. J. (1992). Gas exchange, circulation, and energy metabolism in arachnids. In: Physiological Adaptations in Vertebrates (eds. S. C. Wood, R. E. Weber, A. R. Hargens, and R. W. Millard), pp. 169–97. Marcel Dekker, New York.

Paul, R. J., Bergner, B., Pfeffer-Seidl, A., Decker, H., Efinger, R. and Storz, H. (1994b). Gas transport in the haemolymph of Arachnids I. Oxygen transport and the physiological role of haemocyanins. Journal of Experimental Biology 188, 25–46.

Paul, R. J. and Bihlmayer, S. (1995). Circulatory physiology of a tarantula (*Eurypelma californicum*). Zoology-Analysis of Complex Systems 98, 69–81.

Paul, R. J., Colmorgen, M., Hüller, S., Tyroller, F. and Zinkler, D. (1997). Circulation and respiratory control in millimeter-sized animals (*Daphnia magna*, *Folsomia candida*) studied by optical methods. Journal of Comparative Physiology B: Biochemical, Systemic, and Environmental Physiology 167, 399–408.

Paul, R. J., Pfeffer-Seidl, A., Efinger, R., Pörtner, H. O. and Storz, H. (1994a). Gas transport in the haemolymph of Arachnids II. Carbon dioxide transport and acid-base balance. Journal of Experimental Biology 188, 47–63.

Paul, R. J., Zahler, S., Werner, R. and Markl, J. (1991). Adaptation of an open circulatory system to the oxidatitive capacity of different muscle cell types. Naturwissenschaften 78, 134–5.

Paul, R. J., Zeis, B., Lamkemeyer, T., Seidl, M. and Pirow, R. (2004). Control of oxygen transport in the microcrustacean *Daphnia*: regulation of haemoglobin expression as central mechanism of adaptation to different oxygen and temperature conditions. Acta Physiologica Scandinavica 182, 259–75.

Paul, R. P. and Pirow, R. (1997). The physiological significance of respiratory proteins in invertebrates. Zoology 100, 298–306.

Peck, L. S. and Chapelle, G. (2003). Reduced oxygen at high altitude limits maximum size. Proceedings of The Royal Society B-Biological Sciences 270, S166–7.

Penteado, C. H. S. and Hebling-Beraldo, M. J. A. (1991). Respiratory responses in a brazilian millipede, *Pseudoannolene tricolor*, to declining oxygen pressures. Physiological Zoology 64, 232–41.

Perry, S. F. (1976). Model of exchange barrier and respiratory surface area in the lung of the tortoise (*Testudo graeca*) and its practical application. Mikroskopie 32, 282–93.

Perry, S. F. (1978). Quantitative anatomy of the lungs of the red-eared turtle, Pseudemys scripta elegans. Respiration Physiology 35, 245–62.

Perry, S. F. (1983). Reptilian Lungs: Functional Anatomy and Evolution. Springer, Berlin.

Perry, S. F. (1991). Gas exchange strategies in reptiles and the origin of the avian lung. In: Physiological adaptations in vertebrates (Lung Biology in Health and Disease Vol. 56) (eds. S. C. Wood R. E. Weber A. R. Hargens and R. W. Millard), pp. 149–67. Marcel Dekker, New York.

Perry, S. F. (1998). Lungs: comparative anatomy, functional morphology, and evolution. In: Biology of the Reptilia, vol. 19, Morphology G (eds. C. Gans and A. S. Gaunt), pp. 1–92. Society for the Study of Amphibians and Reptiles, Ithaca.

Perry, S. F. (2007). Swimbladder-lung homolgy in basal osteichtgyes revisited. In: Fish Respiration and Environment (eds. M. N. Fernandes, F. T. Rantin, M. L. Glass, and B. G. Kapoor), pp. 41–54. Science Publishers, Enfield.

Perry, S. F., Breuer, T. and Pajor, N. (2011). Structure and function of the sauropod respiratory system. In: Saurpod Dinosaurs: Understanding the Life of Giants (eds. K. N. K. Remes C. T. Gee and P. M. Sander), pp. 83–93. Indiana University Press, Bloomington and Indianapolis.

Perry, S. F. and Duncker, H. -R. (1980). Interrelationship of static mechanical factors and anatomical structure in lung evolution. Journal of Comparative Physiology 138, 321–34.

Perry, S. F. and Carrier, D. R. (2006). The coupled evolution of breathing and locomotion as a game of leapfrog. Physiological and Biochemical Zoology 79, 997–9.

Perry, S. F., Dariansmith, C., Alston, J., Limpus, C. J. and Maloney, J. E. (1989). Histological structure of the lungs of the loggerhead turtle, *Caretta caretta*, before and after hatching. Copeia 4, 1000–10.

Perry S. F. and Duncker H. R. (1980). Interrelationship of static mechanical factors and anatomical structure in lung evolution. Journal of Comparative Physiology B: Biochemical, Systemic, and Environmental Physiology 138, 321–34.

Perry, S. F., Euverman, R., Wang, T., Loong, A. M., Chew, S. E., Ip, Y. K. and Gilmour, K. M. (2008). Control of breathing in African lungfish (*Protopterus dolloi*): a comparison of aquatic and cocooned (terrestrialized) animals. Respiratory Physiology & Neurobiology 160, 8–17.

Perry, S. F. and Oliveira, E. S. (2010). Respiration in a changing environment. Respiratory Physiology & Neurobiology 173, S20–5.

Perry, S. F., Similowski, T., Klein, W. and Codd, J. R. (2010). The evolutionary origin of the mammalian diaphragm. Respiratory Physiology & Neurobiology 171, 1–16.

Perry, S. F., Wilson, R. J. A., Straus, C., Harris, M. B. and Remmers, J. E. (2001). Which came first the lung or the breath? Comparative Biochemistry and Physiology – Part A: Molecular & Integrative Physiology 129, 37–47.

Peters, H. M. (1978). Mechanism of air ventilation in anabantoids (Pisces Teleostei). Zoomorphologie 89, 93–123.

Peterson, C. C. and Gomez, D. (2008). Buoyancy regulation in two species of freshwater turtle. Herpetologica 64, 141–8.

Pickard, R. S. and Mill, P. J. (1974). Ventilatory movements of the abdomen and branchial apparatus in dragonfly larvae (Odonata: Anisoptera). Journal of Zoology 174, 23–40.

Pinnow, P., Fabrizius, A., Pick, C. and Burmester, T. (2016). Identification and characterisation of hemocyanin of the fish louse *Argulus* (Crustacea: Branchiura). Journal of Comparative Physiology B –Biochemical Systemic and Environmental Physiology 186, 161–8.

Piper, H. (1902). Die Entwicklung von Magen, Duodenum, Schwimmblase, Leber, Pancreas und Milz bei Amia calva. Archiv für Anatomie und Physiologie, Anatomische Abteilung (Supplement), 1–78.

Pirow, R. and Buchen, I. (2004). The dichotomous oxyregulatory behaviour of the planktonic crustacean *Daphnia magna*. Journal of Experimental Biology 207, 683–96.

Pirow, R., Wollinger, F. and Paul, R. J. (1999). The sites of respiratory gas exchange in the planktonic crustacean *Daphnia magna*: An in vivo study employing blood haemoglobin as an internal oxygen probe. Journal of Experimental Biology 202, 3089–99.

Pisanski, A. and Pagliardini, S. (2018). The parafacial respiratory group and the control of active expiration. Respiratory Physiology & Neurobiology. doi: 10.1016/j.resp.2018.06.010.

Pörtner, H. O., Mark, F. C. and Bock, C. (2004). Oxygen limited thermal tolerance in fish?: answers obtained by nuclear magnetic resonance techniques. Respiratory Physiology & Neurobiology 141, 243–60.

Portugal, S. (2015). Chilled-out iguanas have bird-like lungs. Journal of Experimental Biology 218, 650–1.

Prestwich, K. N. (1983a). The roles of aerobic and anaerobic metabolism in active spiders. Physiological Zoology 56(1), 122–32.

Prestwich, K. N. (1983b). Anaerobic metabolism in spiders. Physiological Zoology 56, 112–21.

Prestwich, K. N. (1988). The constraints on maximal activity in spiders. II. Limitations imposed by phosphagen depletion and anaerobic metabolism. Journal of Comparative Physiology B: Biochemical, Systemic, and Environmental Physiology 158, 449–56.

Prestwich, K. N. (1988). The constraints on maximal activity in spiders. I. Evidence against the fluid insufficiency hypothesis. Journal of Comparative Physiology 158, 437–47.

Prestwich, K. N. (2006). Anaerobic metabolism and maximal running in the scorpion *Centruroides hentzi* (Banks) (Scorpiones, Buthidae). Journal of Arachnology 34, 351–6.

Punt, A. (1944). De gasswisseling van enkele bloedzuigende parasieten van warmbloedige dieren (*Cimex*, *Rhodnius*, *Triatoma*). Onderzoek Physiology Laboratory R.U. Utrecht 8, 122–41.

Punt, A. (1950). The respiration of insects. Physiologia Comparata et Oecologia 2, 59–74.

Punt, A. (1956a). Further investigations on the respiration of insects. Physiologia Comparata et Oecologia 4, 121–31.

Punt, A. (1956b). The influence of carbon dioxide on the respiration of *Carabus nemoralis*. Physiologia Comparata et Oecologia 4, 132–9.

Punt, A., Parser, W. J. and Kuchlein, J. (1957). Oxygen uptake in insects with cyclic CO_2 release. Biological Bulletin 112, 108–19.

Purcell, F. (1895). Note on the development of the lungs, entapophyses, tracheae and genital ducts in spiders. Zoologischer Anzeiger 486, 1–5.

Purcell, W. F. (1909). Development and origin of the respiratory organs in Araneae. Quarterly Journal of Microscopical Science 54, 1–110.

Purcell, W. F. (1910). The phylogeny of tracheae in Araneae. Quarterly Journal of Microscopical Science 54, 519–63.

Purnell, M. A., Aldridge, R. J., Donoghue, P. C. J. and Gabbott, S. E. (1995). Conodonts and the first vertebrates. Endeavour 19, 20–7.

Py, C., Denhoff, M. W., Martina, M., Monette, R., Comas, T., Ahuja, T., Martinez, D., Wingar, S., Caballero, J., Laframboise, S. et al. (2010). A novel silicon patch-clamp chip permits high-fidelity recording of ion channel activity from functionally Defined Neurons. Biotechnology and Bioengineering 107, 593–600.

Quinlan, M. C. and Gibbs, A. G. (2006). Discontinuous gas exchange in insects. Respiratory Physiology & Neurobiology 154, 18–29.

Ragg, N. L. C. and Taylor, H. H. (2006a). Heterogeneous perfusion of the paired gills of the abalone *Haliotis iris* Martyn 1784: an unusual mechanism for respiratory control. Journal of Experimental Biology 209, 475–83.

Ragg, N. L. C. and Taylor, H. H. (2006b). Oxygen uptake, diffusion limitation, and diffusing capacity of the bipectinate gills of the abalone, *Haliotis iris* (Mollusca: Prosobranchia). Comparative Biochemistry and Physiology Part A: Molecular & Integrative Physiology 143, 299–306.

Rahn, H. and Paganelli, C. V. (1968). Gas exchange in gas gills of diving insects. Respiratory Physiology 5, 145–64.

Ramirez, M. J. (2000). Respiratory system morphology and the phylogeny of Haplogyne spiders (Araneae, Araneomorphae). Journal of Arachnology 28, 149–57.

Randall, D., Burggren, W., French, K. (2002). Eckert Animal Physiology, 5th ed. Freeman, New York.

Randall, D. J. and Perry, S. F. (1992). Catecholamines. In: Fish Physiology, vol. XIIB (eds. W. S. Hoar D. J. Randall and A. P. Farrell), pp. 255–300. Academic Press, New York.

Rathke, H. (1825a). Kiemen bei Vögeln. Isis von Oken 18, 1100–1.

Rathke, H. (1825b). Kiemen bei Säugetieren. Isis von Oken 18, 747–9.

Rathke, H. (1828). Über das Dasein von Kiemenandeutungen bei menschlichen Embryonen. Isis von Oken 21, 108–109.

Rees, B. B., Sudradjat, F. A. and Love, J. W. (2001). Acclimation to hypoxia increases survival time of zebrafish, Danio rerio, during lethal hypoxia. Journal of Experimental Zoology 289, 266–72.

Reeves, R. B. (1972). An imidazole alphastat hypothesis for vertebrate acid-base regulation: tissue carbon-dioxide content and body temperature in bullfrogs. Respiration Physiology 14, 219–36.

Regier, J. C. and Shultz, J. W. (1997). Molecular phylogeny of the major arthropod groups indicates polyphyly of Crustaceans and a new hypothesis for the origin of hexapods. Molecular Biology and Evolution 14, 902–13.

Rehm, P., Pick, C., Borner, J., Markl, J. and Burmester, T. (2012). The diversity and evolution of chelicerate hemocyanins. BMC Evolutionary Biology 12, 19.

Reisinger, P. W. M., Focke, P. and Linzen, B. (1990). Lung morphology of the tarantula, *Eurypelma californicum*, Ausserer, 1871 (Araneae: Theraphosidae). Bulletin of the British Arachnological Society 8, 165–70.

Reisinger, P. W. M., Tutter, I. and Welsch, U. (1991). Fine structure of the gills of the horseshoe crabs *Limulus polyphemus* and *Tachypleus tridentatus* and of the book lungs of the spider *Eurypelma californicum*. Zoologische Jahrbücher 121, 331–57.

Richter, D. W., Ballantyne, D. and Remmers, J. E. (1987). The differential organization of medullary post-inspiratory activities. Pflugers Archiv: European Journal of Physiology 410, 420–7.

Romer, A. S. (1972). Skin breathing: primary or secondary. Respiration Physiology 14, 183–92.

Rong, C., Yan, M., Zhen-Zhong, B., Ying-Zhong, Y., Dian-Xiang, L., Qi-Sheng, M., Qing, G., Yin, L. and Ge, R. L. (2012). Cardiac adaptive mechanisms of Tibetan antelope (*Pantholops hodgsonii*) at high altitudes. American Journal of Veterinary Research 73, 809–13.

Roux, E. (2002). Origin and evolution of the respiratory tract in Vertebrates. Revue Des Maladies Respiratoires 19, 601–15.

Roussos, C. and Macklem, P. T. (1985). The Thorax, Part A and B (2 Vols.). Marcel Decker, New York.

Rovainen, C. M. (1996). Feeding and breathing in lampreys. Brain Behavior and Evolution 48, 297–305.

Rovner, J. S. (1987). Nests of terrestrial spiders maintain a physical gill: flooding and the evolution of silk constructions. Journal of Arachnology 14, 327–37.

Ruf, T., Biber, C., Arnold, W. and Milesi, E. (2012). Living in a Seasonal World: Thermoregulatory and Metabolic Adaptations. Springer, Heidelberg.

Sagemehl, M. (1885). Beiträge zur vergleichenden Anatomie der Fische - III. Das Cranium der Characiniden nebst allgemeinen Bemerkungen über die mit dem Weber'schen Apparat versehenen Physostomenfamilien. Morphologisches Jahrbuch 10, 1–119.

Saini, R. S. (1977). Ultrastructural observations on the tracheal gills of *Aeshna* (Anisoptera: Odonata) larvae. Journal of Submicroscopic Cytology 9, 347–54.

Sanchez, D., Ganfornina, M. D., Gutierrez, G. and Bastiani, M. J. (1998). Molecular characterization and phylogenetic relationships of a protein with potential oxygen-binding capabilities in the grasshopper embryo. A hemocyanin in insects? Molecular Biology and Evolution 15, 415–26.

Sanders, R. K. and Farmer, C. G. (2012). The pulmonary anatomy of *Alligator mississippiensis* and its similarity to the avian respiratory system. Anatomical Record – Advances in Integrative Anatomy and Evolutionary Biology 295, 699–714.

Schachner, E. R., Hutchinson, J. R. and Farmer, C. G. (2013). Pulmonary anatomy in the Nile crocodile and the evolution of unidirectional airflow in Archosauria. PeerJ 1, e60.

Schachner, E. R., Sedlmayr, J. C., Schott, R., Lyson, T. R., Sanders, R. K. and Lambertz, M. (2017). Pulmonary anatomy and a case of unilateral aplasia in a common snapping turtle (*Chelydra serpentina*): developmental perspectives on cryptodiran lungs. Journal of Anatomy 231, 835–48.

Scheid, P., Hook, C. and Bridges, C. R. (1982). Diffusion in gas exchange of insects. Federation Proceedings 41, 2143–5.

Scheuermann, D. W. (1987). Morphology and cytochemistry of the endocrine epithelial system in the lung. International Review of Cytology 106, 35–88.

Schmidt, C. and Wägele, J. W. (2001). Morphology and evolution of respiratory structures in the pleopod exopodites of terrestrial Isopoda (Crustacea, Isopoda, Oniscidea). Acta Zoologica 82, 315–30.

Schmidt-Nielsen, K. (1995). Animal Physiology. Cambridge University Press, Cambridge.

Schmitz, A. (2004). Metabolic rates during rest and activity in differently tracheated spiders (Arachnida, Araneae): *Pardosa lugubris* (Lycosidae) and *Marpissa muscosa* (Salticidae). Journal of Comparative Physiology B: Biochemical, Systemic, and Environmental Physiology 174, 519–26.

Schmitz, A. (2005). Spiders on a treadmill: influence of running activity on metabolic rates in *Pardosa lugubris* (Araneae, Lycosidae) and Marpissa muscosa (Araneae, Salticidae). Journal of Experimental Biology 208, 1401–11.

Schmitz, A. (2013). Tracheae in spiders: respiratory organs for special functions. In: Spider Ecophysiology, (ed. W. Nentwig), pp. 29–39. Springer, Heidelberg.

Schmitz, A. (2016). Respiration in spiders (Araneae). Journal of Comparative Physiology B – Biochemical Systemic and Environmental Physiology 186, 403–15.

Schmitz, A., Gemmel, M. and Perry, S. F. (2000). Morphometric partitioning of respiratory surfaces in the Amphioxus (*Branchiostoma lanceolatum* Pallas). Journal of Experimental Biology 203, 3381–90.

Schmitz, A. and Paul, R. J. (2003). Probing of hemocyanin function in araneomorph spiders. XII. International Conference on Invertebrate Dioxygen Binding Proteins, Mainz, 96.

Schmitz, A. and Perry, S. F. (1999). Stereological determination of tracheal volume and diffusing capacity of the tracheal walls in the stick insect *Carausius morosus*. Physiological and Biochemical Zoology 72, 205–18.

Schmitz, A. and Perry, S. F. (2000). Respiratory system of arachnids I: Morphology of the respiratory system of *Salticus scenicus* and *Euophrys lanigera* (Arachnida, Araneae, Salticidae). Arthropod Structure and Development 29, 3–12.

Schmitz, A. and Perry, S. F. (2001). Bimodal breathing in jumping spiders: Morphometric partitioning of lungs and tracheae in *Salticus scenicus* (Arachnida, Araneae, Salticidae). Journal of Experimental Biology 204, 4321–34.

Schmitz, A. and Perry, S. F. (2002a). Respiratory organs in wolf spiders: morphometric analysis of lungs and tracheae in *Pardosa lugubris* (L.) (Arachnida, Araneae, Lycosidae). Arthropod Structure & Development 31, 217–30.

Schmitz, A. and Perry, S. F. (2002b). Morphometric analysis of the tracheal walls of the harvestmen *Nemastoma lugubre* (Arachnida, Opiliones, Nemastomatidae). Arthropod Structure and Development 30, 229–41.

Schmitz, A. and Wasserthal, L. T. (1999). Comparative morphology of the spiracles of the Papilionidae, Sphingidae, and Saturniidae (Insecta: Lepidoptera). International Journal of Insect Morphology and Embryology 28, 13–26.

Schneiderman, H. A. (1960). Discontinuous respiration in insects: role of the spiracles. Biological Bulletin 119, 494–528.

Scholtz, G. and Kamenz, C. (2006). The book lungs of Scorpiones and Tetrapulmonata (Chelicerata, Arachnida): evidence for homology and a single terrestrialisation event of a common arachnid ancestor. Zoology 109, 2–13.

Schwarze, K., Campbell, K. L., Hankeln, T., Storz, J. F., Hoffmann, F. G. and Burmester, T. (2014). The globin gene repertoire of lampreys: convergent evolution of hemoglobin and myoglobin in jawed and jawless vertebrates. Molecular Biology and Evolution 31, 2708–21.

Schwarze, K., Singh, A. and Burmester, T. (2015). The full globin repertoire of turtles provides insights into vertebrate globin evolution and functions. Genome Biology and Evolution 7, 1896–913.

Scott, G. R. and Milsom, W. K. (2007). Control of breathing and adaptation to high altitude in the bar-headed goose. American Journal of Physiology – Regulatory, Integrative and Comparative Physiology 293, R379–91.

Selden, P. A. (1990). Terrestrialization (invertebrates). In: Palaeobiology: A Synthesis (eds. D. E. Briggs and P. P. Crowther), pp. 64–8. Blackwell Scientific Publications, Oxford.

Semenza, G. L. (2004). Hydroxylation of HIF-1: Oxygen sensing at the molecular level. Physiology 19, 176–82.

Seymour, R. S. and Webster, M. E. D. (1975). Gas transport and blood acid-base-balance in diving sea snakes. Journal of Experimental Zoology 191, 169–81.

Shams, H. and Scheid, P. (1989). Efficiency of parabronchial gas-exchange in deep hypoxia: measurements in the resting duck. Respiration Physiology 77, 135–46.

Shear, W. A. and Selden, P. A. (2001). Rustling in the undergrowth: animals in early terrestrial ecosystems. In: Plants Invade the Land: Evolutionary and Environmental Perspectives (eds. P. G. Gensel and D. Edwards). New York: Columbia University Press.

Shear, W. A., Gensel, P. G. and Jeram, A. J. (1996). Fossils of large terrestrial arthropods from the Lower Devonian of Canada. Nature 384, 555–7.

Sidorov, A. V. (2005). Effect of acute temperature change on lung respiration of the mollusc *Lymnaea stagnalis*. Journal of Thermal Biology 30, 163–71.

Simons, R. S., Bennett, W. O. and Brainerd, E. L. (2000). Mechanics of lung ventilation in a post-metamorphic salamander, *Ambystoma tigrinum*. Journal of Experimental Biology 203, 1081–92.

Slama, K. (1988). A new look at insect respiration. Biological Bulletin 175, 289–300.

Slama, K. (1995). Respiratory cycles of *Chelifer cancroides* (Pseudscorpiones) and *Galeodes* sp. (Solifugae). European Journal of Entomology 92, 543–52.

Smith, H. M. (1946). Handbook of Lizards: Lizards of the United States and of Canada. Comstock, Ithaca.

Smith, J. C., Ellenberger, H. H., Ballanyi, K., Richter, D. W. and Feldman, J. L. (1991). Pre-Bötzinger complex: a brainstem region that may generate respiratory rhythm in mammals. Science 254, 726–9.

Soetaert, K., Muthumbi, A. and Heip, C. (2002). Size and shape of ocean margin nematodes: morphological diversity and depth-related patterns. Marine Ecology Progress Series 242, 179–93.

Sohal, R. S., Agarwal, A., Agarwal, S. and Orr, W. C. (1995). Simultaneous overexpression of copper-containing and zinc-containing superoxide-dismutase and catalase retards age-related oxidative damage and increases metabolic potential in *Drosophila melanogaster*. Journal of Biological Chemistry 270, 15671–4.

Soivio, A. and Tuurala, H. (1981). Structural and circulatory responses to hypoxia in the secondary lamellae of *Salmo gairdneri* gills at two temperatures. Journal of Comparative Physiology 145, 37–43.

Sokolova, I. M. and Portner, H. O. (2003). Metabolic plasticity and critical temperatures for aerobic scope in a eurythermal marine invertebrate (*Littorina saxatilis*, Gastropoda: Littorinidae) from different latitudes. Journal of Experimental Biology 206, 195–207.

Sollid, J., De Angelis, P., Gundersen, K. and Nilsson, G. E. (2003). Hypoxia induces adaptive and reversible gross morphological changes in crucian carp gills. Journal of Experimental Biology 206, 3667–73.

Somme, L. (1989). Adaptations of terrestrial arthropods to the alpine environment. Biological Reviews of the Cambridge Philosophical Society 64, 367–407.

St. John, W. M. (2009). Noeud vital for breathing in the brainstem: gasping – yes, eupnoea – doubtful. Philosophical Transactions of the Royal Society of London B: Biological Sciences 364, 2625–33.

Stevens, E. D. (1972). Some aspects of gas-exchange in tuna. Journal of Experimental Biology 56, 809–23.

Stewart, T. C. and Woodring, J. P. (1973). Anatomical and physiological studies of water balance in the millipedes *Pachydesmus crassicutis* (Polydesmida) and *Orthoporus texicolens* (Spirobolida). Comparative Biochemistry and Physiology – Part A: Molecular & Integrative Physiology 44, 735–50.

Stokes, M. D. (1997). Larval locomotion of the lancet *Branchiostoma floridae*. Journal of Experimental Biology 200, 1661–80.

Storey, K. B. and Storey, J. M. (2013). Molecular biology of freezing tolerance. Comprehensive Physiology 3, 1283–308.

Straus, C., Vasilakos, K., Wilson, R. J. A., Oshima, T., Zelter, M., Derenne, J. P., Similowski, T. and Whitelaw, W. A. (2003). A phylogenetic hypothesis for the origin of hiccough. Bioessays 25, 182–8.

Strazny, F. and Perry, S. F. (1984). Morphometric diffusing capacity and functional anatomy of the book lungs in the spider *Tegenaria* spp. (Agelenidae). Journal of Morphology 182, 339–54.

Suarez, R. K. (2000). Energy metabolism during insect flight: Biochemical design and physiological performance. Physiological and Biochemical Zoology 73, 765–71.

Suarez, R. K., Lighton, J. R. B., Joos, B., Roberts, S. P. and Harrison, J. F. (1996). Energy metabolism, enzymatic flux capacitiesn and metabolic flux rates in flying honeybees. Proceedings of the National Academy of Sciences of the United States of America 93, 12616–20.

Sverdlova, N. S., Lambertz, M., Witzel, U. and Perry, S. F. (2012). Boundary conditions for heat transfer and evaporative cooling in the trachea and air sac system of the domestic fowl: a two-dimensional CFD analysis. PLoS One 7, e45315.

Swales, L. S., Cournil, I. and Evans, P. D. (1992). The innervation of the closer muscle of the mesothoracic spiracle of the locust. Tissue & Cell 24, 547–58.

Syed, N. I., Bulloch, A. G. M. and Lukowiak, K. (1990). In vitro reconstruction of the respiratory central pattern generator of the mollusk *Lymnaea*. Science 250, 282–5.

Syed, N. I., Harrison, D. and Winlow, W. (1991). Respiratory behavior in the pond snail *Lymnaea stagnalis*. 1. Behavioral-analysis and the identification of motor neurons. Journal of Comparative Physiology A – Sensory Neural and Behavioral Physiology 169, 541–55.

Taylor, B. E. and Lukowiak, K. (2000). The respiratory central pattern generator of *Lymnaea*: a model, measured and malleable. Respiration Physiology 122, 197–207.

Taylor, E. W., Campbell, H. A., Levings, J. J., Young, M. J., Butler, P. J. and Egginton, S. (2006). Coupling of the respiratory rhythm in fish with activity in hypobranchial nerves and with heartbeat. Physiological and Biochemical Zoology 79, 1000–9.

Taylor, E. W., Leite, C. A. C., McKenzie, D. J. and Wang, T. (2010). Control of respiration in fish, amphibians and reptiles. Brazilian Journal of Medical and Biological Research 43, 409–24.

Terblanche, J. S., Klok, C. J., Marais, E. and Chown, S. L. (2004). Metabolic rate in the whip-spider, *Damon annulatipes* (Arachnida: Amblypygi). Journal of Insect Physiology 50, 637–45.

Terwilliger, N. B. (1998). Functional adaptations of oxygen-transport proteins. Journal of Experimental Biology 201, 1085–98.

Terwilliger, R. C. and Terwilliger, N. B. (1985). Molluscan hemoglobins. Comparative Biochemistry and Physiology Part B: Biochemistry & Molecular Biology 81, 255–61.

Thaler, K. (2003). The diversity of high altitude arachnids (Araneae, Opiliones, Pseudoscorpiones) in the alps. In: Ecological Studies, vol. 167 (eds. L. Nagy G. Grabherr C. Körner and D. B. A. Thompson), pp. 281–96. Springer, Berlin.

Thorpe, W. H. and Crisp, D. J. (1947). Studies on plastron respiration I-III. Journal of Experimental Biology 24, 227–328.

Thuesen, E. V., Rutherford, L. D. and Brornmer, P. L. (2005a). The role of aerobic metabolism and intragel oxygen in hypoxia tolerance of three ctenophores: Pleurobrachia bachel, *Bolinopsis infundibulum* and *Mnemiopsis leidyi*. Journal of the Marine Biological Association of the United Kingdom 85, 627–33.

Thuesen, E. V., Rutherford, L. D., Brommer, P. L., Garrison, K., Gutowska, M. A. and Towanda, T. (2005b). Intragel oxygen promotes hypoxia tolerance of scyphomedusae. Journal of Experimental Biology 208, 2475–82.

Ultsch, G. R., Herbert, C. V. and Jackson, D. C. (1984). The comparative physiology of diving in north American freshwater turtles. 1. Submergence tolerance, gas exchange, and acid-base-balance. Physiological Zoology 57, 620–31.

Val, A. L. (1993). Adaptations of fishes to extreme conditions in fresh waters. In: The Vertebrate Gas Transport Cascade: Adaptations to Environment and Mode of Life (ed. J. E. Bicudo), pp. 43–53. CRC Press, Boca Raton.

Val, A. L., Silva, M. N. P. and Almeida-Val, V. M. F. (1998). Hypoxia adaptation in fish of the Amazon: a never-ending task. South African Journal of Zoology 33, 107–14.

van Holde, K. E. (1997). Respiratory proteins of invertebrates: structure, function and evolution. Zoology 100, 287–97.

van Holde, K. E. and Miller, K. I. (1995). Hemocyanins. Advances in Protein Chemistry 47, 1–81.

van Holde, K. E., Miller, K. I. and Decker, H. (2001). Hemocyanins and invertebrate evolution. Journal of Biological Chemistry 276, 15563–6.

Vitalis, T. Z. and Milsom, W. K. (1986). Pulmonary mechanics and the work of breathing in the semi-aquatic turtle, *Pseudemys scripta*. Journal of Experimental Biology 125, 137–55.

Wada, Y., Mogami, Y. and Baba, S. (1997). Modification of ciliary beating in sea urchin larvae induced by neurotransmitters: beat-plane rotation and control of frequency fluctuation. Journal of Experimental Biology 200, 9–18.

Wägele, H., Deusch, O., Handeler, K., Martin, R., Schmitt, V., Christa, G., Pinzger, B., Gould, S. B., Dagan, T., Klussmann-Kolb, A. et al. (2011). Transcriptomic evidence that longevity of acquired plastids in the photosynthetic slugs *Elysia timida* and *Plakobranchus ocellatus* does not entail lateral transfer of algal nuclear genes. Molecular Biology and Evolution 28, 699–706.

Wake, M. H. and Donnelly, M. A. (2010). A new lungless caecilian (Amphibia: Gymnophiona) from Guyana. Proceedings of the Royal Society B 277, 915–22.

Wald, G. and Riggs, A. (1951). The hemoglobin of the sea lamprey, *Petromyzon marinus*. Journal of General Physiology 35, 45–53.

Walentek, P., Bogusch, S., Thumberger, T., Vick, P., Dubaissi, E., Beyer, T., Blum, M. and Schweickert, A. (2014). A novel serotonin-secreting cell type regulates ciliary motility in the mucociliary epidermis of *Xenopus* tadpoles. Development 141, 1526–33.

Wallau, B., Schmitz, A. and Perry, S. F. (2000). Lung morphology in rodents (Mammalia, Rodentia) and its implications for systematics. Journal of Morphology 246, 228–48.

Wallach, V. (1998). The lungs of snakes. In: Biology of the Reptilia: Morphology G, Vol. 19 (eds. C. Gans and A. S. Gaunt), pp. 93–295. Society for the Study of Amphibians and Reptiles, Ithaca.

Ward, P., Labandeira, C., Laurin, M. and Berner, R. A. (2006). Confirmation of Romer's gap as low oxygen interval constraining the timing of initial arthropod and vertebrate terrestrialization. Proceedings of the National Academy of Sciences of the United States of America 103, 16818–22.

Waser, W. P., Schmitz, A., Perry, S. F. and Wobschall, A. (2005). Stereological analysis of blood space and tissue types in the pseudobranch of the rainbow trout (*Oncorhynchus mykiss*). Fish Physiology and Biochemistry 31, 73–82.

Wasserthal, L. T. (1996). Interaction of circulation and tracheal ventilation in holometabolous insects. Advances in Insect Physiology 26, 297–351.

Wassnetzov, W. (1932). Über die Morphologie der Schwimmblase. Zoologische Jahrbücher, Abteilung für Anatomie und Ontogenie der Tiere 56, 1–36.

Weber, R. E. (1995). Hemoglobin adaptations to hypoxia and altitude – the phylogenetic perspective. In: Hypoxia and Brain, Vol. Proc. 9th Int. Hypxia Symp., Lake Louise, Canada (eds. J. R. Sutton C. S. Houston and G. Coates), pp. 31–44. Lake Louise, Canada.

Weber, R. E. (1997). Hemoglobin adaptations in Amazonian and temperate fish, with special reference to hypoxia, allosteric effectors and functional heterogeneity. In: Physiology and Biochemistry of the Fishes of the Amazon (eds. A. L. Val, V. M. F. Almeida-Val, and D. J. Randall), pp. 75–90. Springer, New York.

Weber, R. E. (2007). High-altitude adaptations in vertebrate hemoglobins. Respiratory Physiology & Neurobiology 158, 132–42.

Weber, R. E. and Fago, A. (2004). Functional adaptation and its molecular basis in vertebrate hemoglobins, neuroglobins and cytoglobins. Respiratory Physiology & Neurobiology 144, 141–59.

Weber, R. E. and Vinogradov, S. N. (2001). Nonvertebrate hemoglobins: functions and molecular adaptations. Physiological Reviews 81, 569–628.

Webster, K. A. (2003). Evolution of the coordinate regulation of glycolytic enzyme genes by hypoxia. Journal of Experimental Biology 206, 2911–22.

Wedemeyer, H. and Schild, D. (1995). Chemosensitivity of the osphradium of the pond snail *Lymnaea stagnalis*. Journal of Experimental Biology 198, 1743–54.

Wegener, G. (1987). Insect brain metabolism under normoxic and hypoxic conditions. In: Arthropod Brain: Its Evolution, Development, Structure, and Functions (ed. A. P. Gupta), pp. 87. John Wiley and Sons, New York.

Wegener, G. (1993). Hypoxia and posthypoxic recovery in insects: physiological and metabolic aspects. In: Surviving hypoxia: mechanisms of control and adaptation (eds. P. W. Hochachka P. L. Lutz T. Sick M. Rosenthal and G. van den Thillart), pp. 417–34. CRC Press, Boca Raton.

Wegener, G. (1996). Flying insects: model systems in exercise physiology. Experientia 52, 404–12.

Wegner, N. C., Lai, N. C., Bull, K. B. and Graham, J. B. (2012). Oxygen utilization and the branchial pressure gradient during ram ventilation of the shortfin mako, *Isurus oxyrinchus*: is lamnid shark-tuna convergence constrained by elasmobranch gill morphology? Journal of Experimental Biology 215, 22–8.

Weihe, E., Lucassen, M. and Abele, D. (2009). The hypoxia inducible factor (HIF-1) in the Antarctic limpet *Nacella concinna*. In: 2nd International Congress of Respiratory Science – Abstracts & Scientific Program (eds. S. F.

Perry, S. Morris, T. Breuer, N. Pajor, and M. Lambertz), pp. 107–8. Tharax, Hildesheim.

Weibel, E. R. and Taylor, C. R. (1981). Design of the mammalian respiratory system. I. Problem and strategy. Respiration Physiology 44, 1–10.

Weibel, E. R., Taylor, C. R. and Hoppeler, H. (1991). The concept of symmorphosis: a testable hypothesis of structure-function relationship. Proceedings of the National Academy of Sciences of the United States of America 88, 10357–61.

Weidemann, A. and Johnson, R. S. (2008). Biology of HIF-1 alpha. Cell Death and Differentiation 15, 621–7.

Weis-Fogh, T. (1964a). Functional design of the tracheal system of flying insects as compared with the avian lung. Journal of Experimental Biology 41, 207–28.

Weis-Fogh, T. (1964b). Diffusion in insect wing muscle. The most active tissue known. Journal of Experimental Biology 41, 229–56.

Wells, M. J., Odor, R. K., Mangold, K. and Wells, J. (1983). Oxygen-consumption in movement by octopus. Marine Behaviour and Physiology 9, 289–303.

Wells, M. J. and Wells, J. (1985). Ventilation frequencies and stroke volumes in acute-hypoxia in octopus. Journal of Experimental Biology 118, 445–8.

Wells, M. J. and Wells, J. (1995). The control of ventilatory and cardiac responses to changes in ambient oxygen-tension and oxygen-demand in octopus. Journal of Experimental Biology 198, 1717–27.

Wells, R. M. G. and Dales, R. P. (1974). Oxygenational properties of haemerythrin in blood of *Magelona papillicornis* Muller (Polychaeta: Magelonidae). Comparative Biochemistry and Physiology 49, 57–64.

West, J. B. (ed.) (1996). Respiratory Physiology: People and Ideas. Oxford University Press/American Physiological Society, New York.

West, J. B. (2006). Human responses to extreme altitudes. Integrative and Comparative Biology 2, 25–34.

West, J. B. (2011). History of respiratory gas exchange. Comprehensive Physiology 1, 1509–23.

West, J. B. (2012). History of respiratory mechanics prior to World War II. Comprehensive Physiology 2, 609–19.

West, J. B. (2015). Essays on the History of Respiratory Physiology. American Physiological Society/Springer, Washington, D.C.

West, J. B. (2016). Early history of high-altitude physiology. Annals of the New York Academy of Sciences 1365, 33–42.

West, J. B. and Luks, A. M. (2016). Respiratory Physiology: The Essentials, 10th ed. Wolters Kluwer, Alphen aan den Rijn.

Westneat, M. W., Betz, O., Blob, R. W., Fezzaa, K., Cooper, W. J. and Lee, W. K. (2003). Tracheal respiration in insects visualized with synchroton X-ray imaging. Science 299, 558–60.

White, C. R., Blackburn, T. M., Terblanche, J. S., Marais, E., Gibernau, M. and Chown, S. L. (2007). Evolutionary responses of discontinuous gas exchange in insects. Proceedings of the National Academy of Sciences of the United States of America 104, 8357–61.

Whiting, M. F., Bradler, S. and Maxwell, T. (2003). Loss and recovery of wings in stick insects. Nature 421, 264–7.

Wichard, W. (1979). Zur Feinstruktur der abdominalen Tracheenkiemen von Larven der Kleinlibellen-Art *Epallage fatime* (Odonata: Zygoptera: Euphaidae). Entomology Generalis 5, 129–34.

Wichard, W. and Komnick, H. (1971). Zur Feinstruktur der Tracheenkiemen von *Glyphotaelius pellucidus* Retz. (Insecta, Trichoptera). Cytobiologie 3, 106–10.

Wichard, W. and Komnick, H. (1974). Zur Feinstruktur der rektalen Tracheenkiemen von anisopteren Libellenlarven. I. Das Respiratorische Epithel Odonatologica 3, 121–7.

Wiedersheim, R. (1909). Vergleichende Anatomie der Wirbeltiere. Gustav Fischer Verlag, Jena.

Wigglesworth, V. B. (1930). A theory of tracheal respiration in insects. Royal Society of London Proceedings 106, 229–50.

Wigglesworth, V. B. (1953, 1972). The Principles of Insect Physiology. Chapman Hall, London.

Wigglesworth, V. B. (1983). The physiology of insect tracheoles. Advances in Insect Physiology 17, 85–148.

Wilkinson, M. and Nussbaum, R. A. (1997). Comparative morphology and evolution of the lungless caecilian *Atretochoana eiselti* (Taylor) (Amphibia: Gymnophiona: Typhlonectidae). Biological Journal of the Linnean Society 62, 39–109.

Wilson, R. J. A., Harris, M. B., Remmers, J. E. and Perry, S. E. (2000). Evolution of air-breathing and central CO_2/H+ respiratory chemosensitivity: New insights from an old fish? Journal of Experimental Biology 203, 3505–12.

Winkler, D. (1955). Das Tracheensystem der Dysderidae. Mitteilungen aus dem Zoologischen Museum in Berlin 31, 25–43.

Withers, P. C. (1981). The effects of ambient air pressure on oxygen consumption of resting and hovering honeybees. Journal of Comparative Physiology 141, 433–7.

Wittenberg, J. B. and Wittenberg, B. A. (2003). Myoglobin function reassessed. Journal of Experimental Biology 206, 2011–20.

Wohlgemuth, S. E., Taylor, A. C. and Grieshaber, M. K. (2000). Ventilatory and metabolic responses to hypoxia and sulphide in the lugworm *Arenicola marina* (L.). Journal of Experimental Biology 203, 3177–88.

Wood, C. M., Wilson, R. W., Gonzalez, R. J., Patrick, M. L., Bergman, H. L., Narahara, A. and Val, A. L. (1998). Responses of an Amazonian teleost, the tambaqui (*Colossoma macropomum*), to low pH in extremely soft water. Physiological Zoology 71, 658–70.

Woodman, J. D., Cooper, P. D. and Haritos, V. S. (2007a). Effects of temperature and oxygen availability on water loss and carbon dioxide release in two sympatric saproxylic invertebrates. Comparative Biochemistry and Physiology Part A: Molecular & Integrative Physiology 147, 514–20.

Woodman, J. D., Cooper, P. D. and Haritos, V. S. (2007b). Cyclic gas exchange in the giant burrowing cockroach, *Macropanesthia rhinoceros*: effect of oxygen tension and temperature. Journal of Insect Physiology 53, 497–504.

Woodman, J. D., Cooper, P. D. and Haritos, V. S. (2008). Neural regulation of discontinuous gas exchange in *Periplaneta americana*. Journal of Insect Physiology 54, 472–80.

Wright, J. C. and Machin, J. (1990). Water vapour absorption in terrestrial isopods. Journal of Experimental Biology 154, 13–30.

Wright, J. C. and Machin, J. (1993). Atmospheric water absorption and the water budget of terrestrial isopods (Crustacea, Isopoda, Oniscidea). Biological Bulletin 184, 243–53.

York, J. M., Chua, B. A., Ivy, C. M., Alza, L., Cheek, R., Scott, G. R., McCracken, K. G., Frappell, P. B., Dawson, N. J., Lague, S. L. et al. (2017). Respiratory mechanics of eleven avian species resident at high and low altitude. Journal of Experimental Biology 220, 1079–89.

Yoshihiro, M., Keiko, W., Chieko, O., Akemi, K. and Baba, S. A. (1992). Regulation of ciliary movement in sea urchin embryos: Dopamine and 5-HT change the swimming behaviour. CComparative Biochemistry and Physiology Part C: Comparative Pharmacology 101, 251–4.

Youson, J. H. and Freeman, P. A. (1976). Morphology of gills of larval and parasitic adult sea lamprey, *Petromyzon marinus* L. Journal of Morphology 149, 73–103.

Zebe, E., Salge, U., Wiemann, C. and Wilps, H. (1981). The energy metabolism of the leech *Hirudo medicinalis* in anoxia and muscular work. Journal of Experimental Zoology 218, 157–63.

Zinkler, D. and Rüssbeck, R. (1986). Ecophysiological adaptations of Collembola to low oxygen concentrations. 2nd International Seminar on Apterygota, pp. 195–8. Siena, Italy, 4–6 September 1986.

Index

Tables and figures are indicated by an italic *t* and *f* following the page number.

D

T

U

V

W

X

Y

Z

Printed and bound by CPI Group (UK) Ltd, Croydon, CR0 4YY